UNITEXT for Physics

UNITEXT for Physics series publishes textbooks in physics and astronomy, characterized by a didactic style and comprehensiveness. The books are addressed to upper-undergraduate and graduate students, but also to scientists and researchers as important resources for their education, knowledge, and teaching.

Sergio Cecotti

Statistical Mechanics

A Concise Advanced Textbook

 Springer

Sergio Cecotti
String Theory, Hefangkou Village
Beijing Institute of Mathematical Science
Beijing, China

ISSN 2198-7882　　　　　　ISSN 2198-7890　(electronic)
UNITEXT for Physics
ISBN 978-3-031-67876-9　　　ISBN 978-3-031-67874-5　(eBook)
https://doi.org/10.1007/978-3-031-67874-5

This Springer imprint is published by the registered company Springer Nature Switzerland AG
The registered company address is: Gewerbestrasse 11, 6330 Cham, Switzerland

If disposing of this product, please recycle the paper.

To my teacher Pietro Menotti

Preface

This is the second textbook in the series of Lecture Notes from the classes the author taught in Qiuzhen College (Tsinghua University), the Chinese *elite* Institution for the very talented students in mathematics. The intended audience explains the character of the book: while it introduces Statistical Mechanics from the very scratch and is fully self-contained, the exposition tries to be more mathematically precise than in typical physics textbooks, paying attention to distinguish rigorous results from controlled approximations or physical pictures of what is going on. While concise (as suited for a one semester basic course) the book contains several topics not usually covered in introductory textbooks, such as the information-theoretic interpretation of entropy by Shannon, the gauge approach to order-disorder duality in the Ising model, the Yang-Lee theory, the quantum dissipation-fluctuation theorem, frustrated and quenched systems, including an introduction to the celebrated Parisi solution of the Sherrington-Kirkpatrick model of spin glasses.

The path integral formalism is introduced and discussed from different perspectives suited for the different applications to statistical problems. In Chap. 3 path integrals are seen from the side of the Feymann-Kac formula, second quantization, and quantum statistics. In Chap. 5 we look at them from the viewpoint of the effective field theories *á la* Landau-Ginzburg, while in Chap. 6 we study them in relation with the Brownian motion, Langevin stochastic differential equations, and Fokker-Planck diffusion PDEs. In the last direction we elaborate on the connection between stochastic processes and supersymmetry. Several techniques to compute path integrals (and in particular functional determinants) are introduced: in each relevant chapter, we explain the computing tools which are the most convenient ones for the given application.

We have divided the book in two parts. Part II two contains fundamental but more specialized materials which may be omitted in a quick introductory course.

<table>
<tr><td>Beijing, China</td><td align="right">Sergio Cecotti</td></tr>
<tr><td>June 2024</td><td></td></tr>
</table>

Contents

Part I
Basic Theory

The material of this concise textbook has been divided in two parts. The first part, which encompasses the first five chapters, contains the basic theory which should be covered in any introductory course in Statistical Mechanics. Teachers willing to give a shorter and more elementary presentation may limit themselves to this first part.

We tried our best to use the same language and notation as in the first volume of this series: *Analytic Mechanics. A Concise Textbook* (Springer, 2024).

Chapter 1
Thermodynamical Formalism

Therefore the deep impression which classical thermodynamics made upon me. It is the only physical theory of universal content, which I am convinced that, within the framework of applicability of its basic concepts, will never be overthrown.

Albert Einstein

If someone points out to you that your pet theory of the universe is in disagreement with Maxwell's equations – then so much the worse for Maxwell's equations. If it is found to be contradicted by observation – well, these experimentalists do bungle things sometimes. But if your theory is found to be against the second law of thermodynamics I can give you no hope; **there is nothing for it but to collapse in deepest humiliation.**

Arthur S. Eddington in [1] page 74

1.1   The Unique Role of Thermodynamics in Physics

We assume the student has already encountered Thermodynamics in a General Physics course. The purpose of this chapter is to review the aspects of the theory which are relevant for Statistical Mechanics and Theoretical Physics in a convenient mathematical language. The focus is on the universal concepts rather than on the phenomenology of specific thermal systems.

Thermodynamics has a unique double-faced role in Physics. From one point of view it is a "phenomenological" macroscopic description of complex systems with a huge number of degrees of freedom—typically $O(10^{25})$—in terms of a handful of collective variables (such as volume, pressure, temperature, etc) called *control parameters*.[1] From this point of view the fundamental theory is the one which describes the detailed dynamics of the underlying microscopic system. However, given the huge number of its degrees of freedom, a statistical description of its physical behavior is well justified: the statistical analysis of the dynamics of the

---

[1] They are the variables that an experimentalist may *control* i.e. tune to desired values.

microscopic degrees of freedom is the defining task of Statistical Mechanics. In this first interpretation Thermodynamics is *deduced* from the more fundamental Statistical Mechanics which in turn is rooted in the microscopic dynamical theory.

On the other hand, as the two opening quotations of this chapter emphasize, Thermodynamics is *a lot more* than a phenomenological theory: its Laws have universal validity and Thermodynamics may be interpreted as a *meta-theory,* that is, not merely a theory which describes a specific class of physical phenomena, but rather a collection of fundamental principles which should be obeyed by *all* meaningful physical theories:

> *A mathematical model is entitled to call itself a "physical theory" iff it agrees with the Laws of Thermodynamics*

Of course, a theory may be consistent with Thermodynamics and nevertheless be *totally wrong,* but it would be wrong with some degree of dignity, without *collapsing in deepest humiliation* (to use Eddington's opening words).

As suggested by Lorentz in 1903 and stressed by Einstein in 1905, Classical Physics is an example of a theory which is as good as it can possibly be as a mathematical construct, but is inconsistent with the second Law and hence—as a physical theory—had collapsed, leaving the stage to Quantum Physics (see Sect. 2.8, for the proof of inconsistency of Classical Physics).

There are other meta-theories in Physics, namely the Quantum Principle and the Relativity Principle. These principles are most conveniently framed as *formalisms.* Assuming these principles, *all physical theory should be formulable in the corresponding formal language.*[2] For instance: the Quantum Principle requires all physical theory to be formulable in the language of Hilbert spaces of states, while the Relativity Principle requires the dynamical laws to be expressed by covariant tensor equations. The same kind of formal universality holds for the Laws of Thermodynamics. They produce a Thermodynamical Formalism which applies to all physical theories which have some degree of dignity.

In the closing section of this chapter, Sect. 1.12, we illustrate the power of Thermodynamics as a meta-theory by reviewing one of its most dramatic successes: the proof by Boltzmann in 1884 of the Stefan law for the black body radiation [2].

---

[2] In special cases there may be other formalisms, but they should ultimately be equivalent—under the specific circumstances of the particular theory—with the universal ones.

1.2 Zeroth Law of Thermodynamics

Thermal Systems

Thermodynamics, as all Physics, is concerned with a class of systems [3]. The ones of interest in Thermodynamics are called *thermal systems*; sometimes we loosely call them *bodies*. A typical thermal system consists of a macroscopic quantity of "matter", with finite mass and energy, contained in a finite (but "macroscopic") volume V. We write "matter" between quotes because, in addition to ordinary matter (ultimately made of atoms and molecules, that is, of protons, neutrons, electrons,...), our system may consist of electromagnetic radiation (photons) or anything else that can carry energy or information: gravitational waves, neutrinos, dark matter,...*just anything*—Thermodynamics is truly universal! We shall loosely refer to the content of our system as "matter" (or "particles") by abuse of language. The meaning of the adjective "macroscopic" will be clarified momentarily. Thus any bounded "macroscopic" physical system defines a thermal system, while all thermal systems have an underlying fully-fledged physical system with, typically, a huge number of degrees of freedom. We refer to the degrees of freedom of the underlying system as the *microscopic* degrees of freedom, to distinguish them from the *macroscopic* (global) quantities studied in Thermodynamics.

A thermal system is *isolated* (or *closed*) if it cannot exchange "matter", energy (in any form), or any other additively conserved quantity, with other thermal systems. A quantity A is *additively conserved* if its variation ΔA in our system is equal to minus its variation in all other systems, that is,

$$\sum_{s \,\in\{\text{systems}\}} (\Delta A)_s = 0. \tag{1.1}$$

Energy is an additive conserved quantity for all systems which are independent of time, a condition that we assume throughout the book but in Chap. 6. We say that two systems are *in contact* if they can exchange such conserved quantities, in particular if they can freely exchange energy. When convenient, we think of two bodies in contact as a single thermal system containing two subsystems. Only the energy of the compound system is then conserved: this assumes that the microscopic degrees of freedom of the two subsystems are coupled by some kind of interaction. We think of this interaction as "very weak", so that it makes sense to describe the compound system in terms of the individual properties of the two component subsystems. The set of systems with which our system is in contact is called its *surroundings*.

Thermal States

The reader has already encountered the notion of *state* in Analytic Mechanics [3]. In any physical theory the state at a given time t is the set of (consistent) informations about the system at time t which are needed and suffice to determine the dynamical behavior of the system at all subsequent times $t' > t$. The notion of *thermal state* of a thermal system follows this general pattern: it consists of the specification of all parameters that are necessary and sufficient to predict the later *thermal behavior* of the system. We write $\mathfrak{T}$ for the set of thermal states of our thermal system.

We refer to the global *control* parameters which specify a thermal state $\theta \in \mathfrak{T}$ as the *macroscopic* quantities, while the parameters which specify the state $w \in \mathfrak{W}$ of the underlying dynamical system (the coordinates of its "phase space"[3] $\mathfrak{W}$) are called *microscopic*. A thermal state θ is obtained from the microscopic state $w \in \mathfrak{W}$ of the underlying physical system by a kind of "forgetful functor"

$$\mathsf{F}\colon \mathfrak{W} \to \mathfrak{T} \qquad w \mapsto \theta, \tag{1.2}$$

that is, by forgetting all but a few global properties of the underlying physical state $w \in \mathfrak{W}$. Think, for instance, of a gas composited of $O(10^{24})$ molecules. The state $w \in \mathfrak{W}$ of the underlying dynamical system is (classically!) specified by giving the positions, velocities, and rotation angles of each molecule individually, requiring $O(10^{25})$ real numbers. The macroscopic description requires much less information.

Thermodynamics is mostly concerned with a special class of thermal states, where the macroscopic description simplifies even further.

Equilibrium Thermal States
A thermal state $\theta \in \mathfrak{T}$ is a *state of equilibrium* if it does not evolve with time, that is, if the values of the parameters which specify it remain *macroscopically* constant. In a state of equilibrium $\theta \in \mathfrak{T}$ the microstate w of the underlying system will evolve with time, but remaining confined in the fiber $\mathsf{F}^{-1}(\theta) \subset \mathfrak{W}$ over θ.

We are mainly interested in the Thermodynamics of equilibrium states which produces the deepest and most universal theory. A posteriori it turns out that this theory sheds light also on the non-equilibrium Physics. All thermal states we shall consider are equilibrium states, unless explicitly stated otherwise. We write

$$\mathcal{E} \subset \mathfrak{T} \tag{1.3}$$

for the *subset of equilibrium states*. For analytic purposes it is convenient to think of the set of equilibrium thermal states as a smooth *manifold* $\mathcal{E}$ whose local coordinates are the control parameters which specify the particular equilibrium thermal state. While this is obviously an idealization, it is well justified for the typical thermal systems: for instance, the energy of the underlying system may be quantized, but when the system has a huge number of degrees of freedom, the number of quanta is so large that the energy behaves effectively as a smooth function. We also think of the space $\mathfrak{T}$ of *all* thermal states as a "manifold", but its geometry will be described only in an informal way.

---

[3] Here we use "phase space" in quotes since the term is appropriate when the underlying system is a classic mechanical system with many degrees of freedom. For other physical systems "phase space" should be understood as the appropriate space of states.

Underlying the definition of *equilibrium thermal state* there is the physical intuition of the *thermalization* process:[4] the intuition is that, if we wait long enough, eventually our system will end up in an equilibrium state, that is, as one says, it *will thermalize.* Most real systems do thermalize, but *there are exceptions* to the rule (cf. footnote 4).

The dimension $\dim \mathcal{E}$ of the submanifold $\mathcal{E}$ of equilibrium states is of order 1, and typically small (see below). A system is *macroscopic* if the number of microscopic degrees of freedom (say the dimension of the underlying "phase space" $\mathfrak{W}$) is huge with respect to $\dim \mathcal{E}$

$$\dim \mathcal{E} \lll \dim \mathfrak{W}. \tag{1.4}$$

This entails that to a typical equilibrium state $\theta \in \mathcal{E}$ there corresponds a huge number of distinct states of the underlying system called *microstates.*

Two systems in given thermal states are in *thermal equilibrium* with respect to each other *in the restrict sense* iff putting them in contact *in such a way that they can exchange **only energy*** (but not "matter", "volume", or other quantities), the combined system is also in a state of equilibrium, that is, iff there is no net flow of energy between the two systems. We consider the set $\mathfrak{E} = \amalg_s \mathcal{E}_s$ of the totality of equilibrium states for *all* thermal systems. Then

The Zeroth Law of Thermodynamics *Thermal equilibrium (in the above restrict sense!) is an equivalence relation $\sim$ in $\mathfrak{E}$.*

In plain English: if A is equilibrium with B and B is in equilibrium with C (in the restrict sense!), A is in equilibrium with C. $\mathfrak{E}$ is the disjoint union of $\sim$ equivalence classes. A $\sim$ equivalence class is called a *temperature.*

This is an unsatisfactory definition of temperature on several counts. For analytic and experimental purposes we need a notion of temperature as an *observable quantity* to which we may assign a *numerical value*, not as an abstract equivalence class. Later we shall give a fully canonical and deep definition of temperature *as an observable quantity* and check that indeed it yields a one-to-one parametrization of the space $\mathfrak{E}/\sim$ of $\sim$ equivalence classes. For the moment we content ourselves with a *conventional scale* of temperature. The obvious procedure to get a conventional definition of temperature as a numerical quantity is to choose a convenient representative in each $\sim$ equivalence class. For instance, we can take a fixed quantity of mercury in a cylindric glass with a free surface. Then the pressure on the system is fixed to be the atmospheric one, and the states of the system are

---

[4] The physical idea is that the state of an isolated system reaches an "equilibrium state" after a "sufficiently long time", which is yet another vague notion. For most systems the thermalization time is not very long on a human scale, but there are materials—such as *glass,* a non-equilibrium state of *silicium*—which can remain in their non-equilibrium state for very long times. In the case of glass at least for a few millennia since archeology museums contain ancient glass artifacts. The thermodynamics of glasses is a very subtle topic for which Giorgio Parisi got the Nobel Prize, see Chap. 7.

parametrized by the volume of mercury, i.e. by the level of the mercury column in the cylinder. Each mercury level corresponds to an unique $\sim$ equivalence class. We claim that all equivalence classes in some domain of $\mathfrak{E}/\sim$ have a representative of this form. This is easily established: putting our cylinder filled of mercury in contact with our system, and waiting a long time to allow the combined system to thermalize, the mercury will eventually reach a thermal equilibrium with our system at some level inside the tube, unless the temperature is so extreme that the mercury freezes, or evaporates, or the glass cylinder melts. Then in the appropriate domain $\subset \mathfrak{E}/\sim$ we use the level of mercury in the cylinder as a numerical parametrization of temperature. This is not quite correct since in the process the mercury will exchange energy with our system changing the thermal state of the latter (including its "temperature"): as we say, the measure instrument *back reacts on the system*. The formal definition of temperature requires the mercury to be put in contact with an infinite union of identical copies of our system so that, no matter how much energy the mercury cedes to (or acquires from) the infinite collection, the state of each copy will not change. This conventional definition of temperature is, of course, the usual thermometer.

The problem with the back reaction prompts the

Definition 1.1 The surroundings of a thermal system are a *thermal bath* iff they are in equilibrium and the back reaction of the thermal system on them vanishes.

It is a consequence of the Zeroth Law that the notion of thermal bath is well posed, and independent of the system/state chosen, as long as it belongs to the appropriate $\sim$ equivalence class. A good choice is the *Gibbs one:* infinitely many copies of the thermal system under study [4].

Remark 1.1 The only purpose of the Zeroth Law is to state a priori that "temperature is well-defined". This leads to an ugly statement in terms of an unnaturally *restricted notion* of mutual thermal equilibrium.

Thermodynamical Properties
A *thermodynamical property* (or *quantity*) is a function[5] $G \in \Omega^0(\mathcal{E})$ of the equilibrium thermal state of our system. A thermodynamical quantity is called *extensive* if it is additive under the union of identical copies of the system. For instance: volume V and energy[6] U are extensive. A quantity is called *intensive* if it invariant under the union of identical copies of the same system. A stronger notion (sometimes called a *field*) applies when the quantity is invariant under the union of *arbitrary* systems at mutual equilibrium. The Zeroth Law says that temperature T is a field (hence intensive). The pressure is also a field in isotropic systems.

---

[5] Throughout the book $\Omega^k(M)$ stands for the vector space of smooth k-forms on the smooth manifold M (same notation as in [3]). In particular $\Omega^0(M)$ is the space of zero-forms, i.e. the vector space of smooth functions on M.

[6] In Thermodynamics it is more common to call U the *internal energy*. We shall use *energy* and *internal energy* interchangeably.

The number of *functionally independent* thermodynamical quantities is $\dim \mathcal{E}$ (the dimension of the manifold $\mathcal{E}$ of equilibrium states). Many issues become trivial when $\dim \mathcal{E} = 1$, so the simplest class of interesting thermal systems are the ones with $\dim \mathcal{E} = 2$.

Definition 1.2 A thermodynamical system is called *simple* iff its manifold of equilibrium states is 2-dimensional, $\dim \mathcal{E} = 2$, i.e. if the number of functionally independent thermodynamical quantities at equilibrium is *two*.

For instance, for an inert gas in a containing vessel (so that it cannot exchange molecules with its surroundings), specification of its volume V and pressure P will suffice to determine its equilibrium thermal state. For didactical reasons we develop the Thermodynamical formalism for simple systems and then present its straightforward extension to arbitrary systems with $\dim \mathcal{E} > 2$.

Processes. Reversible Processes

A *process*[7] is a change in the thermal state from an initial equilibrium state $\theta_1 \in \mathcal{E}$ to a final equilibrium state $\theta_2 \in \mathcal{E}$. A typical thermal process is not confined to the submanifold $\mathcal{E}$ of equilibrium states. Indeed, to start the process we must perturb the initial thermal state away from equilibrium, and then the system will evolve through general thermal states, and will reach an equilibrium state only eventually at late times through thermalization. Such general processes are *irreversible*, meaning that we cannot run them backwards, going back from the final equilibrium state to the initial one. We illustrate the issue in a simple example.

Example 1.1 Consider two vessels containing water at two different temperatures $T_1 > T_2$. Pour all water in the same container. The resulting system will eventually reach an equilibrium at some intermediate temperature $T_2 < T < T_1$. Now try to revert this process getting back the two samples of water at the two distinct temperatures T_1, T_2 *without supplying energy to the system from the outside*. You will never succeed! This reverse process is impossible: the mixing of two liquids of different temperatures is an *irreversible process*.

A special class of process are the *quasi-static* ones, where the change in the state of the system happens so slowly that it has plenty of time to thermalize at each intermediate step of the process which is then an equilibrium state. Thus a quasi-static process is mathematically a curve

$$\gamma : [0, 1] \to \mathcal{E}, \quad \gamma(0) = \theta_1, \quad \gamma(1) = \theta_2. \tag{1.5}$$

We see γ as a 1-chain in $\mathcal{E}$ with boundary $\partial \gamma = \theta_2 - \theta_1$.

---

[7] A priori one may consider processes where the initial and final states are not in equilibrium. However, as $t \to +\infty$ the final state will eventually thermalize into an equilibrium state, and, dually, the initial state was in equilibrium as $t \to -\infty$. So all processes may be completed to transformations starting and ending in equilibrium states.

We are interested in a special class of *ideal* quasi-static processes, called *reversible* where both 1-chains γ and $-\gamma$ are allowed thermal processes, that is, we can go back from the final to the initial state along the same curve γ in $\mathcal{E}$. A reversible process is in particular quasi-static. The notion of reversible processes is an idealization—real processes are irreversible—but a very useful one which allows to construct a deep theory which afterwards can be applied to irreversible processes as well.

The reversible process γ is called a *(thermal) cycle* if the 1-chain γ is a cycle in the sense of singular homology, i.e. $\partial\gamma = 0$, which means that the initial and final states coincide $\theta_2 \equiv \theta_1$. In Thermodynamics one assumes $H_1(\mathcal{E}, \mathbb{Z}) = 0$, so that all cycles are boundaries.

1.3   First Law of Thermodynamics

Work

Let our *simple* thermal system occupy a bounded convex region B in the space $\mathbb{R}^3$ of volume V (see Fig. 1.1) whose boundary ∂B is diffeomorphic to S^2. In polar coordinates, centered at some point $0 \in B$, the region is given by

$$B = \{r \leq f(\theta, \phi)\} \subset \mathbb{R}^3, \tag{1.6}$$

for some function $f(\theta, \phi)$ on S^2 which specifies the shape of the boundary $\partial B \simeq S^2$ of the system in $\mathbb{R}^3$. The volume of B is

$$V = \int_B r^2 \mathrm{d}r \wedge \sin\theta\, \mathrm{d}\theta \wedge \mathrm{d}\phi = \frac{1}{3} \int_{S^2} f(\theta, \phi)^3 \sin\theta\, \mathrm{d}\theta \wedge \mathrm{d}\phi \tag{1.7}$$

We think $\mathbb{R}^3$ as filled by some medium which plays the role of *surroundings* for our system. The body B produces a force on the medium through its boundary which (for isotropic media) is normal to the boundary, directed outward, and has a surface intensity equal to the pressure P per unit area of the boundary. Thus the intensity of the normal force at the point (θ, ϕ) of the boundary is given by the 2-form

$$P\,\mathrm{d}\Sigma = P\, f(\theta, \phi)^2 \sin\theta\, \mathrm{d}\theta \wedge \mathrm{d}\phi. \tag{1.8}$$

Fig. 1.1 A body B which increases its volume makes work on its surroundings

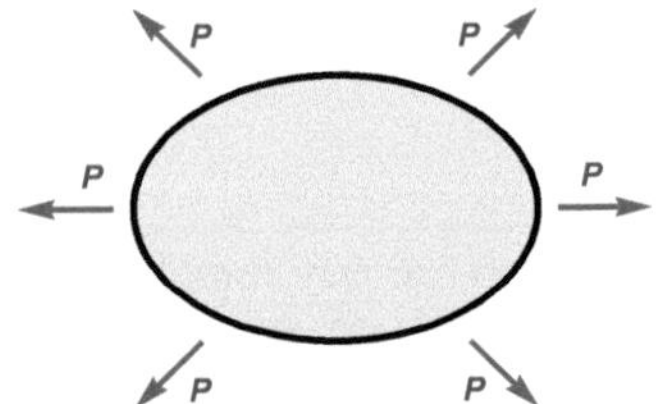

Except for the non-negativity of the pressure, $P \geq 0$, all these statements are consequences of the principle of virtual works [3, Chap. 1].

Consider a process of expansion of the system, where the boundary $f(\theta, \phi)$ moves outward. The virtual work done in an infinitesimal change $\delta f(\theta, \phi)$ of the boundary shape is the integral over S^2 of the density of force (1.8) at the boundary point (θ, ϕ) times its local infinitesimal displacement $\delta f(\theta, \phi)$. The total virtual work is then

$$\delta W = \int_{S^2} \delta f(\theta, \phi)\, P\, f(\theta, \phi)^2\, \sin\theta\, d\theta \wedge d\phi = P\, dV. \tag{1.9}$$

The equality $\delta W = P\, dV$ remains true for all isotropic homogeneous bodies (which have the property that at equilibrium the pressure P is equal everywhere) for any topology of their boundary ∂B which needs not to be diffeomorphic to a sphere.

We focus on a *reversible* process $\gamma : [0, 1] \to \mathcal{E}$ of our simple thermal system. The total work done by the system in the process γ is[8]

$$W(\gamma) = \int_0^1 \gamma^* P\, dV. \tag{1.10}$$

We stress that the work depends on the process γ and not only on its initial and final states $\partial\gamma = \theta_2 - \theta_1$: indeed

$$d(\gamma^* P\, dV) = \gamma^*(dP \wedge dV) \neq 0, \tag{1.11}$$

because, in a simple system P, V are independent coordinates on $\mathcal{E}$. Equation (1.11) holds a fortiori for non-simple systems where the number of independent parameters is greater than 2. In particular, when γ is a cycle

$$W(\text{cycle}) = \int_\gamma P\, dV = \int_\Sigma dP \wedge dV \tag{1.12}$$

where $\Sigma \subset \mathcal{E}$ is a 2-chain with $\partial\Sigma = \gamma$. We see from (1.12) that the closed 2-form $dP \wedge dV$ has a physical meaning, hence it is natural to see it as a geometric structure on the equilibrium manifold $\mathcal{E}$ of the thermal system.

Corollary 1.1 *The manifold of equilibrium states of a* simple *thermodynamical system is a* symplectic manifold *equipped with the natural symplectic form* $dP \wedge dV$.

Symplectic manifolds are reviewed in Appendix 1: the geometric notions we use are defined there. Note that one Darboux coordinate, P, is intensive and the other one, V, is extensive.

---

[8] The notation $\gamma^* P\, dV$ stands for the pull-back of the 1-form $P\, dV$ by the smooth map γ cf. [3].

On the other hand the (internal) energy U of the system is a globally defined function on the manifold $\mathcal{E}$, so its variation in the process γ is simply

$$\Delta U = \int_0^1 \gamma^* dU = \int_\gamma dU = \int_{\partial \gamma} U \equiv U(\theta_2) - U(\theta_1), \tag{1.13}$$

and depends only on the initial and final states of γ. The First Law of Thermodynamics states that *energy is conserved,* where "energy" is the sum of all possible kinds of energies: mechanic, thermic, electric, etc. Conservation of the *total* energy is a general property of all time-invariant[9] physical system. Consider the quantity

$$\Delta U + W \equiv \int_0^1 \gamma^* (dU + P\, dV), \tag{1.14}$$

which is the variation of the energy of the system plus the work the system has done, that is, the amount of mechanic energy the body has transferred to its surroundings. If the mechanic energy was the *only* form of energy, we would conclude from the conservation law that the quantity (1.14) should be zero *for all* 1-chain γ, i.e.

$$dU + P\, dV \overset{?}{=} 0 \quad \text{as an equality in } \Omega^1(\mathcal{E}), \tag{1.15}$$

but this conclusion would lead to a contradiction: indeed

$$d(dU + P\, dV) = dP \wedge dV \neq 0 \tag{1.16}$$

because $\mathcal{E}$ is symplectic for a simple system. We are forced to conclude that, during the process γ, the surroundings have supplied a quantity $\Delta U + W$ of energy to the system *by means different from mechanical work.* We call *heat* this "new" form of energy which is required to make the conservation of energy satisfied. The First Law of Thermodynamics then says that the quantity of heat supplied to the system during the process γ is

$$Q_\gamma = \Delta U + W = \int_0^1 \gamma^* (dU + P\, dV), \tag{1.17}$$

or, in differential form,

$$dQ = \gamma^* P\, dV + d(\gamma^* U) \quad \text{equality in } \Omega^1([0, 1]) \tag{1.18}$$

usually written (a bit abusively) as

$$\delta Q = P\, dV + dU \tag{1.19}$$

---

[9] Cf. the Noether theorem [3].

where the notation is meant to stress that the 1-form $\delta Q \in \Omega^1(\mathcal{E})$ is **not** the differential of any function $Q \in \Omega^0(\mathcal{E})$. Of course, δQ becomes exact when pulled back to $[0, 1]$ because all 1-forms are exact in $\Omega^1([0, 1])$.

The situation with non-simple systems is similar. Besides mechanical energy and heat, a system may exchange various kinds of additive (extensive) conserved quantities a_i with its surroundings. The quantity of energy transferred to the surroundings by a variation da_i of the extensive quantity a_i has the form $b_i\, da_i$ for some intensive quantity b_i "dual" to a_i as we shall see in Sect. 1.10. Then the First Law reads in general

$$\delta Q = \mathrm{d}U + P\,\mathrm{d}V + \sum_i b_i\, \mathrm{d}a_i. \tag{1.20}$$

1.4 Second Law of Thermodynamics

Traditional Nineteenth Century Statements
There are two classical statements of the Second Law of Thermodynamics that we recall to make contact with the history of the subject.

Second Law: Clausius Formulation *No cyclic process can exist whose only effect is to transfer heat from a body at a temperature T_2 to a body at temperature $T_1 > T_2$.*

Second Law: Kelvin Formulation *A cyclic process whose only effect is the transformation of heat taken from a body at fixed temperature T into work is impossible.*

The two statements are equivalent.

Clausius Implies Kelvin If we could produce work from a single source at temperature T_2 we can use that work to heat up (say by friction) a body at any temperature T_1. Choosing $T_1 > T_2$ we get a contradiction with the Clausius principle. $\square$

Kelvin Implies Clausius To show the opposite implication Kelvin $\Rightarrow$ Clausius, we use the *Carnot cycle,* defined as a reversible cycle $\gamma \subset \mathcal{E}$ along which the system exchanges heat with the surroundings at *two* different temperatures $T_1 > T_2$. We shall see later that Carnot cycles exist. At the higher temperature the system absorbs a quantity Q_1 of heat, while at the lower temperature it releases a quantity Q_2 of heat. The mechanical work produced in the cycle is then $Q_1 - Q_2$ by conservation of energy. If we can transfer the heat Q_2 from the body at the lower temperature T_2 to the body at temperature T_1 we would produce work out of heat extracted only at the fixed temperature T_1, getting a contradiction with Kelvin's principle. $\square$

Carnot Theorem
The *thermal efficiency* η of a thermal machine working in cycle between the two temperatures T_1, T_2 is the ratio of the work W produced in a cycle to the heat Q_1 supplied to the system at the higher temperature T_1 (the heat Q_2 released at the lower temperature T_2 is a total loss). Since the internal energy U does not change in

a cycle, the First Law gives $W = Q_1 - Q_2$, and

$$\eta \equiv \frac{W}{Q_1} = \frac{Q_1 - Q_2}{Q_1} < 1. \qquad (1.21)$$

Remark 1.2 A Carnot cycle is composed by the sequence of four *reversible* processes: the first one is isothermal at the fixed temperature T_1, in the second process the temperature is lowered from T_1 to T_2 *adiabatically* i.e. without exchange of heat with the surroundings, the third process is isothermal at temperature T_2, and finally the temperature is raised back to T_1 by a second adiabatic transformation.

Carnot Theorem *The efficiency η of a thermal machine working in cycle between two given temperatures $T_1 > T_2$ is maximal when the thermal machine works in a reversible cycle γ, and is equal for all reversible machines working between the same two temperatures.*

Proof We consider an arbitrary machine of efficiency $\eta < 1$ and a reversible machine of efficiency $\xi < 1$ working in cycle between the same temperatures T_1, T_2. In each cycle the first machine absorbs a quantity Q_1 of heat at the higher temperature T_1, produces a work ηQ_1, and releases the heat $(1 - \eta)Q_1$ at the lower temperature T_2. We run the reversible machine *in reverse* (that is, as a refrigerator): in each cycle we supply to the reversible machine a quantity Q_2 of heat at the lower temperature T_2, and also a positive amount of work $\xi Q_2/(1-\xi)$, so that the machine will release $Q_2/(1-\xi)$ heat at the higher temperature T_1. We tune the two machines so that the heat released by the first machine at T_2 is equal to the heat absorbed at the same temperature by the reversible machine

$$Q_2 = (1 - \eta)Q_1, \qquad (1.22)$$

so that the combined machine does not exchange heat with the surroundings at temperature T_2. The combined machine absorbs at T_1 a net quantity of heat equal to

$$Q_1 - \frac{Q_2}{1-\xi} = Q_1 \left(1 - \frac{1-\eta}{1-\xi} \right) \qquad (1.23)$$

which is equal to the work done in the cycle by conservation of energy. This work cannot be positive by Kelvin's version of the Second Law. Thus

$$\left(1 - \frac{1-\eta}{1-\xi} \right) \leq 0 \quad \Rightarrow \quad \xi - \eta \geq 0, \qquad (1.24)$$

which is the first statement in Carnot's theorem. If both machines are reversible, inverting their role in the argument, we get the two inequalities $\xi - \eta \geq 0$ and $\eta - \xi \geq 0$, so $\eta = \xi$. $\qquad\qquad\square$

Absolute Thermodynamical Temperature
Carnot's theorem states that efficiency ξ of a reversible thermal machine which works between two temperatures T_1, T_2 is *a universal function $\xi(T_1, T_2)$ of the two temperatures.* Suppose we have two reversible machines working, respectively, between temperatures T_1, T_2 and T_2, T_3. We may combine the two machines to make a reversible device working between temperatures T_1 and T_3 by using the heat $(1 - \xi(T_1, T_2))Q_1$ released by the first machine at temperature T_2 to feed the second one which then releases the heat

$$(1 - \xi(T_1, T_2))(1 - \xi(T_2, T_3))Q_1 \tag{1.25}$$

at the lower temperature T_3. Since the compound is again a reversible machine, now working between temperatures T_1 and T_3, the amount of heat released at the lower temperature must be $(1 - \xi(T_1, T_3))Q_1$. Writing $f(T_1, T_2) \equiv 1 - \xi(T_1, T_2)$, we get

$$f(T_1, T_3) = f(T_1, T_2)\, f(T_2, T_3) \quad \Rightarrow \quad f(T_1, T_2) = g(T_2)/g(T_1) \tag{1.26}$$

that is,

$$\xi(T_1, T_2) = \frac{g(T_1) - g(T_2)}{g(T_2)} \tag{1.27}$$

for some *universal* function $g : \mathbb{R} \to \mathbb{R}$. Until now the temperature T was defined in purely conventional terms (by means of a "thermometer") and the temperature scale was unique only up to a reparametrization $T \rightsquigarrow f(T)$ by an arbitrary function $f(\cdot)$. The Carnot theorem yields an absolute definition of temperature by choosing the redefinition $T \rightsquigarrow g(T)$ where g is the above *universal function.*

Definition 1.3 The *absolute thermodynamical temperature* T is defined by the property that the efficiency of *any* reversible thermal machine working between absolute temperatures T_1, T_2 is $(T_1 - T_2)/T_1$.

Note that with this definition the temperature T is *non-negative.* The point $T = 0$ is called the *absolute zero.* From now on the symbol T will always stand for the *absolute thermodynamical temperature.*

Entropy

The Second Law implies the existence of an universal function $S \in \Omega^0(\mathcal{E})$ called *entropy.*

Theorem 1.1

(1) In a general thermal cycle γ we have

$$\oint_\gamma \frac{\delta Q}{T} \leq 0. \tag{1.28}$$

*(2) If $\gamma \subset \mathcal{E}$ is a **reversible** cycle, one has*

$$\oint_\gamma \frac{\delta Q}{T} \equiv \int_0^1 \gamma^* \left(\frac{dU}{T} + \frac{P}{T} dV \right) = 0, \tag{1.29}$$

that is,

$$d \left(\frac{dU}{T} + \frac{P}{T} dV \right) = 0 \quad \text{equality in } \Omega^2(\mathcal{E}). \tag{1.30}$$

Proof Statement **(1)** is true for a cycle working between two temperatures: indeed

$$\eta \equiv \frac{Q_1 - Q_2}{Q_1} \underset{\leq}{\leq} \frac{T_1 - T_2}{T_1} \quad \Rightarrow \quad \frac{Q_1}{T_1} - \frac{Q_2}{T_2} \leq 0. \tag{1.31}$$

Suppose now that our system undergoes a closed cycle γ composed of several isothermal processes at various temperature $T_1, T_2, \ldots, T_s$ connected by adiabatic processes. We assume $T_1 \geq T_i$ for all i with no loss. In the isothermal process at temperature T_i the system exchanges a quantity Q_i of heat with its surroundings (Q_i is positive if the heat is taken in, negative if it is let out). Now we form a thermal machine by adding to our system working in cycle one Carnot machine for each $i \geq 2$ which works between temperatures T_1 and T_i: if Q_i is positive the i-th Carnot machine will work in the direct way, releasing Q_i heat at T_i and absorbing $T_1 Q_i / T_i$ at T_1; if Q_i is negative the i-th Carnot machine will work in the reverse way, taking in a quantity $|Q_i|$ of heat at temperature T_i and releasing a quantity $T_1 |Q_i| / T_i = -T_1 Q_i / T_i$ of heat at the higher temperature T_1.

The compound machine now works at a single temperature T_1, so the amount of work it produces must be non-positive by Kelvin's Law. The work is equal to the total quantity of heat supplied to the combined machine at temperature T_1, so that

$$Q_1 + T_1 \sum_{i=2}^{s} \frac{Q_i}{T_i} = T_1 \sum_{i=1}^{s} \frac{Q_i}{T_i} \leq 0 \tag{1.32}$$

Now **(1)** follows from the fact that all reversible cycles $\gamma \subset \mathcal{E}$ can be approximated, as well as we wish, by a piece-wise sequence of isothermal and adiabatic processes: just subdivide γ in small arcs where the temperature is constant up to order ϵ. **(2)** follows from considering the reversed cycle $-\gamma$, getting the opposite inequality. $\square$

The Poincaré lemma applied to Eq. (1.30) yields

Corollary 1.2 *There exists a function* $S \in \Omega^0(\mathcal{E})$*, called* entropy*, such that*

$$\mathrm{d}S = \frac{1}{T}\left(\mathrm{d}U + P\,\mathrm{d}V\right) \quad \text{equality in } \Omega^1(\mathcal{E}) \tag{1.33}$$

while for a general process (possibly irreversible)

$$\mathrm{d}S \geq \frac{\delta Q}{T}. \tag{1.34}$$

Equation (1.33) *implies*

$$\frac{\partial S(U, V)}{\partial U} = \frac{1}{T}, \qquad \frac{\partial S(U, V)}{\partial V} = \frac{P}{T}. \tag{1.35}$$

Classically S is well defined only up to an additive constant. In Chap. 2 we shall see that at the quantum level the entropy is uniquely determined. The entropy is an extensive quantity, in facts additive, since U, V are extensive and P, T intensive.

In a closed system $\delta Q = 0$, therefore:

Corollary 1.3 *The entropy of a closed system cannot decrease.*

In facts most real physical processes will increase the entropy since the reversible processes are idealizations. In the words of Clausius:

> *The energy of the universe is constant*
> *The entropy of the universe tends to a maximum*

The Principle of Maximal Entropy

Consider an isolated system contained in a rigid "box". The values of all additive *conserved* quantities (including the quantity of "matter" and energy U) and the volume V are fixed. For most systems there is a maximum value of the entropy compatible with the given values of V, U and the other conserved quantities. The system will increase its entropy until it reaches this maximal value, and then should stop there because it cannot go anywhere else by Corollary 1.3. But a state which does not evolve is a state of equilibrium. We conclude

Corollary 1.4 (Principle of Maximal Entropy) *A state of maximal entropy of an isolated system (if it exists) is an equilibrium state.*

This principle vindicates the intuitive idea of *thermalization:* just wait until the entropy gets its maximum, and the system eventually will be in equilibrium.

Back to the Zeroth Law

Consider two bodies which are put in contact so that only the total energy $U = U_1 + U_2$ is conserved while, for simplicity, we keep fixed their volumes. Since the entropy is additive, the entropy of the combined system is

$$S_1(U_1, V_1) + S_2(U - U_1, V_2). \tag{1.36}$$

At equilibrium the energy is distributed between the two bodies in such a way that the entropy is maximal. In particular the first variation with respect to U_1 must vanish:

$$0 = \frac{\partial S_1}{\partial U_1} - \frac{\partial S_2}{\partial U_1} = \frac{1}{T_1} - \frac{1}{T_2}, \tag{1.37}$$

that is, at equilibrium the (thermodynamical) temperatures of the two bodies in contact must be equal, vindicating the Zeroth Law.

Remark 1.3 The principle of maximal entropy will be put on firm math grounds in Sect. 2.4 in the context of the information-theoretical interpretation of entropy.

1.5 The Language of Symplectic Geometry

In this section we state the relations between the various thermal quantities in the math language of *symplectic geometry* introduced in Analytic Mechanics [3]. For the benefit of the reader we have collected the relevant definitions in Appendix 1.

Simple Systems

For a simple thermal system the First and Second Laws combine in the differential relation

$$T\,\mathrm{d}S = \mathrm{d}U + P\,\mathrm{d}V \quad \text{equality in } \Omega^1(\mathcal{E}) \tag{1.38}$$

which gives

$$\mathrm{d}T \wedge \mathrm{d}S = \mathrm{d}P \wedge \mathrm{d}V. \tag{1.39}$$

Corollary 1.5 *The Second Law of thermodynamics (for simple systems) states that the change of coordinates*

$$(V, P) \rightsquigarrow (S, T) \tag{1.40}$$

of the symplectic manifold $\mathcal{E}$ of equilibrium states is a canonical transformation in the sense of Analytic Mechanics [3, Chap. 8].

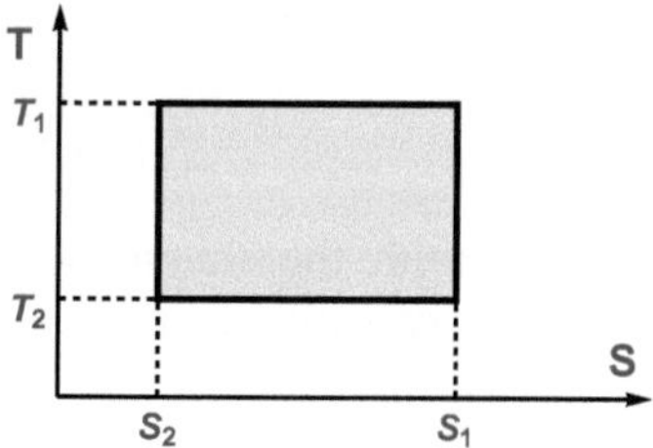

Fig. 1.2 The image of the Carnot cycle under the canonical map $(V, P) \rightsquigarrow (S, T)$. The work produced in a cycle is the area in gray

In the analogy with mechanics we take the extensive Darboux quantities V, S to play the roles of the q's and their dual intensive Darboux quantities P, T to correspond to conjugate momenta. The canonical transformation maps the Carnot cycle in a rectangle in the S-T plane, Fig. 1.2. This figure shows that a cycle $\gamma \subset \mathcal{E}$ realizing the Carnot cycle does exist.

Comparing with the Poincaré-Cartan description of canonical transformations in Mechanics [3, Chap. 8], we identify the internal energy U, written as a function of the old coordinate V and the new coordinate S, with the generating function of the first kind of the canonical transformation $(V, P) \rightsquigarrow (S, T)$. From this identification we get the equations

$$T = \frac{\partial U(S, V)}{\partial S}, \qquad P = -\frac{\partial U(S, V)}{\partial V}, \tag{1.41}$$

for their dual "momenta" which have the same content as the First/Second Laws (1.38).

More intrinsically (cf. [3, Chap. 7]) we can consider the manifold $Y \simeq \mathbb{R}^2$ parametrized by the additive quantities S and V and consider the 4-dimensional symplectic manifold $X = T^*Y$ with the twisted tautological form $\Omega = T\,dS - P\,dV$:

Corollary 1.6 *The manifold $\mathcal{E}$ of equilibrium states is a Lagrange submanifold of $(X, d\Omega)$ which is a section of $T^*Y \to Y$.*

A section whose graph is a Lagrangian submanifold is a closed 1-form on Y.[10] In the case of $\mathcal{E}$ the closed form is $dU(S, V)$. Comparing with [3] we see that the internal energy $U(S, V)$ (written as a function of S and U) is a *generating function* of the first kind for the Lagrangian submanifold $\mathcal{E}$.

As in Mechanics, the ultimate formulation is in terms of contact geometry á la Poincaré-Cartan. For a general thermal system we interpret the First Law (1.20), together with the second one as the definition of the *contact form*

$$\kappa = dU - T\,dS + P\,dV + \sum_i b_i\,da_i \tag{1.42}$$

---

[10] See [3] Proposition 7.1.

in the 'thermal contact manifold' $\mathcal{T}$ with Darboux coordinates

$$(U, S, V, a_1, \ldots, a_s, T, P, b_1, \ldots, b_s), \tag{1.43}$$

together with the statement that the submanifold of equilibrium states $\mathcal{E}$ is a *Legendre submanifold* $\iota\colon \mathcal{E} \hookrightarrow \mathcal{T}$ for the contact structure κ defined by the two Laws (see Appendix 1).

The Equations of State

The physics of a thermal system is specified by its *equations of state,* that is, by the functional relations between the several thermal quantities which hold for all equilibrium states. In mathematical terms

Definition 1.4 The *equations of state* are the equations which describe the Legendre submanifold $\iota\colon \mathcal{E} \hookrightarrow \mathcal{T}$ of the contact form κ.

Traditionally by "equation of state" (for a simple system) one means the expression of $T \in \Omega^0(\mathcal{E})$ as a function of P and V, i.e. an equation of the form $T = T(P, V)$. But this information may not be enough to recover $\mathcal{E}$, even if we use the fact that it is Legendrian.

Example 1.2 The "traditional" equation of state for a ideal gas is

$$T = \rho\, P\, V \tag{1.44}$$

where ρ is a constant which depends on the particular sample of gas. Then the Lagrangian condition (1.30) yields

$$0 = \mathrm{d}T^{-1} \wedge dU + \mathrm{d}\left(\frac{P}{T}\right) \wedge \mathrm{d}V = \mathrm{d}T^{-1} \wedge \mathrm{d}U + \rho\, \mathrm{d}V^{-1} \wedge \mathrm{d}V = \mathrm{d}T^{-1} \wedge \mathrm{d}U \tag{1.45}$$

which says that U is a function of T only. But it does not says which function of T, a datum which is a major part of the thermal physics of the system. Instead, if we give the generating function of the canonical transformation

$$U(V, S) = \alpha\, e^{S/\alpha}\, V^{-\beta} \tag{1.46}$$

(α, β positive constants which depend on the particular ideal gas), we get

$$T = \frac{\partial U}{\partial S} = \frac{1}{\alpha} U(V, S), \qquad P = -\frac{\partial U}{\partial V} = \beta \frac{U(V, S)}{V} \tag{1.47}$$

two equations of state which fully determine $\mathcal{E}$ and the physics

$$P\, V = \alpha\beta\, T, \qquad U = \alpha\, T. \tag{1.48}$$

1.6 Thermodynamical Potentials (Simple Systems)

In the most general sense, a *thermodynamical potential* Φ is a thermal quantity, written as a function of certain *natural variables* $y_1, \ldots, y_s$ ($s = \dim \mathcal{E}$), from which we can read the thermal physics at equilibrium, i.e. determine the equations of state, that is, the submanifold $\mathcal{E}$ of equilibrium states. The energy $U(S, V)$ as a function of entropy and volume is a first example of thermodynamical potential (for a simple system). Most crucially, the thermodynamical potentials provide a *variational characterization* of the equilibrium states in the bigger space $\mathcal{T}$ of all thermal states as the *ones which extremize the potential*.[11] A very fundamental property of the thermodynamical potentials is that they are "convex" for "sound" thermal systems which do thermalize to equilibrium states. In this context "convex" refers to the fact that the second derivatives of the thermodynamical potential satisfy an inequality which may be of the form ≥ 0 (convex) or of the form ≤ 0 (concave) depending on the sign conventions[12] we use in the definition of Φ. Another general property of the potentials is that a priori *they are well defined only up to the addition of a constant,* just as the mechanical potentials.

We start from the most "classical" potentials usually referred to as "the *four* thermodynamical potentials*".

The Four *Thermodynamical Potentials*

In Mechanics a generic Lagrangian submanifold may be equivalently described by four generating functions called of the first, second, third, and fourth kind, each one written in terms of its natural variables [3, §. 8.6]. We have already seen that the internal energy as a function of entropy and volume, $U(S, V)$, is a generating function of the *first kind*. The generating functions of the other three kinds are related to $U(S, V)$ by a Legendre transform [3, Chaps. 6, 7 and 8] and contain exactly the same physical information—that is, they describe the same equations of state $\mathcal{E}$— but may be more convenient for the particular application at hand.

Definition 1.5 [For a simple system] the four generating functions of the Lagrangian submanifold $\mathcal{E}$ are called *the four thermodynamical potentials:*

First kind: $U(S, V)$ *internal energy* as a function of S and V

Second kind: $F(T, V)$ *free energy*[13] as a function of T and V

---

[11] This is the historical reason for the term "potential". In mechanics a configuration which extremizes the potential is stationary, and this property is shared by the thermodynamical potentials. This is not mere analogy: the path integral formulation identifies the (Euclidean) effective quantum potential with a thermodynamic potential.

[12] As almost everybody, we adhere to the *historical* sign conventions which look pretty bizarre to the modern eye.

[13] Also called the *Helmholtz free energy* to distinguish it from the Gibbs free energy.

Third kind: $H(S, P)$ *enthalpy* as a function of S and P
Fourth kind: $G(T, P)$ the *Gibbs free energy* as a function of T and P.

As in Mechanics, one passes from one potential to another one by a Legendre transform in the appropriate variable(s). Thus

$$F(T, V) = U(S(T, V), V) - T S(T, V), \quad T \equiv \left.\frac{\partial U(S, V)}{\partial S}\right|_{S=S(T,V)} \tag{1.49}$$

$$H(S, P) = U(S, V(S, P)) + P V(S, P), \quad P \equiv -\left.\frac{\partial U(S, V)}{\partial V}\right|_{V=V(S,P)} \tag{1.50}$$

$$G(T, P) = H(S(T, P), P) - T S(T, P) \quad T \equiv \left.\frac{\partial H(S, P)}{\partial S}\right|_{S=S(T,P)} \tag{1.51}$$

which entail the differential relations

$$dU = T\, dS - P\, dV \quad \Rightarrow \quad T = \left(\frac{\partial U}{\partial S}\right)_V \qquad P = -\left(\frac{\partial U}{\partial V}\right)_S \tag{1.52}$$

$$dF = -S\, dT - P\, dV \quad \Rightarrow \quad S = -\left(\frac{\partial F}{\partial T}\right)_V \qquad P = -\left(\frac{\partial F}{\partial V}\right)_T \tag{1.53}$$

$$dH = T\, dS + V\, dP \quad \Rightarrow \quad T = \left(\frac{\partial U}{\partial S}\right)_P \qquad V = \left(\frac{\partial U}{\partial P}\right)_S \tag{1.54}$$

$$dG = -S\, dT + V\, dP \quad \Rightarrow \quad S = -\left(\frac{\partial G}{\partial T}\right)_P \qquad V = \left(\frac{\partial G}{\partial P}\right)_T \tag{1.55}$$

where we used the traditional thermodynamical notation for derivatives:

$$\left(\frac{\partial A}{\partial B}\right)_C \quad \text{for} \quad \frac{\partial A(B, C)}{\partial B} \tag{1.56}$$

standing for the derivative of the function A with respect to the variable B while keeping C constant. Equations (1.52)–(1.55) show the

Fact 1.2 *One can read the thermal physics of the system (that is, recover the equations of state $\mathcal{E}$) from any one of the thermal potentials written in its natural variables.*

More Precise Math Definitions
We need to make the Legendre definition of the four thermodynamical potentials more precise and in particular *globally* well-defined in the space of thermal parameters. As we shall see momentarily, their precise definition is *variational* in nature, and the variational interpretation of the potentials is quite useful for analytic purposes.

We start with some refinements of the Legendre transform. In [3] §.6.1 we defined the Legendre transform to act on convex stable functions $f : V \to \mathbb{R}$, where V is a real vector space, to return a convex stable function $g : V^\vee \to \mathbb{R}$, where $V^\vee$ is the vector space dual to V. In sound physical systems, thermal quantities like T, V, P, U, S, do not take value in a real vector space: rather each of them takes value in the positive half-line[14] $\mathbb{R}_{\geq 0}$ and a set of k such quantities "in involution"[15] takes value in the 2^k-tant $(\mathbb{R}_{\geq 0})^k$. Therefore we need a more general notion of Legendre transform.

Definition 1.6

(1) A *convex cone* $C \subset V$ in the real vector space V is a subset such that

$$x_1, x_2 \in C \quad \Rightarrow \quad a\,x_1 + b\,x_2 \in C \text{ for all } a, b \geq 0. \tag{1.57}$$

The convex cone is called *strict* iff $x \in C$ and $x \neq 0$ implies $-x \notin C$. For instance 2^k-tants $(\mathbb{R}_{\geq 0})^k$ are strict convex cones.

(2) if C is a *strict* convex cone, the dual cone $C^\vee \subset C$ is the strict convex cone

$$C^\vee = \left\{ y \in V^\vee : \langle y, x \rangle \geq 0 \ \forall\, x \in C \right\}. \tag{1.58}$$

For instance $(\mathbb{R}_{\geq 0})^k$ is its own dual.

(3) A convex function f on a strict convex cone C is *stable* if the function $\langle y, x \rangle - f(x)$ has a unique maximum for all $y \in C^\vee$.

(4) If $f : C \to \mathbb{R}$ is a convex stable function defined on the strict convex cone C, its *Legendre transform* $g : C^\vee \to \mathbb{R}$ is the (globally defined) convex stable function on the dual convex cone $C^\vee$

$$g(y) = \max_{x \in C} \left(\langle y, x \rangle - f(x) \right). \tag{1.59}$$

Remark 1.4 The Legendre transform for a dual pair of vector spaces (resp. for a dual pair of strict convex cones) is the asymptotic limit of the Fourier (resp. Laplace) transform, see Sect. 2.5 for details.

The thermodynamical potentials of a simple system are functions on the quadrant $\mathbb{R}^2_{\geq 0}$ which are stable convex or concave (depending on sign conventions) and

---

[14] When defined with the appropriate additive constant.

[15] Two thermodynamical quantities y_1, y_2 are *in involution* iff they are in involution with respect to the symplectic form $d\kappa|_{\mathcal{E}}$. We stress that the set of natural variables on which a potential depends are always in mutual involution.

related by Legendre transforms whose precise definitions are

$$F(T, V) \stackrel{\text{def}}{=} \min_{S}(U(S, V) - TS) \tag{1.60}$$

$$H(S, P) \stackrel{\text{def}}{=} \min_{V}(U(S, V) + PV) \tag{1.61}$$

$$G(T, P) \stackrel{\text{def}}{=} \min_{S,V}(U(S, V) + PV - TS) \tag{1.62}$$

These variational characterizations look very natural from the mechanical analogy
(cf. Hamilton's and Maupertuis' principles [3]). In Chap. 2 we explain the variational characterization of the potentials from first principles in the "microscopic"
description. Here we present a traditional thermodynamic argument for their
validity.

Physical Justification We have to show that in a "good" thermal system, that is, *in
a system capable of thermalizing into an equilibrium state,* the internal energy U at
equilibrium is a convex function of S and V. We start from the dual of Corollary 1.4.

Corollary 1.7 (Principle of Minimal Energy) *Suppose we have a thermal system
which is kept at fixed entropy and volume but can exchange energy with its
surroundings. If the energy U is minimal for the given values of S and V, the system
is in equilibrium.*

Indeed from the second principle we have

$$\frac{\delta Q}{T} \leq \Delta S = 0, \tag{1.63}$$

so the only allowed processes are the ones in which the system transfers energy to
its surroundings. If the energy has a minimal value, the system cannot give away
any of its energy, and its state is stationary.

Theorem 1.3 *In a thermal system which is capable to thermalize into an equilibrium state when put in contact with any thermal bath, the internal energy $U(S, V)$ is
a convex function of its natural variables, in facts a stable function in the Legendre
sense.*[16]

Proof Take two identical copies of the system in the same equilibrium state. Put
them in contact through a movable boundary (so that only the total volume $V =
V_1 + V_2$ is fixed) in a way that the total entropy $S = S_1 + S_2$ is kept constant. The
energy of the total system is

$$U = U(S_1, V_1) + U(S - S_1, V - V_1). \tag{1.64}$$

---

[16] That is, the function $f_T(S, V) \equiv U(S, V) - TS$ has a minimum for all $T > 0$, cf. [3] §. 6.1.

By Corollary 1.7, at equilibrium the distribution of volume and entropy between the two copies minimizes the energy U. The first variation of U vanishes when $S - S_1 = S_1$, $V - V_1 = V_1$ corresponding to the fact that when the two identical systems are in the same equilibrium state, their compound state is also in equilibrium. The condition that this equilibrium state is a minimum of the free energy implies that the Hessian

$$\begin{pmatrix} \frac{\partial^2 U}{\partial S^2} & \frac{\partial^2 U}{\partial S \partial V} \\ \frac{\partial^2 U}{\partial S \partial V} & \frac{\partial^2 U}{\partial V^2} \end{pmatrix} \tag{1.65}$$

is a positive definite matrix, that is, we have the three inequalities

$$\frac{\partial^2 U}{\partial S^2} \geq 0, \qquad \frac{\partial^2 U}{\partial V^2} \geq 0, \qquad \frac{\partial^2 U}{\partial V^2} \frac{\partial^2 U}{\partial S^2} \geq \left(\frac{\partial^2 U}{\partial S \partial V} \right)^2 . \tag{1.66}$$

$\square$

The *Twelve* Potentials (Simple Systems)

As already mentioned, it is more natural to look to Thermodynamics from the viewpoint of contact geometry *á la* Poincaré-Cartan. For simple systems, the first and second Laws together say that the manifold of equilibrium states $\mathcal{E}$ is a *Legendre submanifold* of the space parametrized by the five thermal quantities (U, T, S, P, V) with respect to the contact form

$$\kappa = \mathrm{d}U + P\,\mathrm{d}V - T\,\mathrm{d}S. \tag{1.67}$$

The thermal quantities (U, T, S, P, V) are Darboux coordinates of the space of thermal quantities in the sense of contact geometry (see Appendix 1). This entails that the restriction to $\mathcal{E}$ of the three functions U, V, S are functionally dependent

$$F(U, V, S) = 0 \quad \text{on } \mathcal{E}, \tag{1.68}$$

a relation which may be made explicit in three different ways

$$U = U(S, V), \quad S = S(U, V), \quad \text{or } V = V(S, U). \tag{1.69}$$

Each one of these functions (written in terms of its natural variables) allows to reconstruct the equation of $\mathcal{E}$, i.e. fully determines the thermal physics: hence they also are thermodynamic potentials. For instance, the entropy written as a function of energy and volume $S(U, V)$ satisfies

$$\frac{\partial S(U, V)}{\partial U} = \left(\frac{\partial U(S, V)}{\partial S} \right)^{-1} = \frac{1}{T} \tag{1.70}$$

$$\frac{\partial S(U, V)}{\partial V} = -\left(\frac{\partial U}{\partial S}\right)^{-1} \frac{\partial U}{\partial V} = \frac{P}{T} \tag{1.71}$$

which corresponds to the new contact form—equivalent to κ in the sense of contact geometry—

$$\mathrm{d}S - \frac{\mathrm{d}U}{T} - \frac{P}{T}\,\mathrm{d}V \equiv -\frac{1}{T}\,\kappa. \tag{1.72}$$

Example 1.3 For the ideal gas of Example 1.2

$$S(U, V) = \alpha \log(U/\alpha) + \alpha\beta \log V \tag{1.73}$$

from which we read the equations of state

$$T^{-1} = \frac{\partial S}{\partial U} = \frac{\alpha}{U}, \qquad \frac{P}{T} = \frac{\partial S}{\partial V} = \frac{\alpha\beta}{V}. \tag{1.74}$$

In this way we get 12 thermodynamical potentials for the simple systems each one endowed with its natural variables

$$
\begin{array}{ccc}
\hline
U(S, V) & S(U, V) & V(U, S) \\
F(V, T) & T(F, V) & V(F, T) \\
H(P, S) & P(H, S) & S(H, P) \\
G(P, T) & T(G, P) & P(G, T) \\
\hline
\end{array} \tag{1.75}
$$

When restricted to the submanifold $\mathcal{E}$, they satisfy the differential relations

$$
\begin{array}{lll}
\hline
\mathrm{d}U = T\,\mathrm{d}S - P\,\mathrm{d}V & \mathrm{d}S = \frac{1}{T}\mathrm{d}U + \frac{P}{T}\mathrm{d}V & \mathrm{d}V = \frac{T}{P}\mathrm{d}V - \frac{\mathrm{d}U}{T} \\
\mathrm{d}F = -S\,\mathrm{d}T - P\,\mathrm{d}V & \mathrm{d}T = -\frac{\mathrm{d}F}{S} - \frac{P}{S}\mathrm{d}V & \mathrm{d}V = -\frac{\mathrm{d}F}{P} - \frac{S}{P}\mathrm{d}T \\
\mathrm{d}H = T\,\mathrm{d}S + V\,\mathrm{d}P & \mathrm{d}P = \frac{\mathrm{d}H}{V} - \frac{T}{V}\mathrm{d}S & \mathrm{d}S = \frac{\mathrm{d}H}{T} - \frac{V}{T}\mathrm{d}P \\
\mathrm{d}G = -S\,\mathrm{d}T + V\,\mathrm{d}P & \mathrm{d}T = -\frac{\mathrm{d}G}{S} + \frac{V}{S}\mathrm{d}P & \mathrm{d}P = \frac{\mathrm{d}G}{V} + \frac{S}{V}\mathrm{d}T \\
\hline
\end{array}
$$

$$\tag{1.76}$$

The 12 thermodynamical potentials do not exhaust the list of the functions $\Phi(y_1, \ldots, y_s)$ satisfying our definition of *thermodynamic potential* even for the class of *simple thermal systems*. We shall present some other important ones at the end of next section.

1.7 Thermodynamical Potentials: Convexity Properties

We know that the Legendre transform (in the precise variational version of Eqs. (1.60)–(1.62)) is well-defined when the function is convex (and stable), see [3] §. 6.1. For instance, to have a well-defined free energy

$$F(T, V) \stackrel{\text{def}}{=} \min_{S}(U(S, V) - TS) \qquad (1.77)$$

the internal energy $U(S, V)$ should satisfy the inequality

$$\frac{\partial^2 U}{\partial S^2} \geq 0. \qquad (1.78)$$

In Theorem 1.3 we have already seen that such inequality follows from the minimum energy principle (or the maximal entropy principle). In this section we wish to understand the physical meaning of this *convexity condition* more in detail.

Heat Capacity at Fixed Volume We note the equality

$$\frac{\partial U}{\partial S} = T \quad \Rightarrow \quad \frac{\partial^2 U}{\partial S^2} = \left(\frac{\partial T}{\partial S}\right)_V. \qquad (1.79)$$

Definition 1.7 Consider a reversible (infinitesimal) process where the volume V is kept fixed while the body absorbs a quantity δQ of heat. As a result the temperature of the body increases by dT. The *heat* (or thermal) *capacity at fixed volume C_V* of the body is the ratio

$$C_V \equiv \frac{\delta Q}{dT} = \left(\frac{\partial U}{\partial T}\right)_V \qquad (1.80)$$

where we used that at constant V, $\delta Q = dU$ by the First Law. The function $C_V(T, V)$ is a characteristic of the particular body and an extensive quantity.

When $C_V > 0$ heating up the body (i.e. supplying heat to it) increases its temperature. When $C_V < 0$ heating up the body *decreases* its temperature: then the body will absorb more heat by Clausius' principle, so its temperature will decrease even further, and then it will absorb even more heat…and so on, without ever reaching an equilibrium state. From Eq. (1.79) we get

$$C_V \equiv \left(\frac{\partial U}{\partial T}\right)_V = T\left(\frac{\partial S}{\partial T}\right)_V = T\left(\frac{\partial T}{\partial S}\right)_V^{-1} = T\left(\frac{\partial^2 U}{\partial S^2}\right)_V^{-1} \qquad (1.81)$$

comparing with (1.78) we conclude:

Theorem 1.4 *The free energy $F(T, V)$ of a thermal system is well defined if and only if its heat capacity at fixed volume is positive, $C_V > 0$, i.e.* if heating up the body its temperature increases.

Remark 1.5 There exist systems with negative thermal capacity. They are *extremely* rare and their physics is counterintuitive. The most important example is a black hole in asymptotically flat space time [5] which has

$$S = \alpha \, U^2 \tag{1.82}$$

(α a constant) so that

$$\frac{1}{T} = \frac{\partial S}{\partial U} = 2\alpha \, U \quad \Rightarrow \quad C_V = \left(\frac{\partial U}{\partial T}\right)_V \equiv \left(\frac{\partial}{\partial T}\frac{1}{2\alpha \, T}\right)_V = -\frac{1}{2\alpha \, T^2} \tag{1.83}$$

Clearly this system cannot reach an equilibrium at any given T. The physical mechanism underlying a negative thermal capacity will be clarified in Chap. 2. The simple answer is that the thermal capacity is negative when entropy grows too fast as a function of energy. If the entropy scales with energy as

$$S = \alpha \, U^\beta \qquad \alpha > 0, \ \beta \in \mathbb{R} \ \text{constants} \tag{1.84}$$

the condition of positive heat capacity requires $\beta < 1$. This observation gives a crucial bound on entropy which is valid for all systems with a positive heat capacity

$$S = o(U) \quad \text{as} \ U \to \infty. \tag{1.85}$$

We point out the useful formula

$$C_V = T \left(\frac{\partial S}{\partial T}\right)_V = -T \left(\frac{\partial^2 F}{\partial T^2}\right)_V. \tag{1.86}$$

Adiabatic Compressibility Analogously, the enthalpy $H(S, P)$ is well defined iff

$$\frac{\partial^2 U}{\partial V^2} \geq 0. \tag{1.87}$$

Definition 1.8 The *adiabatic (or isoentropic) compressibility*

$$\kappa_S = -\frac{1}{V}\left(\frac{\partial V}{\partial P}\right)_S, \tag{1.88}$$

measures how much the volume of the system shrinks when compressing it with extra pressure in a reversible adiabatic process (i.e. without exchange of heat). If compressibility is positive the body will contract when compressed, whereas if κ_S is negative the body *will expand under compression.*

Now

$$\frac{\partial^2 U}{\partial V^2} = -\left(\frac{\partial P}{\partial V}\right)_S = -\left(\frac{\partial V}{\partial P}\right)_S^{-1} = \frac{1}{\kappa_S V}, \tag{1.89}$$

hence

Theorem 1.5

(1) The enthalpy of a body is well defined if and only if its adiabatic compressibility is positive, $\kappa_S > 0$, i.e. compressing the body adiabatically will make its volume smaller.

(2) The Gibbs free energy is well defined, if and only if the body has both $C_V > 0$ and $\kappa_S > 0$.

Remark 1.6 In Eq. (1.66) we saw that convexity of U requires a third inequality

$$\frac{\partial^2 U}{\partial V^2}\frac{\partial^2 U}{\partial S^2} \geq \left(\frac{\partial^2 U}{\partial S\,\partial V}\right)^2 \tag{1.90}$$

besides the positivity of $\partial^2 U/\partial S^2$ and $\partial^2 U/\partial V^2$. On the other hand, the above argument shows that these two conditions are sufficient to guarantee the existence of the Gibbs free energy. Therefore condition (1.90) must hold automatically when $C_V > 0$ and $\kappa_S > 0$ (this is true: for a detailed proof see [6]).

The same token also shows that when all four potentials are well-defined the following inequalities hold

$$\frac{\partial^2 H(S, P)}{\partial S^2} \geq 0, \qquad \frac{\partial^2 F(T, V)}{\partial V^2} \geq 0. \tag{1.91}$$

To explain their physical interpretation, we give two definitions:

Definition 1.9

(1) The *heat* (or thermal) *capacity at fixed pressure*, C_P is the ratio of the heat absorbed by the body and the variation of its temperature in an (infinitesimal) process at fixed pressure P

$$C_P = \left.\frac{\delta Q}{\delta T}\right|_P = \left(\frac{\partial H}{\partial T}\right)_P = T\left(\frac{\partial S}{\partial T}\right)_P = T\left(\frac{\partial T}{\partial S}\right)_P^{-1} = T\left(\frac{\partial^2 H}{\partial^2 S}\right)_P^{-1} \tag{1.92}$$

(2) The *isothermal compressibility* κ_T is

$$\kappa_T = -\frac{1}{V}\left(\frac{\partial V}{\partial P}\right)_T = -\frac{1}{V}\left(\frac{\partial P}{\partial V}\right)_T^{-1} = \frac{1}{V}\left(\frac{\partial^2 F}{\partial V^2}\right)_T^{-1} \tag{1.93}$$

Corollary 1.8 *When the four thermodynamics potential are well defined, the thermal capacity C_P at fixed P and the isothermal compressibility κ_T are positive.*

The fixed P counterpart to Eq. (1.86) is

$$C_P = -T\left(\frac{\partial^2 G}{\partial T^2}\right)_P. \tag{1.94}$$

We shall show below that $C_P > C_V$. In Appendix 2 it is shown that

$$1 < \frac{C_P}{C_V} = \frac{\kappa_T}{\kappa_S}. \tag{1.95}$$

From the general properties of the Legendre transform [3, Lemma 6.1] we get that, when $U(S, V)$ is convex, the second derivatives of all four thermodynamic potentials with respect to their natural variables are either ≥ 0 or ≤ 0. More precisely:

Fundamental Principle **(Convexity of Thermodynamic Potentials)** *Let Φ be one of the **four** potentials and y one of its natural variables. Then*

$$\frac{\partial^2 \Phi}{\partial y^2}\begin{cases} \geq 0 & y \text{ is extensive} \\ \leq 0 & y \text{ is intensive.} \end{cases} \tag{1.96}$$

In Table 1.1 we give the physical interpretation of each inequality.

Table 1.1 Second derivatives of the four potentials with respect to their natural variables

Physical interpretation of second derivative	Sign	See Equation
$\left(\frac{\partial^2 U}{\partial S^2}\right)_V = \frac{T}{C_V}$	>0	(1.81)
$\left(\frac{\partial^2 U}{\partial V^2}\right)_S = \frac{1}{\kappa_S V}$	>0	(1.89)
$\left(\frac{\partial^2 F}{\partial T^2}\right)_V = -\left(\frac{\partial S}{\partial T}\right)_V = -\frac{C_V}{T}$	<0	(1.81), (1.86)
$\left(\frac{\partial^2 F}{\partial V^2}\right)_T = -\left(\frac{\partial P}{\partial V}\right)_T = -\left(\frac{\partial V}{\partial P}\right)_V^{-1} = \frac{1}{\kappa_T V}$	>0	(1.93)
$\left(\frac{\partial^2 H}{\partial S^2}\right)_P = \frac{T}{C_P}$	>0	(1.92)
$\left(\frac{\partial^2 H}{\partial P^2}\right)_S = \left(\frac{\partial V}{\partial P}\right)_S = -V\kappa_S$	<0	(1.88)
$\left(\frac{\partial^2 G}{\partial T^2}\right)_P = -\left(\frac{\partial S}{\partial T}\right)_P = -\frac{C_P}{T}$	<0	(1.92),(1.94)
$\left(\frac{\partial^2 G}{\partial P^2}\right)_T = \left(\frac{\partial V}{\partial P}\right)_T = -V\kappa_T$	<0	(1.93)

More Thermodynamical Potentials

As already stated, there are other thermal quantities which satisfy the definition of thermodynamical potential and are useful in our applications. We present one example which is central in Statistical Mechanics—also called *free energy*—

$$\mathcal{F}(T^{-1}, V) \overset{\text{def}}{=} T^{-1} F(T^{-1}, V) \tag{1.97}$$

that is, the Helmholtz free energy F as a function of the inverse temperature T^{-1} times T^{-1}. $\mathcal{F}$ satisfies the differential relation

$$d\mathcal{F} = U\, dT^{-1} - \frac{P}{T}\, dV. \tag{1.98}$$

Comparing with Eq. (1.76) we see that $\mathcal{F}$ is minus the Legendre transform of the entropy $S(U, V)$ with respect to U. Indeed

$$dS(U, V) = \frac{1}{T}\, dU + \frac{P}{T}\, dV, \qquad \frac{\partial S(U, V)}{\partial U} = \frac{1}{T} \tag{1.99}$$

and the Legendre transform is

$$S(U(T^{-1}, V), V) - T^{-1} U(T^{-1}, V) = -\mathcal{F}(T^{-1}, V). \tag{1.100}$$

To be precise, we notice that as a function of U the entropy S is *concave*

$$\left(\frac{\partial^2 S}{\partial U^2}\right)_V = \left(\frac{\partial}{\partial U}\frac{1}{T}\right)_V = -\frac{1}{T^2}\left(\frac{\partial T}{\partial U}\right)_V = -\frac{1}{T^2}\left(\frac{\partial U}{\partial T}\right)_V^{-1} = -\frac{1}{C_V T^2} < 0, \tag{1.101}$$

so that $\mathcal{F}$ is the Legendre transform of the convex potential $-S$:

$$\mathcal{F} = \min_U \left(T^{-1} U - S\right). \tag{1.102}$$

Other thermodynamical potentials may be constructed in the same fashion. One we shall need in Chap. 2 is the Legendre transform of $\mathcal{F}$ with respect to volume

$$\mathcal{G}(T^{-1}, P) = \mathcal{F} - V\frac{\partial \mathcal{F}}{\partial V} = \frac{F + PV}{T} = T^{-1} G(T, P) \tag{1.103}$$

where $G(T, P)$ is the Gibbs potential. While the potentials F, G are the traditional ones, and are the most convenient functions in Chemical Physics, from a fundamental physical point of view the basic potentials at fixed T, V, and respectively fixed T, P, are $\mathcal{F}(T^{-1}, V)$ and $\mathcal{G}(T^{-1}, P)$, see Chap. 2.

Maxwell Relations

When acting on smooth functions the partial derivatives commute. In view of the differential relations satisfied by the thermodynamical potentials, these commutation relations yield identities between the derivatives of thermal quantities known collectively as *Maxwell relations*. For instance

$$\left(\frac{\partial}{\partial V}\left(\frac{\partial U}{\partial S}\right)_V\right)_S = \left(\frac{\partial}{\partial S}\left(\frac{\partial U}{\partial V}\right)_S\right)_V \tag{1.104}$$

gives the Maxwell relation

$$\left(\frac{\partial T}{\partial V}\right)_S = -\left(\frac{\partial P}{\partial S}\right)_V \tag{1.105}$$

or, inverting both sides,

$$\left(\frac{\partial V}{\partial T}\right)_S = -\left(\frac{\partial S}{\partial P}\right)_V \tag{1.106}$$

which says that the *coefficient of volume dilation* when we increase the temperature adiabatically

$$\rho \overset{\text{def}}{=} \frac{1}{V}\left(\frac{\partial V}{\partial T}\right)_S \tag{1.107}$$

is equal to the coefficient of entropy grow (per unit volume) when we decrease the pressure at constant volume. We collect the Maxwell relations from the second derivatives of the four classical thermodynamical potentials in the following table:

Potential	Maxwell Relation	
$U(S, V)$	$\left(\frac{\partial T}{\partial V}\right)_S = -\left(\frac{\partial P}{\partial S}\right)_V$	
$F(T, V)$	$\left(\frac{\partial S}{\partial V}\right)_T = \left(\frac{\partial P}{\partial T}\right)_V$	(1.108)
$H(S, P)$	$\left(\frac{\partial T}{\partial P}\right)_S = \left(\frac{\partial V}{\partial S}\right)_P$	
$G(S, V)$	$\left(\frac{\partial S}{\partial P}\right)_T = -\left(\frac{\partial V}{\partial T}\right)_P$	

Next we present some examples of applications of the Maxwell relations. The *coefficient of thermal expansion* is defined to be

$$\beta \overset{\text{def}}{=} \frac{1}{V}\left(\frac{\partial V}{\partial T}\right)_P \tag{1.109}$$

From table (1.108) we see that β is equal to minus the variation of entropy with pressure at fixed temperature

$$\beta = -\frac{1}{V}\left(\frac{\partial S}{\partial P}\right)_T.\tag{1.110}$$

As a second application of the Maxwell relations let us prove our claim that $C_P > C_V$ in any "sound" thermal system. We have

$$C_P = T\left(\frac{\partial S}{\partial T}\right)_P = T\left(\frac{\partial S}{\partial T}\right)_V + T\left(\frac{\partial S}{\partial V}\right)_T\left(\frac{\partial V}{\partial T}\right)_P \equiv C_V + T\left(\frac{\partial S}{\partial V}\right)_T\left(\frac{\partial V}{\partial T}\right)_P\tag{1.111}$$

Using the second Maxwell relation this becomes

$$C_P - C_V = T\left(\frac{\partial P}{\partial T}\right)_V\left(\frac{\partial V}{\partial T}\right)_P.\tag{1.112}$$

The formula for the derivative of implicit functions[17] yields the identity

$$\left(\frac{\partial P}{\partial T}\right)_V = -\left(\frac{\partial P}{\partial V}\right)_T\left(\frac{\partial V}{\partial T}\right)_P\tag{1.113}$$

which inserted back in (1.112) yields

$$C_P - C_V = -\left(\frac{\partial P}{\partial V}\right)_T T\left(\frac{\partial V}{\partial T}\right)_P^2 = \frac{V}{\kappa_T}T\beta^2 > 0.\tag{1.114}$$

As a third application in Appendix 2 we show Eq. (1.95), i.e.

$$\frac{\kappa_T}{\kappa_S} = \frac{C_P}{C_V}.\tag{1.115}$$

1.8 Third Law of Thermodynamics

The Third Law is less general than the first two, but it applies to all "usual" thermal systems. Its statement is that we can define the entropy so that it vanishes at (absolute) temperature zero, $S|_{T=0} = 0$. Since entropy was defined up to an additive constant, this amounts to making the redefinition $S \rightsquigarrow S - S|_{T=0}$, and the statement

---

[17] See Eq. (1.169) below.

may sound empty. However what it really means it that zero-temperature is at *finite distance* in entropy with respect to a typical equilibrium state, that is,

Third Law (Nernst) *Let* $\gamma : [0, 1] \rightarrow \mathcal{E}$ *be any process (which we assume reversible for simplicity) with* $\gamma(0)$ *a zero temperature state and* $\gamma(1)$ *a state at finite temperature. Then the integral*

$$\int_0^1 \gamma^* \mathrm{d}S < \infty \qquad converges. \tag{1.116}$$

To get a flavor of the implications of the Third Law (when it applies), suppose the process γ is at fixed volume V or, respectively, fixed pressure P: then

$$\int_0^T \left(\frac{\partial S}{\partial T}\right)_V \mathrm{d}T \equiv \int_0^T \frac{C_V(T, V)}{T} \mathrm{d}T < \infty \tag{1.117}$$

$$\int_0^T \left(\frac{\partial S}{\partial T}\right)_P \mathrm{d}T \equiv \int_0^T \frac{C_P(T, P)}{T} \mathrm{d}T < \infty \tag{1.118}$$

which say that the thermal capacities $C_V(T)$, $C_P(T)$ should go to zero as $T \rightarrow 0$ rapidly enough to make the integrals converge.

1.9 Non-simple Systems: The Grand Potential

Now it is time to consider general systems, that is, *non-simple* ones with $\dim \mathcal{E} > 2$. We start from the most classical situation.

The Quantity of "Matter" N as a Thermal Variable

We consider a thermal system which can trade with its surrounding not just energy, entropy, and volume (by expanding or contracting) but also "matter". In other words, the "matter" content of our system is no longer kept constant. For instance: if our system is a gas, we can measure the amount of "matter" in the system either macroscopically in terms of the number n of moles or microscopically using the number N of "particles"; the two quantities differ only in overall normalization

$$N = n \, N_A, \tag{1.119}$$

where $N_A \approx 6.022 \, 10^{23}$ is the *Avogadro number*. We shall use the symbol N for the quantity of "matter" in the system,[18] its precise physical meaning will depend on the system and the chosen normalization of the quantity itself. The crucial property that we must require is that N is *additively conserved* in the sense that the variation

---

[18] By abuse of language we shall sometimes refer to N as the "number of particles".

$(\Delta N)_{\text{syst}}$ of the "quantity of matter" in the system is equal to minus its variation in the surroundings

$$(\Delta N)_{\text{syst}} + (\Delta N)_{\text{surrou}} = 0. \tag{1.120}$$

The internal energy of the system, viewed as a thermodynamical potential, is now function of *three* natural variables

$$U = U(S, V, N), \tag{1.121}$$

and the First/Second Laws take the extended form

$$dU = T\,dS - P\,dV + \mu\,dN, \tag{1.122}$$

where μ is a new *intensive* thermal quantity, canonically conjugate to N, called the *chemical potential*, which is defined as

$$\mu \stackrel{\text{def}}{=} \left(\frac{\partial U}{\partial N}\right)_{S,V} \tag{1.123}$$

From the properties of the Legendre transform [3, Lemma 6.1(3)] we get

$$dF(T, V, N) = -S\,dT - P\,dV + \mu\,dN \tag{1.124}$$

$$dH(S, p, N) = T\,dS + V\,dP + \mu\,dN \tag{1.125}$$

$$dG(T, p, N) = -S\,dT + V\,dP + \mu\,dN, \tag{1.126}$$

so that

$$\mu = \left(\frac{\partial U}{\partial N}\right)_{S,V} = \left(\frac{\partial F}{\partial N}\right)_{T,V} = \left(\frac{\partial H}{\partial N}\right)_{S,P} = \left(\frac{\partial G}{\partial N}\right)_{T,P} \tag{1.127}$$

Putting together m copies of our system, all in equilibrium at the same temperature and pressure, we get a system with $N' = mN$ particles, and all extensive quantities become m times bigger. In particular for the Gibbs potential

$$G(T, P, mN) = m\,G(T, P, N) \quad \Rightarrow \quad G(T, P, N) = N\,g(T, P) \tag{1.128}$$

for some function $g(T, P)$. Now

$$\mu(T, P) = \left(\frac{\partial G}{\partial N}\right)_{T,P} = g(T, P). \tag{1.129}$$

That is: *the chemical potential $\mu(T, P)$ is the Gibbs free energy per particle, $g(T, P)$, for the system at fixed temperature T and pressure P*

The Grand Potential

When studying the simple systems, we got new thermodynamical potentials by taking the Legendre transform of old ones with respect to some of their natural variables. In doing this we exploited the "convexity" properties of the several potentials. It is natural to try to do the same with the new extensive natural variable N. However, recall from Sect. 1.6 that the Legendre transform makes sense only when the function is *convex* (and stable in its convex cone of definition), and this condition is not satisfied in general by the dependence on N. For instance the Gibbs potential $G(T, P, N)$ is linear in N, Eq. (1.128), and its Legendre transform would be meaningless. However,

Lemma 1.1 *The free energy $F(T, V, N)$ is a convex function of N.*

Proof One has

$$F(T, V, N) = G(T, P, N) - PV = Ng(T, P) - PV \qquad (1.130)$$

where

$$V = \frac{\partial G}{\partial P} = N \frac{\partial g(T, P)}{\partial P}, \qquad (1.131)$$

and the Legendre transform of $G(T, P, N)$ with respect to the pressure is

$$F(T, V, N) = N f(T, V/N) \qquad (1.132)$$

where $f(T, v)$ is the Legendre transform of $g(T, P)$ with respect to P, that is, the free energy *per particle* in the normalized volume $v \equiv V/N$. One has

$$\frac{\partial F}{\partial N} = f(T, V/N) - \frac{V}{N} \frac{\partial f}{\partial v} \qquad (1.133)$$

and

$$\frac{\partial^2 F}{\partial N^2} = \frac{V^2}{N^3} \frac{\partial^2 f}{\partial v^2} \geq 0 \qquad (1.134)$$

by Eq. (1.93) and $\kappa_T > 0$ (in "sound" systems). $\qquad\qquad\square$

Therefore it makes perfect sense to take the Legendre transform of the free energy $F(T, V, N)$ with respect to N. The resulting thermodynamical potential $\Phi(T, V, \mu)$ is called the *Grand potential* $\Phi(T, V, \mu)$

$$\Phi(T, V, \mu) = F(T, V, N) - N\frac{\partial F(T, V, N)}{\partial N} = F - N\mu \qquad (1.135)$$

where N is written as a function of μ, T, V by inverting the relation

$$\mu = \frac{\partial F(T, V, N)}{\partial N}. \qquad (1.136)$$

In terms of the Grand potential the First/Second Laws read

$$d\Phi = -S\,dT - P\,dV - N\,d\mu \qquad (1.137)$$

that is,

$$S = -\left(\frac{\partial \Phi}{\partial T}\right)_{V,\mu}, \quad P = -\left(\frac{\partial \Phi}{\partial V}\right)_{T,\mu}, \quad N = -\left(\frac{\partial \Phi}{\partial \mu}\right)_{T,V}, \qquad (1.138)$$

from which we can read the equations of state. In *value* the Grand potential is simply

$$\Phi = F - \mu N = F - G = -PV, \qquad (1.139)$$

but recall that what matters is not the value of a thermal potential but its *functional expression* in terms of its natural variables (in this case T, V, μ).

1.10 General Thermodynamical Formalism

A general system has a number of additive conserved quantities

$$U, V, a_1, \ldots, a_s. \qquad (1.140)$$

For instance our system may be a mixture of gases composed of stable particles of different species and $a_i \equiv N_i$ is the number of particles of the i-th species. Or the a_i's can be electric/magnetic charges or Noether charges associated to some symmetry of the problem.

Independently of the nature and physical interpretation of the additive quantities a_i, the basic differential relations are

$$dU(T, V, a_1, \ldots, a_s) = T\,dS - P\,dV + \sum_i b_i\,da_i \qquad (1.141)$$

$$dF(T, V, a_1, \ldots, a_s) = -S\,dT - P\,dV + \sum_i b_i\,da_i \tag{1.142}$$

$$dH(S, p, a_1, \ldots, a_s) = T\,dS + V\,dP + \sum_i b_i\,da_i \tag{1.143}$$

$$dG(T, p, a_1, \ldots, a_s) = -S\,dT + V\,dP + \sum_i b_i\,da_i, \tag{1.144}$$

where the intensive quantity b_i is the *chemical potential* associated (or dual) to the i-th conserved quantity a_i. When a_i are particle numbers, the b_i are chemical potentials in the sense of the previous section.

The Legendre transform with respect to the a_i's is the *Grand potential*

$$\Phi(T, V, b_1, \ldots, b_s) = F - \sum_i a_i\,b_i. \tag{1.145}$$

It makes sense iff the Hessian matrix

$$\frac{\partial^2 F}{\partial a_i\,\partial a_j} \tag{1.146}$$

is positive-definite.

The Gibbs-Duhem Relation

Let us consider the important case where $a_i = N_i$ is the number of particles of the i-th species. Clearly the Gibbs potential $G(T, P, N_1, \ldots, N_s)$ is a homogeneous function in the N_i's of degree 1. By the Euler formula for the derivative of homogeneous functions

$$G = \sum_i N_i\,\frac{\partial G}{\partial N_i} = \sum_i N_i\,\mu_i, \tag{1.147}$$

and hence

$$dG = \sum_i (\mu_i\,dN_i + N_i\,d\mu_i). \tag{1.148}$$

Subtracting (1.144) we get the Gibbs-Duhem relation

$$\sum_i N_i\,d\mu_i = -S\,dT + V\,dP. \tag{1.149}$$

1.11   Coexistence of Two Phases

We consider a closed system where coexist two phases of the same substance, say, liquid and gas. The total internal energy has the form

$$U = U_1(S_1, V_1, N_1) + U_2(S - S_1, V - V_1, N - N_1) \tag{1.150}$$

where N is the total number of molecules and N_1 is the fraction of molecules in the gas phase. U_1 and U_2 are, respectively, the internal energy functions in the gas and liquid phase. At equilibrium the energy is minimal, so that

$$\frac{\partial U}{\partial S_1} = 0 \quad \Rightarrow \quad T_1 - T_2 = 0 \tag{1.151}$$

$$\frac{\partial U}{\partial V_1} = 0 \quad \Rightarrow \quad P_1 - P_2 = 0 \tag{1.152}$$

$$\frac{\partial U}{\partial N_1} = 0 \quad \Rightarrow \quad \mu_1 - \mu_2 = 0. \tag{1.153}$$

Hence at equilibrium $T_1 = T_2 = T$ and $P_1 = P_2 = P$ while

$$\mu_1(T, P) = \mu_2(T, P). \tag{1.154}$$

Since the chemical potentials in the two phases are given by analytically different functions of P, T, (1.154) is the equation of *a curve* in the P-T plane: the so-called *curve of coexistence of the two phases*. Away from this curve only one phase is possible, and the coexistence curve is the boundary separating the region in the P-T plane where the first phase is realized from the region where the second phase takes place. On the coexistence curve both phases may be simultaneously present. For instance, the point

$$(T, P) = (0\,^\circ\mathrm{C}, 1\ \mathrm{atm}) \tag{1.155}$$

belongs to the coexistence curve for water and ice, and it is a common experience that at this point we can have any mixture of them: for any $0 \le x \le 1$ a fraction x of H_2O molecules can be liquid water while a fraction $1 - x$ of them are solid ice.

Using the formula for the derivative of an implicit function (cf. Eq. (1.169)), the derivative of the pressure with respect to the temperature along the coexistence curve becomes

$$\frac{\partial P}{\partial T} = -\frac{\frac{\partial(\mu_1 - \mu_2)}{\partial T}}{\frac{\partial(\mu_1 - \mu_2)}{\partial P}} = \frac{s_1 - s_2}{v_1 - v_2} \tag{1.156}$$

where (v_a, s_a) are the volume and entropy per particle (or mole) in the a-th phase. Writing $L = T(s_1 - s_2)$ for the *latent heat* of the phase transition per particle (or mole) we get the *Clausius-Clapeyron equation*

$$\frac{\partial P}{\partial T} = \frac{L}{T(v_1 - v_2)},$$

(1.157)

which yields the shape of the coexistence curve in terms of the latent heat L and the variation of specific volumes $v_1 - v_2$ between the two phases.

1.12   EPILOGUE: Boltzmann's Proof of the Stefan Law

We started this chapter by emphasizing the unique role of Thermodynamics as a meta-theory of universal principles that all scientific theories should obey. This implies that we can make exact predictions about physical phenomena in the *most absolute ignorance* of the dynamical theory which governs them, just by requiring consistency with the Laws of Thermodynamics. The best example of this wonder is the proof by Boltzmann of the Stefan law, a turning point in the history of science.

The Stefan law says that the total intensity of the black body radiation, which is essentially the energy density u of the electromagnetic field inside an empty box with reflecting walls, is proportional to the fourth power of the temperature

$$u(T) = a\,T^4$$

(1.158)

where a is an universal constant. Stefan had found his law empirically, but Boltzmann in 1884 gave a rigorous math proof of it [2]. The reason why this result is amazing is that in 1884 nobody—including the great Ludwig Boltzmann—had the vaguest clue about the correct theory of the black body radiation, and the only then existing theory (classical physics) predicted that $u(T)$ was *linear* in T (with a divergent coefficient) not *quartic!* However, until the 1905 paper by Einstein on the photoelectric effect, nobody computed the classical answer, so the discrepancy went unnoticed. (The classical computation giving the wrong answer is reviewed in Sect. 2.8).

Yet, *in total ignorance of the underlying theory*, Boltzmann not only got the *correct* result that follows from the quantum theory (to be introduced more than twenty years later), but he also gave a *mathematically fully rigorous proof* of it which stands even today to our learned scrutiny. How could he do that? He simply looked for the most general function $u(T)$ which is consistent with the Second Law.

To make a thermodynamical study of a particular physical system we need only one piece of information about it: its equation of state. In the case of the electromagnetic radiation it suffices to know that *light travels at the speed of light.*

This implies that the radiation pressure P and the energy density u are related by the equation

$$P = \frac{1}{3}u. \tag{1.159}$$

We shall deduce this equation in Chap. 3 from special relativity. Boltzmann did not know special relativity in 1884, but he knew the Maxwell equations from which one proves the *Poynting theorem* which implies (1.159). Armed with the equation of state (1.159) we proceed to reproduce Boltzmann's proof of Stefan law.

Energy is an extensive quantity proportional to the volume, so

$$U = V\,u(T). \tag{1.160}$$

We plug the last two equations in (1.33) getting

$$dS \equiv \frac{dU}{T} + \frac{P}{T}dV = \frac{d(u\,V)}{T} + \frac{1}{3}\frac{u}{T}\,dV = \frac{V}{T}du + \frac{4}{3}\frac{u}{T}dV. \tag{1.161}$$

Next we take the differential of the two sides, using Poincaré identity $d^2 = 0$

$$0 = d^2 S = \frac{u'(T)}{T}dV \wedge dT - \frac{4}{3}\left(\frac{u'(T)}{T} - \frac{u(T)}{T^2}\right)dV \wedge dT, \tag{1.162}$$

We got a differential equation for the dependence of $u(T)$ on the temperature T

$$-\frac{u'(T)}{T} + 4\frac{u(T)}{T^2} = 0 \quad \Rightarrow \quad T\frac{\partial \log u(T)}{\partial T} = 4 \tag{1.163}$$

whose general integral is Eq. (1.158) with a the integration constant. We are entitled to rewrite the overall integration constant a in a fancy way, parametrizing it in terms of a new constant $\hbar$ in the form

$$u(T) = a(\hbar)\,T^4 = \frac{\pi^2 k^4}{15c^3\hbar^3}\,T^4 \tag{1.164}$$

where $\hbar$ is an integration constant still to be determined, while k is Boltzmann's constant, and c is the speed of light. The new fundamental constant $\hbar$—whose existence is predicted by the Second Law—is the celebrated *quantum* Planck constant.

Boltzmann's argument is amazingly simple, elegant, and valid whatever the fundamental theory is (as long as light does travel at the speed of light!). We stress that Planck's constant $\hbar$ emerges from the meta-theoretic Boltzmann's analysis as a "mere" integration constant. Classical Physics corresponds to $\hbar = 0$, so classically the energy density $u(T)$ diverges for all positive temperatures!

We return to the quotation by Eddington which opened this chapter:

> Classical Physics is not consistent with the second principle of Thermodynamics and hence **there is nothing for it but to collapse in deepest humiliation**

Appendix 1: Symplectic and Contact Geometries

We recall here the basic geometric definitions used in the main text. Besides Ref. [3], readers looking for background may have a look to [7–9].

Definition 1.10 A *symplectic manifold* is a pair (M, Ω) where M is a smooth manifold and $\Omega \in \Omega^2(M)$ is a smooth 2-form on M with two properties:[19]

(a) $\Omega = \Omega_{ij}\, \mathrm{d}x^i \wedge \mathrm{d}x^j$ is *nowhere singular,* i.e. $\det \Omega_{ij} \neq 0$ everywhere in M. This requires the dimension of M to be even;
(b) Ω is *closed:* $\mathrm{d}\Omega = 0$.

An important example of symplectic manifold is the total space T^*M of the cotangent bundle of any smooth manifold M. T^*M has the canonical symplectic 2-form $\Omega = \mathrm{d}(p_i\, \mathrm{d}q^i) = \mathrm{d}p_i \wedge \mathrm{d}q^i$ where the q^i's are local coordinates on the base M and the p_i's their dual coordinates along the fiber.

Definition 1.11 Let (M, Ω) be a symplectic manifold of dimension $2m$. A *Lagrangian submanifold* $L \subset M$ is a submanifold of dimension m such that $\Omega|_L = 0$.

A map $f \colon M_1 \to M_2$ between two symplectic manifolds is a *symplectomorphism* ($\equiv$ a diffeomorphism which preserves the symplectic form) if and only if its graph is a Lagrangian submanifold of the symplectic manifold $(M_1 \times M_2, \pi_1^*\Omega_1 - \pi_2^*\Omega_2)$.

The *Darboux theorem* states that in any symplectic manifold we may find local coordinates q^i, p_j (called canonical or Darboux coordinates) such that $\Omega = \mathrm{d}p_i \wedge \mathrm{d}q^i$.

The odd dimensional counterpart to symplectic geometry is *contact geometry.*

Definition 1.12 A *(strict) contact manifold* is a pair (X, κ) where X is a $(2m + 1)$-dimensional manifold and $\kappa \in \Omega^1(X)$ is a one-form on X such that the $(2m + 1)$-form

$$\kappa \wedge (\mathrm{d}\kappa)^m \neq 0 \quad \text{everywhere in } X. \tag{1.165}$$

---

[19] Here and below Einstein's convention on the sum over repeated indices is in force.

The contact version of the Darboux theorem states that we can find local coordinates $w, q^1, \ldots, q^m, p_1, \ldots, p_m$ such that the contact form κ has the local form

$$\kappa = \mathrm{d}w + p_i \, \mathrm{d}q^i. \tag{1.166}$$

Definition 1.13 A *Legendre submanifold* $\mathcal{L} \subset X$ of a $(2m+1)$-dimensional contact manifold (X, κ) is a m-dimensional submanifold of X such that $\kappa|_{\mathcal{L}} = 0$.

In local Darboux coordinates a (generic) Legendre submanifold $\mathcal{L}$ has the form

$$\left(w(q), q^1, \ldots, q^m, -\frac{\partial w(q)}{q^1}, \ldots, -\frac{\partial w(q)}{q^m} \right) \subset X \tag{1.167}$$

for some function $w(q) \equiv q(q^1, \ldots, q^m)$ called the *generating function of* $\mathcal{L}$.

Appendix 2: The Ratio κ_T/κ_S

We prove the relation

$$\frac{C_P}{C_V} = \frac{\kappa_T}{\kappa_S}. \tag{1.168}$$

The proof is based on the differential identity which holds whenever three functions x, y and z are functionally dependent, i.e. they satisfy a relation $F(x, y, z) = 0$:

$$\left(\frac{\partial x}{\partial y} \right)_z \left(\frac{\partial z}{\partial x} \right)_y \left(\frac{\partial y}{\partial z} \right)_x = -1. \tag{1.169}$$

We write $\overset{\mathrm{M}}{=}$ iff the two sides are equal by a Maxwell relation (1.108). We have

$$\frac{C_V}{T} = \left(\frac{\partial S}{\partial T} \right)_V = -\left(\frac{\partial S}{\partial V} \right)_T \left(\frac{\partial V}{\partial T} \right)_S \overset{\mathrm{M}}{=} -\left(\frac{\partial P}{\partial T} \right)_V \left(\frac{\partial V}{\partial T} \right)_S =$$
$$= \left(\frac{\partial P}{\partial V} \right)_T \left(\frac{\partial V}{\partial T} \right)_P \left(\frac{\partial V}{\partial T} \right)_S \tag{1.170}$$

and multiplying both sides by $T\kappa_T$

$$C_V \kappa_T = -\frac{T}{V} \left(\frac{\partial S}{\partial T} \right)_V \left(\frac{\partial V}{\partial P} \right)_T = -\frac{T}{V} \left(\frac{\partial V}{\partial T} \right)_P \left(\frac{\partial V}{\partial T} \right)_S. \tag{1.171}$$

The corresponding computation for C_P is

$$\frac{C_P}{T} = \left(\frac{\partial S}{\partial T}\right)_P = -\left(\frac{\partial S}{\partial P}\right)_T \left(\frac{\partial p}{\partial T}\right)_S \overset{M}{=} \left(\frac{\partial V}{\partial T}\right)_P \left(\frac{\partial S}{\partial V}\right)_P =$$
$$= -\left(\frac{\partial V}{\partial T}\right)_P \left(\frac{\partial P}{\partial V}\right)_S \left(\frac{\partial S}{\partial P}\right)_V$$

(1.172)

and then

$$C_P \, \kappa_S = -\frac{T}{V}\left(\frac{\partial S}{\partial T}\right)_P \left(\frac{\partial V}{\partial P}\right)_S = \frac{T}{V}\left(\frac{\partial V}{\partial T}\right)_P \left(\frac{\partial S}{\partial P}\right)_V$$

(1.173)

which is equal to (1.171) by the Maxwell relation

$$\left(\frac{\partial S}{\partial P}\right)_V \overset{M}{=} -\left(\frac{\partial V}{\partial T}\right)_S .$$

(1.174)

Problems

1.1 Consider the thermal system described Sect. 1.12 (black body radiation):

(a) compute its (Helmholtz) free energy;
(b) find the relation between temperature T and volume at constant entropy S.

1.2 A substance has enthalpy $H = a\,S^2 \log(P/b)$ (a, b constants). Find C_V.

1.3 The free energy for a particular gas is

$$F = a\,T\exp(bT) - cT[\log(V - d) + e/V]$$

(a, b, c, d, e constants). Find the equation of state and the internal energy.

1.4 The manifold of equilibrium states $\mathcal{E}$ has equation

$$(S - a)^2 = b\,VU^2$$

where $a, b > 0$ are constants.

(a) Find the Gibbs energy;
(b) Compute C_P.

1.5 Compute β and κ_T for an ideal gas.

1.6 A system has $\kappa_T = aT^3/P^2$ and $\beta = bT^2/P$ (a, b constants). Is there any relation between the constants a and b? Find the equation of state.

References

1. A.S. Eddington, *The Nature of the Physical World* (Cambridge University Press, 1928)
2. L. Boltzmann, Ableitung des Stefan'schen Gesetzes, betreffend die Abhängigkeit der Wärmestrahlung von der Temperatur aus der electromagnetischen Lichttheorie [Derivation of Stefan's law, concerning the dependency of heat radiation on temperature from the electromagnetic theory of light]. Ann. Phys. Chem. **258**, 291–294 (1884)
3. S. Cecotti, *Analytic Mechanics. A Concise Textbook* (Springer, 2024)
4. E. Schröedinger, *Statistical Thermodynamics* (Dover, 1989)
5. D. Grumiller, M.M. Sheikh-Jabbari, *Black Hole Physics. From Collapse to Evaporation.* Graduate Texts in Physics (Springer, 2022)
6. A.J. Berlinsky, A.B. Harris, *Statistical Mechanics. An Introductory Graduate Course* (Springer, 2019)
7. A. Cannas da Silva, *Lectures on Symplectic Geometry.* Lectures Notes in Mathematics, vol. 1764 (Springer, 2008)
8. R. Berndt, *An Introduction to Symplectic Geometry.* Graduate Studies in Mathematics, vol. 26 (AMS, 2001)
9. J.-L. Koszul, Y.M. Zou, *Introduction to Symplectic Geometry* (Science Press/Springer, Beijing, 2019)

Chapter 2
Equilibrium Ensembles

Thermodynamics gives us a wonderful web of thermal quantities, and relations between them, which are pretty universal, that is, valid in whatever sound "macroscopic" physical system. However we are interested in the detailed physics of important particular systems not just in universal statements which, being completely general, cannot say much on specific issues which are relevant for the real-world applications. The *detailed* thermal physics of a system is encoded in its *equations of state,* and we need effective methods to find out and understand the equations of state of any system of interest. Studying the Thermodynamical Formalism we learned that the equations of state are determined once we know any one of the thermodynamic potentials discussed in Chap. 1 *written in its natural variables.*

Statistical Mechanics is the theory (methodology) which allows us to compute and analyze the equations of state for any "reasonable" physical system and check that they satisfy all the properties predicted by the Laws of Thermodynamics, thus "proving" the Laws and explaining the physical meaning of each thermodynamic potential—such as the entropy $S(U, V)$—from the very first principles of the underlying microscopic dynamics.

The main tool one uses to compute the thermodynamic potentials are the *equilibrium ensembles* which are the topic of this chapter.

2.1 Preliminaries

Our problem is to compute the thermodynamical potentials, as functions of their natural variables, from a statistical treatment of the dynamics of the microscopic degrees of freedom. The basic idea is that a statistical approach is well justified because the number of degrees of freedom is huge (typically $O(10^{25})$). From the

S. Cecotti, *Statistical Mechanics*, UNITEXT for Physics,
https://doi.org/10.1007/978-3-031-67874-5_2

thermodynamical potentials we can read the equations of state of the particular system, that is, recover its thermal physics at equilibrium. We focus on thermodynamic potentials which are *extensive,* a class which includes the *four potentials* discussed at length in Chap. 1 and a few others.

Thermal Baths. Thermodynamical Limit

To compute a particular thermodynamical potential we study the statistical dynamics of the microsystem *at fixed values of its natural variables* $y_1, \ldots, y_r, z_1, \ldots z_s$ where the y_i are *extensive* and the z_j *intensive.* For instance, if we want to compute the free energy $F(T, V)$ for a simple system, we keep fixed temperature and volume (and the quantity of "matter"), but allow our system to exchange energy (heat) and entropy with its surroundings to be able to thermalize itself into an equilibrium state with the given values T, V of temperature and volume. If the system has to reach equilibrium at the fixed values $y_1, \ldots, y_r, z_1, \ldots z_s$ of the natural variables, its surroundings should enjoy two properties:

S1   they should not be allowed to exchange the additive natural quantities y_i with our system;

S2   the intensive quantities z_j should remain fixed at their prescribed values independently of the exchanges going on with our system.

In the example of $F(T, V)$, the system should be enclosed in a rigid box, whose volume cannot change with time, while the surroundings are kept at the fixed temperature T. In addition, the surroundings should be large enough so that their temperature is not modified by the back-reaction of our system. In other words: the surroundings should be a *thermal bath* at fixed natural variables in the sense of Chap. 1. The detailed properties of the thermal bath are immaterial: as long as it has the correct values of the z_j's and cannot exchange y_i's with the system, all thermal baths are equivalent for the purpose of computing the thermodynamical potential of our system at equilibrium. This universality property of the thermal baths is the ultimate version of the Zeroth Law.

To avoid the technical nuisance of having to mathematically model the thermal bath in addition to the system, one uses the so-called *Gibbs' trick:* as thermal bath one takes infinitely many copies of the system under consideration. Thus it suffices to model the system of interest.

Let us see how the trick works in practice. Let $\Phi(y_i, z_j)$ be the (additive) thermodynamical potential of interest for our system in presence of the appropriate thermal bath. The thermodynamical potential $\Phi_N(y_i, z_j)_{\text{no-bath}}$ for $N \gg 1$ identical copies of our system *in absence of a bath* is

$$\Phi_N(y_i, z_j)_{\text{no-bath}} \approx N\, \Phi(y_i, z_j), \tag{2.1}$$

while

$$\Phi_N(y_i, z_j)_{\text{no-bath}} = \Phi(N y_i, z_j)_{\text{no-bath}}, \tag{2.2}$$

and these formulae become exact in the limit $N \to \infty$, when the combined system can be interpreted as our original system in presence of a Gibbs' trick thermal bath. Hence we conclude

$$\Phi(y_i, z_j) = \lim_{N \to \infty} \frac{\Phi(N y_i, z_j)_{\text{no-bath}}}{N}. \tag{2.3}$$

The limit in the RHS is called the *thermodynamic limit*, since it is in this infinite limit that the thermodynamical behavior becomes exactly valid. A fundamental problem of Statistical Mechanics is to establish the *existence of the thermodynamic limit* for the ensemble of interest. If the thermodynamic limit does **not** exist, the system has no equilibrium thermal physics at fixed values of the corresponding natural variables.

The thermodynamical limit has other important properties in sound physical systems. First of all, for a given thermodynamical potential $\Phi(y_i, z_j)$ there is no single prescription for the corresponding ensemble[1] but the potential $\Phi(y_i, z_j)$ *in the thermodynamic limit* must be independent of the choices *up to an immaterial additive constant*. This property is called *universality of the thermodynamical limit*.

When the limit exists, but it is not unique (i.e. it depends on the detailed way we define the limit[2]), we should look at the result with a pinch of salt: the non-uniqueness may be not a signal of inconsistencies but of new *observable physical phenomena* which are of the utmost interest (see next subsection). In this situation universality of the thermodynamic limit is re-established once the relevant phenomena are duly taken into account.

Another crucial property is the independence of the state equations from the particular ensemble we use to compute them. This property is known as *the equivalence of ensembles* in the thermodynamic limit. Again, when it fails one should look for subtle physical phenomena and take them into due account.

The last general property is *convexity of the thermodynamical potential:* the second derivatives with respect to the natural variables of the function $\Phi(y_i, z_j)$ produced by the thermodynamical limit should have the sound physical sign discussed in Chap. 1. Proving that the thermodynamical potentials of interest *do are* "convex" is one of the main math problems arising from Statistical Physics.

Non-uniqueness of the Thermodynamical Limit
As mentioned above, the thermodynamical limit may be *non-unique* even for well-behaved systems which satisfy all the required thermodynamic conditions.

As a basic example consider a system composed by $O(10^{24})$ molecules of H_2O (water) in a thermal bath at temperature T and pressure P. Depending on T and P the water may be in a liquid, or solid (ice), or gaseous state (vapor). These

---

[1] Of course, some prescriptions are better than others.

[2] For instance there may be an order of limits issue. We shall encounter examples of this situation and see how subtle this issue may be in some class of systems. See in particular Chap. 7.

three phases have clearly very different equations of state, hence thermodynamical potentials with rather different functional forms. At generic T and P only one phase is in equilibrium, but along the coexistence curve,

$$\mu(T, P)_{\text{liquid}} = \mu(T, P)_{\text{gas}}, \tag{2.4}$$

both the liquid and the gas are at equilibrium, while at the triple point in the P-T plane

$$\mu(T, P)_{\text{liquid}} = \mu(T, P)_{\text{gas}} = \mu(T, P)_{\text{ice}} \tag{2.5}$$

all three phases coexist, cf. Sect. 1.11. This means that the thermodynamic limit of an ensemble of H_2O molecules at fixed T, P laying on a coexistence curve *is not unique*, but we have different limits which produce the inequivalent thermodynamical potentials describing pure liquid water, pure ice, or any mixture with a fraction x of ice coexisting with a fraction $(1 - x)$ of liquid water. How, why, and when this non-uniqueness appears is the subject of the theory of *phase transitions* which is one of the deepest topics in Statistical Mechanics. An introduction to the subject will be given in Chaps. 4 and 5.

Microsystem Classes

The underlying microscopic system of our thermal system may be of different kinds. Just to name a few typical examples:

μS1 it may be a *continuous classical system* with a phase space $\mathfrak{W}$ parametrized by $(\mathbf{q}_i, \mathbf{p}_j) \in \mathbb{R}^{3N}$ ($N = O(10^{23})$) and an "usual" Hamiltonian of the form

$$H = \sum_{i=1}^{N} \frac{\mathbf{p}_i^2}{2m} + \sum_{i<j} V(|\mathbf{q}_i - \mathbf{q}_j|) \tag{2.6}$$

μS2 it may be a *continuous quantum system* with a separable Hilbert space $\mathcal{H}$ of states and a quantum Hamiltonian H of the form (2.6), where now H is a Hermitian operator acting on $\mathcal{H}$ with a spectrum bounded below;

μS3 it may be a *field theory*—e.g. the electromagnetic field—treated at either the classical or quantum level;

μS4 it may be a *classical discrete system*, usually called a *lattice* system. The prototypical example is the Ising model in d dimensions which models a ferromagnetic material. We have a box K of size $L \in \mathbb{N}$ in $\mathbb{R}^n$ and at each point i of the finite lattice $\mathbb{Z}^n \cap K$ there is a "spin", σ_i that is, a classical variable which may take the two values ± 1. The energy of a spin configuration is

$$H = -J \sum_{\langle i,j \rangle} \sigma_i \sigma_j \tag{2.7}$$

where the notation $\langle i, j \rangle$ means that we sum only over nearest neighbor pairs of lattice points, and $J > 0$ is a coupling constant. The Ising model will be our main "hero" in Chaps. 4 and 5;

μS5 it may be a *quantum discrete system* (or a *quantum lattice system*) where now each "spin" σ_i is an operator acting on a finite-dimensional Hilbert space $\mathcal{H}_i$, so that the states of the full system belongs to the "huge" Hilbert space[3]

$$\mathcal{H}_N \equiv \bigotimes_{i=1}^{N} \mathcal{H}_i \tag{2.8}$$

which becomes quite subtle in the thermodynamic limit $N \to \infty$.

Remark 2.1 Classical discrete systems can be seen as special instance of quantum systems where the degrees of freedom $\{\sigma_i\}$ are quantum operators which commute, so that they can be all simultaneously diagonalized. Written in the diagonal basis, the Hamiltonian is just a classical function of the eigenvalues of the operators σ_i. Therefore all general results proved for the continuous quantum case apply equally to the discrete systems (classical or quantum).

Thermodynamical Potentials vs. Partition Functions

An *ensemble* is a statistical distribution of *microstates* ($\equiv$ states of the underlying microsystem) satisfying the constraint that the relevant natural (macroscopic) quantities y_j, z_i have their prescribed values.

For a classical continuous system (2.6), an *ensemble* is a *non-negative measure*

$$\phi = \rho \, \mathrm{d}^m q \, \mathrm{d}^m p, \qquad \rho \geq 0 \tag{2.9}$$

on the phase space $\mathfrak{W}$ of the microscopical system ($\dim \mathfrak{W} \equiv m = O(10^{24})$). In the quantum set-up an ensemble is a *density matrix*, i.e. a Hermitian non-negative operator ϱ. The *total mass* of the measure (resp. the *trace of the density matrix*) is called the *partition function Z*

$$Z = \int_{\mathfrak{W}} \rho(q, p) \, \mathrm{d}^m q \, \mathrm{d}^m p \quad \text{classical}$$
$$Z = \mathrm{Tr}\, \varrho \qquad\qquad\qquad \text{quantum} \tag{2.10}$$

where $\mathrm{Tr}(\cdot)$ stands for the trace over the Hilbert space $\mathcal{H}$.

---

[3] N is the number of "spin" degrees of freedom in the system.

The *normalized measure*

$$\frac{\rho}{Z}\, \mathrm{d}^m q\, \mathrm{d}^m p, \tag{2.11}$$

is then a *probability distribution.* The *normalized density matrix*

$$\frac{\varrho}{Z}, \tag{2.12}$$

has eigenvalues $0 \leq p_n \leq 1$ satisfying $\sum_n p_n = 1$ and hence can also be seen as a probability distribution. This allows to define thermal expectation values and correlation functions associated to each given ensemble

$$\langle O_1 \ldots O_r \rangle = \begin{cases} \frac{1}{Z}\int O_1 \cdots O_r\, \rho\, \mathrm{d}^m q\, \mathrm{d}^m p & \text{classical} \\ \frac{1}{Z}\mathrm{Tr}[O_1 \cdots O_r\, \varrho] & \text{quantum,} \end{cases} \tag{2.13}$$

where in the classical case the O_a are functions in the phase space $\mathfrak{W}$ while in the quantum set-up they are operators acting on the Hilbert space $\mathcal{H}$. The correlation functions (2.13) are essential tools to reconstruct the physics of the system, in particular to determine the reaction of the system to modifications of the external conditions (*response theory,* see Chap. 6).

The partition function Z_N of N identical copies of the system is Z^N. Therefore

$$\log Z \tag{2.14}$$

is an *additive* function (an extensive thermal quantity) which is a function of the natural variables y_j, z_i which are kept fixed in the definition of the ensemble. $\log Z(y_j, z_i)$ then looks like a natural candidate for the role of the thermodynamic potential associated to the ensemble in consideration. Using the Gibbs' trick and performing the thermodynamic limit, we get the proposal for the *thermodynamical potential whose natural variables are* (y_j, z_i):

$$\Phi(y_j, z_i) = \lim_{N \to \infty} \frac{1}{N} \log Z(N y_j, z_i). \tag{2.15}$$

Below we shall consider each standard statistical ensemble in turn and check that the log of its partition function becomes indeed a good thermodynamical potential in the limit (2.15). In particular we will verify its "convexity" properties. The standard ensembles are in one-to-one correspondence with the *four* classical thermodynamical potentials introduced in Chap. 1. They are *all physically equivalent* in the sense that they produce the same equations of state. The use of one or another ensemble in the study of a particular system is a matter of convenience not of principle.

2.2 REVIEW: Classical and Quantum Liouville Equation

The physical foundation of the statistical ensembles lays in the *Liouville equation*
that we review in this section. We start by recalling the Liouville equation of
classical mechanics [1, §. 6.6]. Let[4]

$$\mu = \frac{\Omega^m}{m!} \tag{2.16}$$

be the *Liouville volume $2m$-form* of the phase space $\mathfrak{W}$ (here $2m \equiv \dim \mathfrak{W}$). Suppose
we have a distribution of states in the classical phase space $\mathfrak{W}$ given by the non-
negative $2m$-form

$$\rho(q, p, t)\, \mu \tag{2.17}$$

where $\rho(q, p, t)$ is a (possibly time-dependent) *non-negative* function on the phase
space $\mathfrak{W}$. Often it is convenient to normalize the distribution so that its total mass is
1: then the distribution can be interpreted as a *probability measure*

$$p(q, p, t) = \frac{\rho(q, p, t)}{Z} \quad \text{where } Z \overset{\text{def}}{=} \int_{\mathfrak{W}} \rho\, \mu. \tag{2.18}$$

Z is the *partition function*, cf. Eq. (2.10). The continuity equation together with the
Liouville theorem yields the *classical Liouville equation* [1, §. 6.6][5]

$$\frac{\partial \rho}{\partial t} + [\rho, H]_{\text{PB}} = 0. \tag{2.19}$$

von Neumann introduced in Quantum Mechanics an operator ϱ—called the
density matrix—which is the quantum counterpart to the classical density function
ρ. The operator ϱ satisfies the quantum version of the Liouville equation.

We consider[6] a quantum dynamical system which at a given time t_0 is in one
between several possible states according to a probability law. For simplicity we
assume that the possible states form a discrete set of orthogonal states labeled by
a positive integer n. $|n\rangle$ stands for a normalized vector representing the n-th state.
Let p_n be the probability that our system is in the state $|n\rangle$. We define the *density
operator ϱ* to be

$$\varrho = \sum_n |n\rangle p_n \langle n|, \qquad 0 \le p_n \le 1, \qquad \sum_n p_n = 1. \tag{2.20}$$

---

[4] As always Ω is the canonical symplectic 2-form on the phase space $\mathfrak{W}$ (Appendix 1 of Chap. 1).

[5] $[A, B]_{\text{PB}}$ is the Poisson bracket of the two functions $A, B \in \Omega^0(\mathfrak{W})$ [1].

[6] The following discussion follows Dirac [2] §. 33.

ϱ is a Hermitian operator acting on the Hilbert space[7] $\mathcal{H}$ whose eigenvalues are non-negative. Since the $|n\rangle$'s are orthonormal (i.e. $\langle n|n'\rangle = \delta_{n,n'}$)

$$\operatorname{Tr}\varrho = \sum_n p_n = 1, \tag{2.21}$$

where, as before, $\operatorname{Tr}(\cdot)$ is the trace over the Hilbert space $\mathcal{H}$. Note that

$$\operatorname{Tr}\varrho^2 = \sum_n p_n^2 \leq 1, \tag{2.22}$$

so that ϱ is a Hilbert-Schmidt operator [3] acting on the separable Hilbert space $\mathcal{H}$, hence a *compact* operator.

Next we study the time-evolution of the operator ϱ. In Eq. (2.20) the states $|n\rangle$ evolve according to the Schrödinger equation

$$i\hbar \frac{d}{dt}|n\rangle = H|n\rangle, \tag{2.23}$$

while the p_n's remain constant because the system cannot jump from a state solving the Schrödinger equation to a different state. Then

$$i\hbar \frac{d\varrho}{dt} = i\hbar \sum_n \left(\frac{d|n\rangle}{dt} p_n \langle n| + |n\rangle p_n \frac{d\langle n|}{dt} \right) =$$
$$= \sum_n \left(H|n\rangle p_n \langle n| - |n\rangle p_n \langle n|H \right) = H\varrho - \varrho H \tag{2.24}$$

which is the *quantum Liouville equation* which reduces to the classical one (2.19) as $\hbar \to 0$ since in this limit the commutator of quantum operators becomes the Poisson bracket of the corresponding classical quantities

$$\left[A, B\right] \rightsquigarrow i\hbar \left[A_{\text{class.}}, B_{\text{class.}}\right]_{\text{PB}} \qquad \text{as } \hbar \to 0. \tag{2.25}$$

A density matrix ϱ *is in equilibrium* iff it does not evolve with time, that is, iff

$$[\varrho, H] = 0 \tag{2.26}$$

where the bracket is the quantum commutator (resp. the Poisson bracket) in the quantum (resp. classical) case.

---

[7] All our Hilbert spaces have at most *numerable* dimension.

If O is a Hermitian operator ($\equiv$ observable) of the quantum system, its expectation value on the quantum distribution ϱ is

$$\langle O \rangle = \mathrm{Tr}[O\,\varrho]. \tag{2.27}$$

The classical limit as $\hbar \to 0$ of the Hilbert-space trace of a quantum operator O with a smooth classical limit O_{class} is

$$\mathrm{Tr}\,O \rightsquigarrow \int_{\mathfrak{W}} O_{\mathrm{class}}\,\frac{\mu}{(2\pi\hbar)^m}, \tag{2.28}$$

a formula which expresses the Heisenberg indetermination principle [4]. Using this formula we see that the quantum distribution ϱ has as classical limit a Liouville density ρ. In particular the first line of Eq. (2.13) is the $\hbar \to 0$ limit of its second line.

2.3 Microcanonical Ensemble(s)

For the moment we focus on statistical models which describe *simple* thermal systems. Later we shall present the straightforward extension to arbitrary systems.

The most widely used ensemble is the *canonical* one (Sect. 2.6) because, typically, computations get easier in the canonical set-up. However the *microcanonical ensemble* is simpler to explain and justify conceptually, so, for didactical reasons, we start from the microcanonical ensemble and then deduce the other ones by their equivalence with the microcanonical one in the thermodynamic limit.

Physically speaking, the most obvious variables to keep fixed are the volume V and the energy U which have crystal clear microscopic meanings. For instance, our system may be a gas (classical or quantum) consisting of $O(10^{24})$ molecules closed inside a vessel which is *rigid* (so that its volume V is fixed) and *thermally isolating* (so that the system cannot exchange heat with its surroundings). In the general case we have a microscopic system closed in a finite box of volume V with fixed total energy U. The class of ensembles defined by the condition of fixed volume V and energy U are called *microcanonical ensemble(s)*.

The thermodynamical potential of the microcanonical ensemble must be an additive ($\equiv$ extensive) quantity whose natural variables are U and V. Comparing with table (1.75), we conclude

Fact 2.1 *The natural microcanonical potential is the* entropy $S(U, V)$.

As in Chap. 1 we have a forgetful map $\mathsf{F}|_{\mathcal{E}} : \mathfrak{W} \to \mathcal{E}$ from microstates to equilibrium thermal states: to a given macroscopic equilibrium state there correspond a huge set of microstates. The most obvious statistical invariant of an equilibrium thermal state $\theta \in \mathcal{E}$—which for our simple system is fully characterized by the fixed values of the volume V and total energy U—is the "number" of the corresponding microstates

$$\Omega(U, V) \stackrel{\text{def}}{=} \left| \mathsf{F}^{-1}(\theta) \right|_{V=V(\theta),\ U=U(\theta)}, \tag{2.29}$$

that is, the *measure* of the fiber $\mathsf{F}^{-1}(\theta)$. "Number" is a rather rough notion, and we may think of various ways of making the definition of $\Omega(U, V)$ mathematically precise for each given system.

We consider the various classes of microsystems $\mu S1$–$\mu S5$ listed in Sect. 2.1 in turn.

$\mu S1$ *Classical Continuous Systems* Consider a classical continuous system (2.6). The set of states with fixed energy U is the set of points in the level hypersurface of the Hamiltonian in the phase space $\mathfrak{W}$

$$\left\{ H(\boldsymbol{p}, \boldsymbol{q}) \equiv H(p_i, q^j) = U \right\} \subset \mathfrak{W}, \qquad i, j = 1, \ldots, m \equiv \tfrac{1}{2} \dim \mathfrak{W}. \tag{2.30}$$

Clearly it makes no sense to count "points" in the level-set. Luckily, the phase space $\mathfrak{W}$ comes with a well-defined canonical measure—the *Liouville volume* that we reviewed in the previous section

$$\mu = \frac{\Omega^m}{m!}. \tag{2.31}$$

It is natural to use the *Liouville total measure* of the energy level-set (suitably defined) as a substitute of the "number of its points" (which equals the total measure for a discrete set). To be meaningful, this definition of the "number" of states should satisfy a number of consistency requirements: in particular the resulting thermodynamical potential cannot depend on the absolute normalization of the Liouville measure (up to inessential additive constants) so that the resulting equations of state are *independent of all choices.*

To make the thermodynamical limit smooth, it is convenient to divide the Liouville measure by $N!$ where N is the number of "equal" particles in our microcanonical ensemble (molecules in the gas example). Morally speaking, we consider equivalent two configurations which differ only by a permutation of two "equal" particles: in the quantum case this is a fundamental physical principle which arises from the quantum indistinguishability of identical particles (see Chap. 3). This normalization of the measure is physically natural also in the classical case where the particles are in principle distinguishable, since it may be thought of as the choice of classical normalization selected by defining the classical system as the limit $\hbar \to 0$ of its quantum parent.[8] However we stress again that the equations

---

[8] As it is clear from Eq. (2.28), the "quantum motivated" normalization contains in addition a power of $\hbar$. A more precise discussion of normalizations will be given momentarily.

of state are *independent* of this natural choice of normalization: this is a matter of convenient choices not of basic principles. Anyhow this "quantum motivated" classical normalization makes things to look nicer in the thermodynamical limit since quantum physics is *by default* more regular than the classical one.

For a *classical continuous system* we may think of the following three definitions of the "number" ($\equiv$ total measure) $\Omega(U, V)$ of states "at fixed energy U"

$$\Omega(U, V)^{(1)} = \mathcal{N} \int_{\mathfrak{W}} \Theta\big(U - H(\boldsymbol{p}, \boldsymbol{q})\big) \frac{\mu}{N!} \tag{2.32}$$

$$\Omega(U, V)^{(2)} = \mathcal{N} \int_{\mathfrak{W}} \Theta\big(U - H(\boldsymbol{p}, \boldsymbol{q})\big) \Theta\big(H(\boldsymbol{p}, \boldsymbol{q}) - U + \delta U\big) \frac{\mu}{N!} \tag{2.33}$$

$$\Omega(U, V)^{(3)} = \mathcal{N} \int_{\mathfrak{W}} \delta\big(U - H(\boldsymbol{p}, \boldsymbol{q})\big) \frac{\mu}{N!} \tag{2.34}$$

where $0 < \delta U \ll U$ is a small "error bar" in the value of energy, and $\Theta(x)$, $\delta(x)$ are, respectively, the *Heaviside step function* and the *Dirac δ-function*

$$\Theta(x) \overset{\text{def}}{=} \begin{cases} 1 & x \geq 0 \\ 0 & x < 0 \end{cases} \qquad \delta(x) \overset{\text{def}}{=} \frac{d}{dx}\Theta(x) \quad \text{(as distributions)}, \tag{2.35}$$

while μ is the Liouville volume form as before. Here and through the book the symbol $\mathcal{N}$ stands for an inessential normalization constant which can be chosen according to convenience (we can always set it to 1).

While for a finite classical continuous system $\Omega(U, V)$ will depend on which one of the three definitions (2.32)–(2.34) we use, in the *thermodynamic limit* (if it exists) all three (and many others) prescriptions should give the same function $\Omega(U, V)$ (up to a constant factor) by universality of the thermodynamic limit.

Similar remarks apply *mutatis mutandis* to the other classes of microsystems.

μS2 Continuous Quantum Systems For quantum continuous systems the integration over the Liouville measure is replaced by a *trace* over the Hilbert space of states $\mathcal{H}$. Now this is an actual counting of states: indeed for a quantum Hamiltonian of the usual form

$$H = \frac{1}{2m} \sum_{i=1}^{N} \boldsymbol{p}^2 + V(\mathbf{r}_1, \ldots, \mathbf{r}_N) \tag{2.36}$$

where the particles are constrained to move inside a *finite box* of volume V and the potential $V(\mathbf{r}_j)$ is bounded below (say by zero), the *energy spectrum is discrete*.

More precisely: the Hermitian operator $1/(H+1)$ is *compact* [3] which implies:

(i) its spectrum

$$\mathsf{Spec}(H+1)^{-1} \equiv \{1/(E_n+1)\} \subset \mathbb{R} \tag{2.37}$$

is *discrete;*
(ii) zero is the only accumulation point of its eigenvalues $1/(E_n+1)$;
(iii) its eigenstates $|n\rangle$ are normalizable;
(iv) all its eigenspaces are finite-dimensional.

Then the space of states with energy $\leq U$ is a linear subspace $\mathcal{H}_{\leq U} \subset \mathcal{H}$ of the (finite box) Hilbert space $\mathcal{H}$ whose dimension as a complex vector space is a *well-defined non-negative integer* which actually counts the number of states with energy $\leq U$.

The orthogonal projector $P_{\leq U}$ onto states of energy $\leq U$

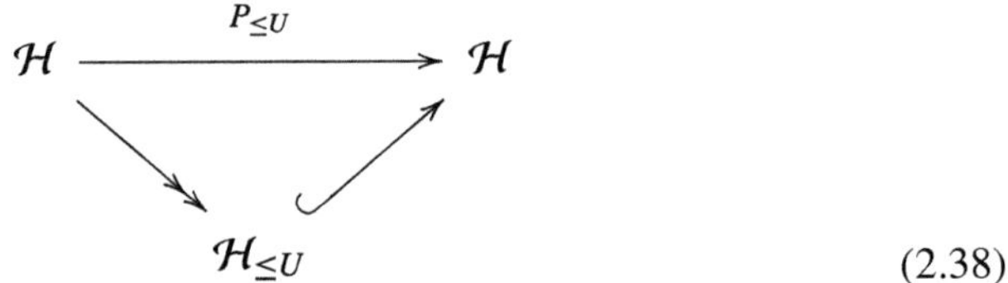

$$\tag{2.38}$$

is the Hermitian operator[9]

$$P_{\leq U} = \Theta(U-H) \equiv \int_{-\infty}^{+\infty} \frac{dt}{2\pi i} \frac{e^{iUt}}{t-i\epsilon} e^{-iHt} \tag{2.39}$$

while the orthogonal projector into the space of states with energy in the range $U - \delta U < E \leq U$ is

$$P_{\leq U}\left(1 - P_{\leq U - \delta U}\right) \equiv P_{\leq U} - P_{\leq(U-\delta U)}. \tag{2.40}$$

---

[9] Recall that if A is a Hermitian operator acting on a Hilbert space $\mathcal{H}$ and $f(x)$ is a real function continuous on its spectrum $\mathsf{Spec}\,A$, $f(A)$ is the Hermitian operator whose eigenstates $|n\rangle$ are the eigenstates of A (which satisfy $A|n\rangle = a_n|n\rangle$, $a_n \in \mathbb{R}$) and whose eigenvalues are the functions $f(a_n)$ of the eigenvalues a_n of A, that is, we have $f(A)|n\rangle = f(a_n)|n\rangle$ with $\{|n\rangle\}$ an orthonormal basis of $\mathcal{H}$. If $f(x) = \sum_k f_k x^k$ is an entire function, $f(A) = \sum_k f_k A^k$. If $f(x) = \int dy\, F(x,y)$ is an integral representation of $f(x)$, we have $f(A) = \int dy\, F(A,y)$.

Then the number of states with energy $E \leq U$ is

$$\Omega(U, V)^{(1)} \equiv \dim \mathcal{H}_{\leq U} = \mathrm{Tr}\left[P_{\leq U}\right] =$$

$$= \mathrm{Tr}\left[\Theta(U - H)\right] = \int_{-\infty}^{+\infty} \frac{dt}{2\pi i} \frac{e^{iUt}}{t - i\epsilon} \, \mathrm{Tr}\left[e^{-iHt}\right], \tag{2.41}$$

while the number of states with energy in the interval $U - \delta U < E \leq U$ is

$$\Omega(U, V)^{(2)} \equiv \dim \mathcal{H}_{\leq U} - \dim \mathcal{H}_{\leq U - \delta U} = \mathrm{Tr}\left[P_{\leq U}\right] - \mathrm{Tr}\left[P_{\leq(U - \delta U)}\right] =$$

$$= \mathrm{Tr}\left[\Theta(U - H)\right] - \mathrm{Tr}\left[\Theta(U - \delta U - H)\right] =$$

$$= \int_{-\infty}^{+\infty} \frac{dt}{2\pi i} \frac{e^{iUt}(1 - e^{-i\delta U t})}{t - i\epsilon} \, \mathrm{Tr}\left[e^{-iHt}\right].$$

$$\tag{2.42}$$

In the semiclassical limit $\hbar \to 0$, we have the correspondence

$$\mathrm{Tr}(O) \rightsquigarrow \int_{\mathfrak{W}} O(\boldsymbol{p}, \boldsymbol{q}) \frac{\mu}{(2\pi\hbar)^m \, N!}, \tag{2.43}$$

where O is any quantum operator with a smooth classical limit

$$O \rightsquigarrow O(\boldsymbol{p}, \boldsymbol{q}) \in \Omega^0(\mathfrak{W}) \quad \text{as } \hbar \to 0, \tag{2.44}$$

such that the LHS of (2.43) is defined (i.e. convergent). Hence in the limit $\hbar \to 0$ the quantum total measure and the classical one $\Omega^{(a)}$ differ only in the overall normalization

$$\Omega^{(a)}(U, V) \rightsquigarrow \frac{1}{(2\pi\hbar)^m \, N} \Omega^{(a)}(U, V)_{\mathrm{class}} \quad \text{as } \hbar \to 0, \quad a = 1, 2, \tag{2.45}$$

a difference which is totally irrelevant since it amounts to a mere constant shift of the thermodynamic potential. Given that the state count is well defined in Quantum Mechanics—where the $\Omega^{(a)}(U, V)$'s are actual number of states!—this formula also explains why the normalization with an extra factor $1/N!$ makes the thermodynamical limit smooth. It may be convenient to choose the normalization factor N in the classical measure to be

$$N = \frac{1}{(2\pi\hbar)^m} \tag{2.46}$$

to make the thermodynamic limit even smoother. Nowadays this quantum-inspired normalization of the classical Liouville measure is the most widely used one.

μS4–μS5 Discrete Systems The set of states of a classical discrete system is discrete by definition, and the counting of states with given energy is defined in the obvious way. If the discrete system is quantum, the counting of states of energy $\leq U$ is again given by the dimension of the subspace $\mathcal{H}_{\leq U} \subset \mathcal{H}$ of the Hilbert space. Using Remark 2.1, we can apply this definition also to classical discrete systems: the Hilbert-theoretic definition coincides with the obvious one in this case. While conceptually straightforward, the actual count of the number of states with given energy may be not easy in presence of a large number of degrees of freedom.

Definition 2.1 The *microcanonical partition function* is the "number of states" ($\equiv$ total measure) $\Omega(U, V)$ at fixed U and V.

We shall check momentarily the independence of the partition function on the detailed counting prescription in the thermodynamic limit (universality of the microcanonical ensemble). We shall also check that this identification is correct in the sense that $\Omega(U, V)$ satisfies all the conditions outlined in Sect. 2.1 for a sound partition function (after taking the thermodynamic limit, of course!).

Microcanonical Probability Distribution
As anticipated in Sect. 2.1, dividing the ensemble distribution by the partition function we get a probability distribution p normalized to 1. For (say) the classical continuous case, the probability distribution for the first definition of the microcanonical ensemble is

$$p^{(1)} = \frac{1}{\Omega^{(1)}(U, V)} \, \Theta(U - H) \, \frac{\mu}{N!}, \tag{2.47}$$

and similar formulae apply to other choices of the measure and other classes of microsystems. Consistency requires that in a sound system the probability measure is totally independent of the detailed choices (in particular the same for all three ensemble measures in Eqs. (2.32)–(2.34)) *after* we take the thermodynamic limit. We shall illustrate this property in an example below.

We stress that the overall normalization of the ensemble measure drops out of the probability measure p, so that, again, the overall normalization of the partition function is a matter of convenience not of principle.

Entropy and Equations of State

The number of states of K identical copies of our system goes roughly as

$$\Omega_K(U, V) \approx \Omega(KU, KV) \approx \Omega(U, V)^K, \tag{2.48}$$

since U and V are both extensive quantities. As advertised in Sect. 2.1, we take the logarithm to get an additive potential which scales linearly with K (for $K \gg 1$), and the thermodynamical limit becomes

$$\Phi(U, V) = \lim_{K \to \infty} \frac{1}{K} \log \Omega(KU, KV). \tag{2.49}$$

From Fact 2.1 we expect the resulting potential to be a function of the entropy

$$\Phi = \Phi(S(U, V)). \tag{2.50}$$

Since both S and Φ are extensive quantities, the function must be homogeneous of degree 1, that is,

Fact 2.2 *The thermodynamical potential* $\Phi(U, V)$ *is* proportional *to the entropy* $S(U, V)$, *that is,*

$$S(U, V) = k\, \Phi(U, V) \approx k \log \Omega(U, V), \tag{2.51}$$

where $k > 0$ *is an* universal *constant. Equation* (2.51) *is called the* Boltzmann equation *and* k *the* Boltzmann constant.

We can compute k in any convenient particular system, and the value we get will apply to all other statistical systems by universality.

To substantiate Fact 2.2 we have to show that $\log \Omega(U, V)$ has the appropriate physical properties to be an *entropy*. In particular: *it should not decrease in any physical process, and be maximal at equilibrium.* The first property requires $\log \Omega(U, V)$ to be *super-additive:* if we put together two systems we must have

$$\log \Omega(U_1 + U_2, V_1 + V_2) \geq \log \Omega(U_1, V_1)_1 + \log \Omega(U_2, V_2)_2. \tag{2.52}$$

But this is a tautology: the set of states of the combined system at energy $U_1 + U_2$ contains all states where the two original subsystems have energies U_1 and U_2, plus other states where some energy has flown between the two subsystems. Therefore the total number of states of the combined system is *not smaller* than the product of the number of states in the separated original systems, that is, $\log \Omega$ is *super-additive.*

In addition $\log \Omega(U, V)$ should be *maximal at equilibrium,* that is, the micro-canonical ensemble should be the distribution with the largest total measure between all distributions with the given energy U. Again this is automatically true since the microcanonical ensemble counts *all* states of the given energy U and any other distribution at fixed U may differ from it only because it misses some of the states with energy U. Such alternative distributions have automatically a smaller $\log \Omega$.

Once we have identified entropy (up to normalization) with the $\log$ of the microcanonical partition function Ω, Eq. (2.51), from the general thermodynamical formalism of Chap. 1 we get

$$
\frac{1}{T} = k \, \frac{\partial \log \Omega(U, V)}{\partial U}, \qquad \frac{P}{T} = k \, \frac{\partial \log \Omega(U, V)}{\partial V}, \tag{2.53}
$$

from which we can read the equations of state once we have computed the microcanonical partition function $\Omega(U, V)$. Moreover

$$
\kappa \, \frac{\partial^2 \log \Omega(U, V)}{\partial U^2} = -\frac{1}{T^2} \left(\frac{\partial T}{\partial U} \right)_V = -\frac{1}{T^2} \left(\frac{\partial U}{\partial T} \right)_V^{-1} = -\frac{1}{T^2 \, C_V} < 0, \tag{2.54}
$$

that is,

Fact 2.3 *In a sound thermal system with* $C_V > 0$, $\log \Omega$ *is a* concave *function of* U.

Example: The Classical Ideal Gas

We work out the example of a classical monoatomic ideal gas for four reasons: to illustrate the technique, to check that we recover the well-known phenomenological equations of state, to verify universality of the ensemble, and to compute the Boltzmann constant k.

At the microscopic level the monoatomic ideal gas consists of N classical "molecules" moving in a 3-dimensional box of volume V. "Monoatomic" means that we model the molecules by Newtonian material points whose only degrees of freedom are their positions inside a cubic box in $\mathbb{R}^3$, without any additional rotational or vibrational degree of freedom. Thus the ideal gas is a classical free system with $3N$ degrees of freedom and Hamiltonian

$$
H = \sum_{i=1}^{N} \frac{\mathbf{p}_i{}^2}{2m}. \tag{2.55}
$$

The region $H \leq U$ in the phase space $\mathfrak{W}$ is then the product of a region of volume V^N in the coordinate space $\mathbb{R}^{3N}$ times a solid sphere (a *ball*) of radius $\sqrt{2m\,U}$ in the momentum space $\simeq \mathbb{R}^{3N}$. We get

$$
\Omega^{(1)} = \frac{V^N}{N!} \, V_{3N}(\sqrt{2mU}) \tag{2.56}
$$

$$
\Omega^{(2)} = \frac{V^N}{N!} \left[V_{3N}(\sqrt{2mU}) - V_{3N}(\sqrt{2m(U - \delta U)}) \right] \tag{2.57}
$$

$$\Omega^{(3)} = \frac{V^N}{N!} \frac{\partial}{\partial U} V_{3N}(\sqrt{2mU})$$
(2.58)

where

$$V_{3N}(r) = \frac{1}{3N} S_{3N} \, r^{3N}$$
(2.59)

is the volume of the ball of radius r in $\mathbb{R}^{3N}$ while S_{3N} is the volume of the unit sphere $S^{3N-1} \subset \mathbb{R}^{3N}$ that we are going to compute.

Volume S_m of the Unit Sphere in $\mathbb{R}^m$
We compare the Gaussian integral in $\mathbb{R}^m$ computed in Cartesian and polar coordinates

$$\sqrt{\pi}^{\,m} = \int_{-\infty}^{+\infty} dx_1 \ldots dx_m \, e^{-(x_1^2 + \cdots + x_m^2)} = S_m \int_0^{\infty} r^{m-1} \, dr \, e^{-r^2} =$$
$$= \frac{1}{2} S_m \int_0^{\infty} s^{m/2-1} \, ds \, e^{-s} = \frac{1}{2} S_m \, \Gamma(m/2).$$
(2.60)

hence

$$S_m = \frac{2\sqrt{\pi}^{\,m}}{\Gamma(m/2)}$$
(2.61)

We return to the computation of the partition function. Equation (2.56) becomes

$$\Omega^{(1)} = \frac{2\sqrt{\pi}^{\,3N} V^N}{N! \, \Gamma(3N/2)} (2m \, U)^{3N/2},$$
(2.62)

and we have corresponding formulae for $\Omega^{(2)}$, $\Omega^{(3)}$. The entropy is then

$$S = kN \log V + \frac{3kN}{2} \log U + C_N,$$
(2.63)

$$\text{where} \quad C_N = k \log \frac{2 \, (2\pi m)^{3N/2}}{\Gamma(N+1) \, \Gamma(3N/2)}.$$
(2.64)

In conclusion, the equations of state of a classical ideal monoatomic gas are

$$\frac{P}{T} = \left(\frac{\partial S}{\partial V}\right)_U = \frac{kN}{V} \quad \Rightarrow \quad PV = kNT, \tag{2.65}$$

and

$$\frac{1}{T} = \left(\frac{\partial S}{\partial U}\right)_V = \frac{3kN}{2}\frac{1}{U} \quad \Rightarrow \quad U = \frac{3kN}{2}T. \tag{2.66}$$

Equation (2.65) has the same form as the well-known "phenomenological" equation of state of an ideal gas (Boyle and Gay-Lussac Laws) which reads

$$PV = nRT, \tag{2.67}$$

where n is the *number of moles* in our sample of gas and R the ideal gas constant

$$R \approx 8.314 \text{ J}/K°\text{mol}. \tag{2.68}$$

Since $N = nN_A$ ($N_A \approx 6.022 \times 10^{23}$ is the Avogadro number), comparing Eqs. (2.66) and (2.67) we get the *Boltzmann constant*

$$k = \frac{R}{N_A} = 1.380 \times 10^{-16} \text{ erg}/K°. \tag{2.69}$$

As first noticed by Planck, it is natural to choose units for temperature such that $k = 1$, that is, to measure temperatures in units of energy. Equation (2.66) becomes

$$U = \frac{3R}{2}nT \equiv C_V T, \tag{2.70}$$

which yields the well-known expression

$$C_V = \frac{3nR}{2} \tag{2.71}$$

for the heat capacity of n moles of a monoatomic ideal gas.

Using the Stirling asymptotic formula for the Γ-function as $x \ggg 1$ [5]

$$\log \Gamma(x) = \left(x - \frac{1}{2}\right)\log x - x + \frac{1}{2}\log(2\pi) + O(1/x) \tag{2.72}$$

we check that with the normalization $1/N!$ the thermodynamic limit is smooth (i.e. we do not need to "renormalize" the additive constant)

$$S = kN\left[\log(V/N) + \frac{3}{2}\log(U/N) + \frac{5}{2} + \frac{3}{2}\log(2m\pi)\right] + O(\log N). \qquad (2.73)$$

Ensemble Universality Next we check *universality*, i.e. independence from the precise definition of the microcanonical ensemble in the thermodynamic limit. The partition function $\Omega^{(2)}$ which counts states in the energy range $U - \delta U < E \le U$ is

$$\Omega^{(2)} = \frac{V^N}{N!} V_{3N}(\sqrt{2mU}) \left(1 - \frac{V_{3N}(\sqrt{2m(U - \delta U)})}{V_{3N}(\sqrt{2mU})}\right) \qquad (2.74)$$

and the factor in the big parenthesis goes to 1 as $N \to \infty$, while

$$\Omega^{(3)} = \frac{3N}{2U}\,\Omega^{(1)}, \qquad (2.75)$$

so that

$$\frac{1}{N}\left(\log\Omega^{(3)} - \log\Omega^{(1)}\right) = \frac{1}{N}\left(\log N + O(1)\right) \to 0 \quad \text{as } N \to \infty. \qquad (2.76)$$

Let us describe the physical mechanism underlying ensemble universality for continuous classical systems. Consider the first ensemble (2.32) where we keep *all* states with energy up to U. Let $\delta U \ll U$, and look at the fraction of states in this ensemble whose energy is in the interval $[U - \delta U, U]$: it is given by

$$1 - \frac{V_{3N}(\sqrt{2m(U - \delta U)})}{V_{3N}(\sqrt{2mU})} = 1 - \left(\frac{U - \delta U}{U}\right)^{3N/2}. \qquad (2.77)$$

In the limit $N \to \infty$ *almost all* states in the ensemble have energy in the range $[U - \delta U, U]$ for *any* positive δU. We express this fact by saying that the *typical state* in the ensemble (2.32) has energy $\approx U$ in the thermodynamical limit.

Definition 2.2 A class of states of an ensemble is said to be *typical* (in the ensemble) iff their probability is 1 in the thermodynamic limit, i.e. iff *non-typical* states have *probability zero* in the thermodynamic limit.

Only typical states matter in computing thermodynamical potentials and equations of states. Since the typical states are the same ones for *all choices* of the microcanonical state measure, we conclude that all microcanonical ensembles are equivalent (provided the thermodynamic limit is well defined).

Remark 2.2 The universality of the microcanonical ensemble for *continuous systems* (in the sense of the equivalence of $\Omega^{(1)}$, $\Omega^{(2)}$, and $\Omega^{(3)}$, see §. 3.3.14 of

[6]) has the following consequence. $\Omega^{(1)}(U)$ counts all states up to energy U so is monotonically non-decreasing as a function of U. Then

$$\frac{1}{kT} \equiv \frac{\partial}{\partial U} \log \Omega^{(1)}(U) \geq 0, \tag{2.78}$$

that is, the absolute temperature T is always non-negative in continuous system. The case of *discrete* systems is subtler, as we are going to see.

Peculiar Phenomena in Discrete Systems

Consider a discrete classical system consisting of N "spins" σ_i, with may take the two values ± 1, coupled to a background magnetic field B via their magnetic dipole moment μ. The Hamiltonian is

$$H = \mu B \sum_{i=1}^{N} \sigma_i + \mu B N \tag{2.79}$$

where the constant term $\mu B N$ is added to normalize the energy so that its minimal value is 0. The energy of a spin configuration depends only on the number N_+ of spins pointing in the positive direction $+1$

$$U = \mu B(N_+ - N_-) + \mu B N = 2\mu B N_+. \tag{2.80}$$

The number of states with energy $U = 2\mu B N_+$ is

$$\Omega^{(3)}(u)_N = \binom{N}{N_+} = \frac{N!}{N_+!\,(N - N_+)!} \approx \frac{N!}{(Nu)^{Nu}(N(1-u))^{N(1-u)}} \tag{2.81}$$

where we set $u \equiv U/(2\mu B N) \equiv N_+/N$ for the rescaled energy per spin and used the Stirling asymptotic formula (2.72) for the factorial. Note that $0 \leq u \leq 1$. The function $\log \Omega^{(3)}(u)$ for $N = 50$ is plotted in Fig. 2.1. Now the entropy per spin in

Fig. 2.1 The function in Eq. (2.81) for $N = 50$

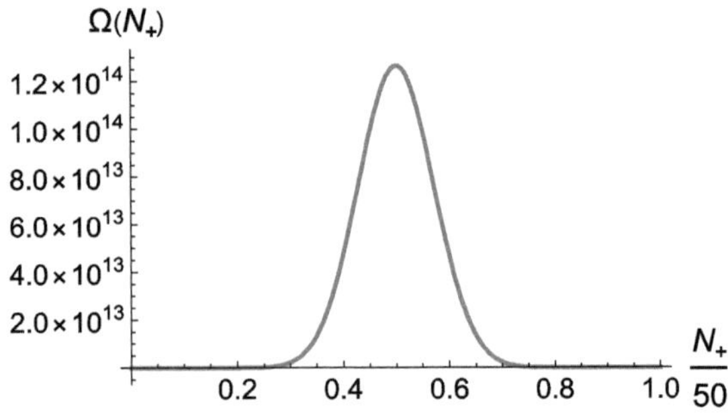

Fig. 2.2 The entropy (2.82) for the decoupled spins in a magnetic field

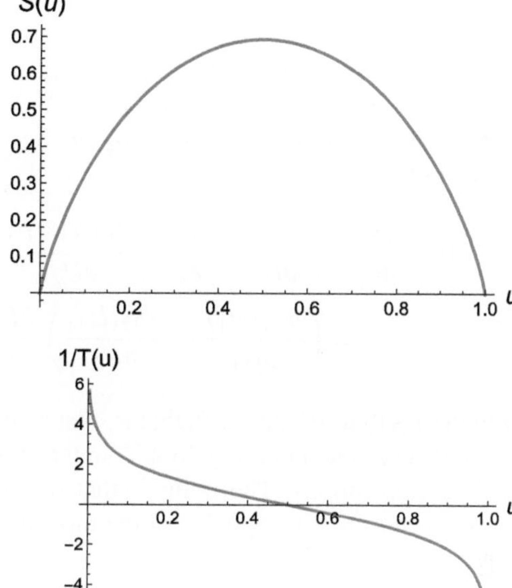

Fig. 2.3 The inverse temperature (2.83) for the decoupled spins in a magnetic field as a function of the (rescaled) energy u

the thermodynamic limit is

$$S(u)/k = \lim_{N\to\infty} \frac{1}{N} \log \Omega(u)_N = -u \log u - (1-u)\log(1-u) + \text{const.} \qquad (2.82)$$

The entropy as a function of u is plotted in Fig. 2.2. Notice that it goes to zero as the energy reaches its maximum at $u \to 1$: not an "usual" behavior for an entropy!

The inverse temperature

$$\frac{1}{kT} \stackrel{?}{=} \frac{1}{k}\left(\frac{\partial S}{\partial u}\right)_V = \log\left(\frac{1-u}{u}\right) \qquad (2.83)$$

is plotted in Fig. 2.3. $1/kT$ is positive when $u < \frac{1}{2}$, but it becomes negative for $u > \frac{1}{2}$, so that at the face of it *we have realized a system with **negative** absolute temperature*. Negative absolute temperature is something which at first looks not to be consistent with the Laws of Thermodynamics. However, according to Ramsey [7], the sacred Laws are not violated at negative temperature provided we formulate them in the proper way [7].

Let us summarize the physicists' common lore on the issue. We start from formal considerations. The first point to stress is that the term "negative temperatures" is rather misleading since these temperatures are *hotter* than all positive temperatures including $T = +\infty$. We say that the temperature T_2 is "hotter" than T_1 if putting in contact two bodies at temperatures T_1, T_2 the heat flows from the body at $T = T_2$ to the one at temperature T_1. That the heat flows from the body at negative temperature to the body at positive temperature is a consequence of the Second Law. Suppose we have a body at temperature $T_1 > 0$ and a body at temperature $T_2 < 0$. We put the two bodies in contact. The total energy $U = U_1 + U_2$ is conserved while the total entropy is $S = S_1(U_1) + S_2(U_2)$. By the Second Law the entropy increases with the time t, so

$$0 < \frac{dS}{dt} = \frac{\partial S_1(U_1)}{\partial U_1} \frac{\partial U_1}{\partial t} + \frac{\partial S_2(U_2)}{\partial U_2} \frac{\partial U_2}{\partial t} =$$
$$= \left(\frac{\partial S_1(U_1)}{\partial U_1} - \frac{\partial S_2(U_2)}{\partial U_2} \right) \frac{\partial U_1}{\partial t} = \left(\frac{1}{T_1} + \frac{1}{|T_2|} \right) \frac{\partial U_1}{\partial t}, \tag{2.84}$$

which says that $\partial U_1/\partial t > 0$, that is, that heat is flowing *into* the body at temperature $T_1 > 0$. The statement may look counterintuitive, but this impression is mainly due to bad conventions. The point is that the usual absolute temperature scale T, as introduced long time ago, is not the most natural one: the natural expression which enters in all equations is

$$y \equiv -\beta \equiv -\frac{1}{kT} \tag{2.85}$$

which starts from the absolute zero at $y = -\infty$ then increases with the temperature until it reaches $y = 0^-$ at $T = +\infty$, next at $y = 0^+$ we have $T = -\infty$ and then y increases to $y = +\infty$ at $T = 0^-$. In terms of the $y \equiv -\beta$ scale the heat always flows from bodies at larger values of y to body at smaller values of y.

There is no logical paradox with negative temperatures, and the Laws of Thermodynamics are still valid (putting some care in the statements [7]) as long as there are no adiabatic reversible process which connects states at negative and positive temperatures. This rules out the possibility of Carnot cycles between temperatures of opposite sign which would have an efficiency greater than 1.

The presence of negative temperature requires the system to have a Hamiltonian which is bounded above and below, as in the example (2.79). Indeed in this case U takes value in a segment, not on the convex cone $\mathbb{R}_{>0}$, and many of the arguments in Chap. 1 no longer apply.

These are the formal statements. *But are negative temperatures for real?*

Can negative temperatures be realized experimentally in the laboratory?

Yes and no. If we define an equilibrium state as a state which remains unchanged forever, no strict equilibrium state of negative temperature may exist in the real world, since sooner or later our sample of negative-temperature material will get in equilibrium with the rest of the laboratory (which is at non-negative temperature). However there are systems, such as LiF crystals, whose spin degrees of freedom thermalize between themselves in 10^{-5} s, while they require the enormous time (on the atomic scale) of 5 min to thermalize with the rest of the world. If one prepares the sample of LiF in a suitable way, by populating the higher energy states, for the atomic eternity of a few minutes the LiF crystal effectively behaves "as if it was" in an equilibrium state at negative temperature. See e.g. the celebrated experiments by Purcell and Pound [8]. After several minutes the sample will thermalize into a positive temperature state. Thus, in the *strict* sense all real equilibrium states are at non-negative temperature, although the math abstraction of "negative temperature" is not inherently contradictory and may be used within its proper limitations.

2.4   Modern Interpretations of Entropy

In Chap. 1 we reviewed the notion of *entropy* as it was originally deduced by Clausius from the Second Law, and in Sect. 2.3 we discussed its first-principle interpretation by Boltzmann as the logarithm of the number of microstates. But there is much more in the entropy concept. *Entropy* is a universal notion that goes well beyond thermal physics: it is the central idea in several different sciences such as probability theory, information theory, geometry, topology, and many other subjects. Arguably the most deep and general definition of entropy is the one in information theory.

Shannon Entropy

Shannon entropy is a fundamental notion in *information theory* which measures the "indeterminacy" (ambiguity) in an information. *Entropy* is used to define and measure the concept of efficient coding of information, see e.g. [9]. In particular the minimal number of bits necessary to construct a code which is "non-ambiguous" is equal to the Shannon entropy of its alphabet (*Shannon's source coding theorem for symbol codes*) [9].

We shall be very schematic. Consider a system which can be in any one of N states labelled by $i = 1, \ldots, N$. The classical probability of being in the i-th state is $p_i \geq 0$, normalized so that the total probability is 1

$$\sum_{i=1}^{N} p_i = 1. \tag{2.86}$$

Definition 2.3 The *Shannon entropy* of the N-state probability distribution $\{p_i\}$ is

$$\mathsf{S}(p_i) \overset{\text{def}}{=} -\sum_{i=1}^{n} p_i \log p_i. \tag{2.87}$$

Remark 2.3 In information theory $\log$ stands for the logarithm in base 2, since $\mathsf{S}(p_i)$ counts the number of "bits" required for an unambiguous code whose symbols have frequencies p_i. In physics we use the formula (2.87) with $\log$ the natural logarithm which amounts to multiply the Shannon entropy by a factor $\log_e 2 \approx 0.693147$.

Up to overall normalization, the Shannon entropy is the only continuous function of the probabilities $\{p_i\}$ with the following four natural properties:

Sh1 $S(p_i)$ is *non-negative*;

Sh2 $\max_{\{p_i\}} S(p_i)$ is a monotonic increasing function of the number N of states;

Sh3 $S(p_i)$ is *additive* under the composition $\{p_i q_j\}_{i,j \in I \times J}$ of two independent probability distributions $\{p_i\}_{i \in I}$ and $\{q_j\}_{j \in J}$, that is:

$$S(p_i q_j) = -\sum_{i,j} p_i q_j \log(p_i q_j) = -\sum_{i,j}(p_i q_j \log p_i + p_i q_j \log q_j)$$

$$= -\sum_i p_i \log p_i - \sum_j q_j \log q_j = S(p_i) + S(q_j);$$

$$(2.88)$$

Sh4 $S(p_i)$ is *intrinsic* in the sense that it is invariant under *coarse graining* of the probability distribution $\{p_i\}$, that is, we can equivalently describe a N-state probability distribution as, say, a distribution into two batches,[10] consisting of k and $N - k$ states respectively. The probabilities that the system belong to either one of the two batches of states are respectively

$$q_1 = \sum_{i=1}^{k} p_i, \qquad q_2 = \sum_{i=k+1}^{N} p_i, \qquad (2.89)$$

while each batch is a probability distribution in its own right with probabilities $\{p_i/q_1\}$ and $\{p_i/q_2\}$, respectively. Naturality requires that the entropy of the original distribution $\{p_1, \ldots, p_N\}$ can be equivalently computed in terms of the single distribution $\{p_i\}$ or using the several batches distributions. In other words: the entropy of the original probability distribution $\{p_i\}$ must be the sum of the entropy of the probability distribution $\{q_1, q_2\}$ in the two batches plus the sum of the expected values of the "internal" entropy in each one of the several batches. This yields a functional equation for the entropy as a function of the probabilities:

$$S(p_1, \ldots, p_N) = S(q_1, q_2) +$$

$$+ q_1 S(p_1/q_1, \ldots, p_k/q_1) + q_2 S(p_{k+1}/q_2, \ldots, p_N/q_2). \qquad (2.90)$$

---

[10] In general we can split our states in n batches. We can get this situation inductively by splitting one batch in two sub-batches at each step. Hence the validity of the functional equation for the splitting in two batches, Eq. (2.90), implies the validity of the corresponding equation for any finite number n of batches.

We leave to the diligent reader to prove that the Shannon entropy (2.87) is the only continuous function of the p_i's with properties **Sh1–Sh4** up to overall normalization. Lazy readers may have a look to §. 11.3 of [10] or §. 7.2 of [11].

Lemma 2.1 *(1) The Shannon entropy is ≥ 0 and zero only if*

$$
p_i = \begin{cases} 1 & i = i_0 \\ 0 & i \neq i_0 \end{cases}
\tag{2.91}
$$

for some $i_0 \in \{1, \ldots, N\}$. (2) The Shannon entropy is maximal when

$$
p_i = 1/N \quad \text{for all } i = 1, \ldots, N,
\tag{2.92}
$$

that is, when the probability is equidistributed between the N states. The maximal value of the entropy for a N-state probability distribution is $\log N$.

Proof (1) Since $p_i \in [0, 1]$, each summand $0 \leq -p_i \log p_i \leq e^{-1}$ while $-p_i \log p_i = 0$ iff $p_i \in \{0, 1\}$. (2) We introduce a Lagrange multiplier to enforce the condition (2.86). The extrema of the entropy are given by

$$
0 = \frac{\partial}{\partial p_i} \left[-\sum_{j=1}^{N} p_j \log p_j + \lambda \left(\sum_{j=1}^{N} p_j - 1 \right) \right] = -\log p_i - 1 + \lambda
\tag{2.93}
$$

which implies $p_i = p_j \equiv 1/N$ for all i, j. $\qquad\square$

From Lemma 2.1 we see that the Shannon entropy $\mathsf{S}(p_i)$ is a measure of the "uncertainty" associated with the probability distribution $\{p_i\}$: if $\mathsf{S}(p_i) = 0$ the system is in some state i_0 with 100% certainty, whereas when the entropy is maximal the system may be in any state with equal probability.

The Principle of Maximal Shannon Entropy

The *indifference principle* in probability theory says that when there are N possible outcomes, about which "a priori we are equally ignorant", namely we have no reason to think that one of them is more likely to occur than another one, we must assign *equal probability* $1/N$ to all outcomes. From Lemma 2.1 we see that the "equally ignorant" distribution is just the one which maximizes the (Shannon) entropy. In other words, a priori ignorance leads to (i.e. justifies) the *principle of maximal entropy* which we deduced in Chap. 1 from thermodynamics at equilibrium.

More generally, we may be in a situation where—while we do not know much about the system—we are not "totally ignorant" about it, but have some partial information. This leads to

The Maximal Entropy Principle *Given partial information about a system, one has to assign to that system the probability distribution with the largest entropy among all the probability distributions consistent with the given partial information.*

A particularly important application of the principle is when we know the expectation value $\langle F \rangle$ of some function of the state $F : \{1, \ldots, N\} \to \mathbb{R}$:

$$\langle F \rangle = \sum_{i=1}^{N} F(i)\, p_i. \tag{2.94}$$

To find the probability distribution that maximizes the entropy in the space of all distributions with the given expectation value $\langle F \rangle$ of F, we use the method of Lagrange multipliers to enforce the two conditions on the probabilities $\{p_i\}$:

$$\begin{aligned}
0 &= \frac{\partial}{\partial p_j}\left(-\sum_i p_i \log p_i - \lambda\left(\sum_i F(i)\, p_i - \langle F \rangle\right) - \mu\left(\sum_i p_i - 1\right)\right) \\
&= -\log p_j - 1 - \lambda\, F(j) - \mu.
\end{aligned} \tag{2.95}$$

Solving this equation we conclude:

The probability distribution $\{p_j\}$ with the maximal Shannon entropy $\mathsf{S}(p_j)$ between all distributions with the given expectation value $\langle F \rangle$ of F is

$$p_j = \frac{e^{-\lambda F(j)}}{Z(\lambda)} \tag{2.96}$$

where $Z(\lambda) \equiv \sum_i e^{-\lambda F(i)}$ is the *partition function* with associated *potential* $-\log Z(\lambda)$. The Lagrange multiplier λ is determined by imposing that the expectation value of $\langle F \rangle$ has the given value, i.e. by inverting the function

$$\lambda \mapsto \langle F \rangle = -\frac{\partial}{\partial \lambda}\log Z(\lambda) \equiv \frac{\sum_j F(j)\, e^{-\lambda F(j)}}{\sum_i e^{-\lambda F(i)}} \tag{2.97}$$

The corresponding maximal value of the Shannon entropy is

$$\mathsf{S} = -\sum_i p_i \log p_i = -\frac{1}{Z}\sum_i e^{-\lambda F(i)}\left(-\lambda F(i) - \log Z\right) = \lambda\langle F \rangle + \log Z, \tag{2.98}$$

that is,

Corollary 2.1 *The maximal value (2.98) of the Shannon entropy* S, *between all probability distributions with given expectation value* $\langle F \rangle$ *of F, is the* Legendre transform *of the potential function*

$$- \log Z(\lambda) \equiv - \log \sum_i e^{-\lambda F(i)} \tag{2.99}$$

with respect to its dual Lagrange multiplier λ. *Cf. Eqs. (2.97) and (2.98).*

We can easily generalize the discussion to the case where we know several expectation values $\langle F_1 \rangle, \ldots, \langle F_s \rangle$. The probability distribution which maximizes entropy between the ones with the given $\langle F_a \rangle$'s is

$$p_j = \frac{\exp\left[- \sum_a \lambda_a F_a(j)\right]}{\sum_i \exp\left[- \sum_a \lambda_a F_a(i)\right]}, \tag{2.100}$$

where the Lagrange multipliers λ_a's are fixed by inverting the map

$$(\lambda_1, \ldots, \lambda_s) \mapsto (\langle F_1 \rangle, \ldots, \langle F_s \rangle). \tag{2.101}$$

We stress that the Hessian of $\log Z$ with respect to the Lagrange multipliers

$$\frac{\partial^2 \log Z}{\partial \lambda_a \, \partial \lambda_b} = \left\langle \left(F_a - \langle F_a \rangle\right)\left(F_b - \langle F_b \rangle\right)\right\rangle \tag{2.102}$$

is the variance of the probability distribution for the variables F_a, and hence is *positive-definite* (assuming no $F_a(i)$ is a constant function). Thus $\log Z$ is *convex* and the Legendre transform in Corollary 2.1 is well defined.

The von Neumann Entropy

In Quantum Mechanics there is a corresponding notion, the *von Neumann entropy* S. Let ϱ be a *normalized* density matrix (cf. Sect. 2.2), i.e. a non-negative (trace-class) Hermitian operator ϱ, acting on the Hilbert space $\mathcal{H}$, normalized such that $\mathrm{Tr}\, \varrho = 1$. The von Neumann entropy is the *physical* entropy in this quantum set-up.

Definition 2.4 The *von Neumann* quantum entropy of the normalized density matrix ϱ (satisfying $\varrho^\dagger = \varrho$, $\varrho \geq 0$, and $\mathrm{Tr}\, \varrho = 1$) is

$$S(\varrho) = -\mathrm{Tr}\left[\varrho \log \varrho\right]. \tag{2.103}$$

The von Neumann quantum entropy is strictly related to the information-theoretic Shannon entropy. Indeed the Hermitian operator ϱ can be diagonalized (in an orthonormal basis)[11]

$$\varrho = \sum_m |m\rangle\, p_m\, \langle m|, \qquad \langle m'|m\rangle = \delta_{m,m'}, \tag{2.104}$$

where the eigenvalues satisfy $0 \leq p_m \leq 1$ and $\sum_m p_m = 1$. Now

$$-\varrho \log \varrho = \sum_m |m\rangle(-p_m \log p_m)\langle m| \tag{2.105}$$

so that

$$S(\varrho) = -\mathrm{Tr}(\varrho \log \varrho) = -\sum_m p_m \log p_m = S(p_m), \tag{2.106}$$

i.e. *the von Neumann quantum entropy is just the Shannon (classical) entropy[12] evaluated on the spectrum $\{p_m\}$ of the normalized density matrix ϱ.* All properties of the Shannon entropy are then shared by the von Neumann one. In particular:

vNE1   the *quantum entropy S is never negative;*
vNE2   the principle of maximal entropy works in the quantum theory exactly as it does in information theory, including when we have partial information on the state of a system. Hence the information-theoretic equations (2.96)–(2.99) are valid also in quantum statistical physics.

The entropy of a *classical system* is best defined as the classical limit of the von Neumann/Shannon quantum entropy. In terms of the normalized Liouville distribution ρ such that

$$\int_{\mathfrak{W}} \rho\, \frac{\mu}{(2\pi\hbar)^m} = 1, \tag{2.107}$$

the classical entropy is given by

$$S(\rho) = -\int_{\mathfrak{W}} \rho\, \log \rho\, \frac{\mu}{(2\pi\hbar)^m}. \tag{2.108}$$

---

[11] For notational simplicity we write the formula for ϱ with a discrete spectrum, which is the case most relevant for the applications we have in mind. The generalization to continuous spectra is obvious and left to the reader.

[12] Historically Shannon was inspired by the work of von Neumann, not the other way around.

The expression in the RHS is not non-negative in general. For instance: let $K \subset \mathfrak{W}$ be a compact region in phase space and

$$\rho(w) = \begin{cases} (2\pi\hbar)^m / \mathsf{Vol}(K) & w \in K \\ 0 & \text{otherwise,} \end{cases} \tag{2.109}$$

then

$$S(\rho) = \log\left(\mathsf{Vol}(K)/(2\pi\hbar)^m\right) \tag{2.110}$$

which is positive iff the Liouville volume of the region $K \subset \mathfrak{W}$ is larger than the volume $(2\pi\hbar)^m$ of the Heisenberg cell. The meaning of this result is obvious: physically it is *impossible* to specify the state with more precision than the maximal one allowed by the Heisenberg indetermination principle. If ρ is physically realizable $\int \rho\mu \geq (2\pi\hbar)^m$, and, the entropy is *non-negative* as it should be.

The Microcanonical Entropy Revisited

Let us check that Boltzmann's statistical definition of entropy

$$S = k \log \Omega(U, V) \tag{2.111}$$

for the microcanonical ensemble—that we inferred in Sect. 2.3 from elementary thermodynamic considerations and deep physical insights—does indeed agree with the von Neumann-Shannon conceptual definitions of entropy. We consider a quantum system; as usual, the continuous classical case can be recovered by taking the classical limit $\hbar \to 0$ and the discrete one by specialization (cf. Remark 2.1).

In Sect. 2.3 we presented different realizations of the microcanonical ensemble (which agree in the thermodynamical limit for sound continuous systems). The corresponding normalized density matrices have the form

$$\varrho^{(a)} = \frac{P^{(a)}}{\Omega^{(a)}}, \quad a = 1, 2 \tag{2.112}$$

where $P^{(1)}$, $P^{(2)}$ are, respectively, the orthogonal projectors into the states with energy $E \leq U$ and $U - \delta U \leq E \leq U$ while $\Omega^{(a)} = \mathrm{Tr}\, P^{(a)}$. Then

$$-\varrho^{(a)} \log \varrho^{(a)} = \frac{P^{(a)}}{\Omega^{(a)}} \log \Omega^{(a)}, \tag{2.113}$$

and the von Neumann entropy is

$$S^{(a)} = -\mathrm{Tr}\left[\varrho^{(a)} \log \varrho^{(a)}\right] = \log \Omega^{(a)} \tag{2.114}$$

in agreement with the result of our elementary thermodynamical analysis i.e. with the original argument by Boltzmann.

2.5 INTERLUDE: Saddle-Point Asymptotic Methods

In physics one often needs to compute integrals (finite-dimensional or functional integrals) in the asymptotic regime where some parameter becomes *large*. We denote this large parameter by N. We want to compute an integral of the general form

$$\int_V e^{NW(x_i)}\, d^m x \tag{2.115}$$

asymptotically for large values of $N \ggg 1$. Here $V \subset \mathbb{R}^m$ while $W \colon V \to \mathbb{R}$ is a real analytic function.[13] We make some simplifying assumptions:

W1 $W(x_i)$ has a unique maximum in $V \subset \mathbb{R}^m$ at the point x_i^0 which is in the *interior* of V;

W2   the Hessian at the critical point x^0

$$\left. \frac{\partial^2 W}{\partial x^i \partial x^j} \right|_{x^0} \tag{2.116}$$

is *negative definite*.

These conditions are generically satisfied in our applications.

The asymptotic limit for $N \ggg 1$ of Eq. (2.115) is computed by a technique known as *(Laplace's) saddle point method*;[14] a closed variant is the *stationary phase method*.

Asymptotics of Integrals

The basic idea is simple: let W_0 be the maximal value of $W(x)$ which is attained at the unique point $x^0 \in V$ (called the *saddle point*[15]); as $N \to \infty$ the function

$$\exp\big[N\big(W(x) - W_0\big)\big] \tag{2.117}$$

goes exponentially to zero for x away from x^0, so in the limit $N \to \infty$ only points at infinitesimal distances from x^0 will contribute to the integral. We expand $W(x)$

---

[13] Weaker regularity conditions will suffice.

[14] We dwell only on the elementary theory which applies under our special assumption of an *unique* (isolated) maximum in the interior of the integration region. Of course a more general theory exists and is quite deep.

[15] The reason for this terminology is as follows. We take $m = 1$ for simplicity. Since $W(x)$ is real analytic, we may consider its extension to a (locally) holomorphic function $W(x + iy)$. We have $\partial_x^2 W = -\partial_y^2 W$ so in the complex domain the graph of the function $\mathrm{Re}\, W$ near the point x^0 is a *saddle*: $\mathrm{Re}\, W(x + iy) \approx W(x^0) + \lambda[y^2 - (x - x^0)^2] + \cdots$ for some $\lambda > 0$.

in power series around the (unique) maximum x^0

$$N\big(W(x) - W_0\big) = \frac{N}{2} \frac{\partial^2 W}{\partial x_i \partial x_j}\Big|_{x^0} (x_i - x_i^0)(x_j - x_j^0)+$$

$$+ \frac{N}{6} \frac{\partial^3 W}{\partial x_i \partial x_j \partial x_k}\Big|_{x^0} (x_i - x_i^0)(x_j - x_j^0)(x_k - x_k^0) + \cdots = \tag{2.118}$$

$$= \frac{1}{2} \frac{\partial^2 W}{\partial x_i \partial x_j}\Big|_{x^0} z_i z_j + \frac{1}{6\sqrt{N}} \frac{\partial^3 W}{\partial x_i \partial x_j \partial x_k}\Big|_{x^0} z_i z_j z_k + O(1/N)$$

where

$$z_i \equiv \sqrt{N}(x_i - x_i^0). \tag{2.119}$$

When written in terms of the rescaled variables z_i, the higher order corrections in the bottom line of (2.118) are suppressed by negative powers of $\sqrt{N}$ with respect to the quadratic term which is non-zero for all $z_i \neq 0$ by assumption **W2**. Therefore these higher terms can be neglected as $N \to \infty$. The integral (2.115) becomes

$$\frac{e^{NW(x_0)}}{N^{m/2}} \int_{V'} \exp\left[\frac{1}{2} \frac{\partial^2 W}{\partial x_i \partial x_j}\Big|_{x_0} z_i z_j + O(1/\sqrt{N})\right] d^m z \tag{2.120}$$

where by **W2** the Hessian matrix is *negative-definite*.

Neglecting the $O(1/\sqrt{N})$ terms in the integrand, and the $O(e^{-aN})$ contributions arising from the boundaries of the integration region V, which do not contribute in the $N \to \infty$ limit, we remain with a Gaussian integral in the z_i's which can be evaluated explicitly

$$e^{NW(x^0)} \left(\frac{2\pi}{N}\right)^{m/2} \left(\det\left[-\frac{\partial^2 W}{\partial x_i \partial x_j}\Big|_{x^0}\right]\right)^{-1/2} \left(1 + O(1/\sqrt{N})\right) \tag{2.121}$$

(since the determinant is positive, its square-root is real). Equation (2.121) is the *saddle point formula*. The corrections to this formula of order $N^{-k/2}$ ($k = 1, 2, \dots$) can be easily computed recursively in k: an efficient method is the Wick theorem [12]. The exponentially small corrections require subtler mathematical techniques as resurgent series and Stokes theory, see e.g. [13]. In our typical applications both kinds of corrections play no role.

Remark 2.4 More generally we may consider integrals of the form

$$\int_V e^{NW(x_i)} \phi(x_i) d^m x \qquad N \gg 1. \tag{2.122}$$

If $\phi(x_i)$ is continuous, its large N limit is $\phi(x_i^0)$ times the expression (2.121).

Relation to the Legendre Transform

We apply the saddle point formula (2.121) to the function

$$W(x) = \Phi(x) - \langle y, x \rangle \tag{2.123}$$

where $\Phi(x)$ is a *stable concave* function on a strict convex cone $C \subset \mathbb{R}^m$ and $y \in C^\vee$ is an element of the dual strict convex cone. Conditions **W1**, **W2** are satisfied. Consider the function $\Psi : C^\vee \to \mathbb{R}$ defined by the following asymptotic limit:

$$\Psi(y) \overset{\text{def}}{=} \lim_{N \to \infty} \frac{1}{N} \log \int_C e^{N[\Phi(x) - \langle y, x \rangle]} \, \mathrm{d}^m x = \max_{x \in C}[\Phi(x) - \langle y, x \rangle] +$$

$$+ \lim_{N \to \infty} \frac{1}{N} \left[-\frac{m}{2} \log \frac{N}{2\pi} - \frac{1}{2} \log \det \left[-\frac{\partial^2 \Phi}{\partial x^i \partial x^j} \Big|_{x_0} \right] + O(1/\sqrt{N}) \right] =$$

$$= \max_{x \in C} \left[\Phi(x) - \langle y, x \rangle \right]. \tag{2.124}$$

We have proven the

Lemma 2.2 *Let $\Phi : C \to \mathbb{R}$ be a **concave** and stable function defined on the strict convex cone C. Its **global** Legendre transform*

$$\Psi : C^\vee \to \mathbb{R}, \qquad \Psi(y) \overset{\text{def}}{=} \max_{x \in C} \left[\Phi(x) - \langle y, x \rangle \right] \tag{2.125}$$

can be written as

$$\Psi(y) = \lim_{N \to \infty} \frac{1}{N} \log \left(\int_C e^{N[\Phi(x) - \langle y, x \rangle]} \, \mathrm{d}^m x \right). \tag{2.126}$$

Comparing with (2.3) we see that the RHS has the formal structure of the *thermodynamic limit* for a statistical ensemble. Thus Lemma 2.2 connects a central notion of Thermodynamics—the Legendre transform of concave/convex thermodynamic potentials—with the fundamental prescription of Statistical Mechanics.

Remark 2.5 The integral in (2.126) is the *Laplace transform* of $\exp[N\Phi(x)]$. We see that the Legendre transform between dual convex cones is the large-N asymptotic form of the Laplace transform.

Subtler Phenomena

In many applications to Statistical Mechanics we have integrals of the form (2.115) where the function $W(x_i; t_a)$ depends on control parameters t_a (such as temperature, pressure, etc.) which take values in some parameter space Y. For generic values

$t_a \in Y$ the critical values $W(x_i^\alpha; t_a)$ at the several *critical points* $x_i^\alpha \in V$ where the differential dW vanishes

$$dW \equiv \frac{\partial W(x_i; t_a)}{\partial x_i} dx_i \bigg|_{x_i = x_i^\alpha} = 0, \qquad (2.127)$$

are pair-wise distinct, hence there is a unique absolute minimum in V (called the *dominant saddle point*) while at the critical points the Hessian is generically non-degenerate: $\det \partial_i \partial_j W|_{x^\alpha} \neq 0$. A smooth function with these two properties is called a *Morse function*. Thus for *generic* t_a's the large N asymptotics of the integral is given by the saddle point formula (2.121) applied to the dominant saddle point.

However, in a codimension-1 locus $C \subset Y$ two things may happen:

C1 two distinct critical points x_i^1 and x_i^2 may have the same critical value $W(x_i^1; t_a^0) = W(x_i^2; t_a^0)$ which is the absolute minimum of the function. In this case for $N \ggg 1$, the integral is given by the sum of the contributions from the two dominating saddle points. The locus C is a "wall" which divides the parameter space Y in two distinct regions Y^1, Y^2: in the first region $x_i^1(t_a)$ is the dominant saddle, while in region Y^2 the dominant saddle is $x_i^2(t_a)$. Along C the two dominant saddles coexist. The function $\Psi(y)$ defined by the "thermodynamic limit" (2.124) will be

$$\Psi(y; t) = \begin{cases} \Phi(x^1(y; t)) - \langle y, x^1(y; t) \rangle & t \in Y^1 \\ \Phi(x^2(y; t)) - \langle y, x^2(y; t) \rangle & t \in Y^2. \end{cases} \qquad (2.128)$$

While the function $\Psi(y; t)$ is continuous, is not analytically: typically some (higher) derivative jumps across C. As we shall see in Chap. 5 this is the basic mechanism underlying phase transitions of the first order: each phase is associated to a saddle point which dominates on one side of C, while along the codimension-1 "wall" C the two phases coexist.

C2 The Hessian $\partial_i \partial_j W$ at the dominating saddle may become singular. In Chap. 5 this situation will be related to phase transitions of the second order.

In higher codimension more complicate phenomena will appear.

2.6 The Canonical Ensemble

The ensemble where we keep fix *temperature* and *volume* is the most widely used and the first one to be formulated historically: it is called the *canonical ensemble*.

Under the condition that the thermodynamic potentials are well-defined and satisfy the appropriate convexity conditions,[16] *in the thermodynamic limit* the canonical ensemble should be physically equivalent to the microcanonical ensemble, that is, it should produce the same equations of state: therefore its thermodynamic potential should be the fixed-temperature Legendre transform of the potential $S(U, V)$ for the fixed-energy microcanonical ensemble. We already encountered a thermodynamical potential of this form in Chap. 1, see Eq. (1.102). We have

$$\mathcal{F} = \min_U \left(T^{-1}U - S \right) = -\max_U \left[S - T^{-1}U \right] \tag{2.129}$$

where $S \in \mathbb{R}_{>0}$ and $T^{-1} \in (\mathbb{R}_{>0})^{\vee}$ belong to dual strict convex cones, while the entropy function $S(U, V)$ is concave as a function of U at fixed V

$$\frac{\partial^2 S}{\partial U^2} = \frac{\partial}{\partial U}\frac{1}{T} = -\frac{1}{T^2}\left(\frac{\partial T}{\partial U}\right)_V = -\frac{1}{T^2}\left(\frac{\partial U}{\partial T}\right)_V^{-1} \equiv -\frac{1}{T^2 C_V} < 0. \tag{2.130}$$

Comparing with Lemma 2.2—where we identify $\Phi = S$, $y = 1/T$, and rescale $N \rightsquigarrow N/k$—we get[17]

$$\frac{F(T, V)}{T} \equiv \mathcal{F}(T^{-1}, V) = -\lim_{N \to \infty}\left(\frac{k}{N} \log \int_0^\infty e^{N/k[S(U,V)-U/T]}\, dU \right) \tag{2.131}$$

for all positive constant k. If we choose k to be the Boltzmann constant, we may rewrite the equation as the asymptotics for large N

$$\exp\left(-NF(T, V)/kT \right) \approx \int_0^\infty \Omega(NU, NV)\, e^{-NU/kT}\, dU. \tag{2.132}$$

We introduce the standard notation

$$\beta = \frac{1}{kT} \tag{2.133}$$

called the *inverse temperature*. In Planck's units where $k = 1$, β is just T^{-1}. By rescaling the system $U \rightsquigarrow NU$, $V \rightsquigarrow NV$, we may rewrite the RHS of (2.132) for a *classical continuous system* as[18]

$$\int dU\, e^{-\beta U} \int_{\mathfrak{W}} \delta(H - U)\frac{\mu}{(2\pi\hbar)^m\, N!} = \int_{\mathfrak{W}} e^{-\beta H(p,q)}\frac{\mu}{(2\pi\hbar)^m\, N!}, \tag{2.134}$$

---

[16] And the absolute temperature is positive (as it should).

[17] Here we assume the energy U to be bounded below, and fix the additive constant in its definition so that its minimum value is zero. Then U takes value in the convex cone $\mathbb{R}_{>0}$.

[18] As always, μ is the Liouville measure on the phase space $\mathfrak{W}$.

and in the *quantum case* as

$$\int dU \, e^{-\beta U} \frac{\partial}{\partial U} \mathrm{Tr}\, P_U = \mathrm{Tr}(e^{-\beta H}), \tag{2.135}$$

where P_U is the orthogonal projector on the subspace of states of energy $\leq U$. When the spectrum of H is discrete, the integral in the LHS of (2.135) is defined in the Stieltjes sense. The semi-classical limit of (2.135) gives back (2.134).

We conclude that in the quantum case the *canonical ensemble* is given by the un-normalized density matrix

$$\varrho = e^{-\beta H} \qquad \begin{array}{l}\text{equality as Hermitian operators}\\ \text{acting on the Hilbert space } \mathcal{H}\end{array} \tag{2.136}$$

while in the classical case it is given by the measure in the phase space $\mathfrak{W}$

$$\rho \equiv \frac{\rho\,\mu}{(2\pi\hbar)^m N!} \equiv e^{-\beta H(p,q)} \frac{\mu}{(2\pi\hbar)^m N!} \in \Omega^{2m}(\mathfrak{W}), \quad \dim \mathfrak{W} = 2m. \tag{2.137}$$

These canonical distributions are obviously at equilibrium, i.e. invariant under the time evolution, because

$$[H, \varrho] = [H, \rho]_{\mathrm{PB}} = 0 \quad \Rightarrow \quad \frac{\partial \varrho}{\partial t} = \frac{\partial \rho}{\partial t} = 0, \tag{2.138}$$

by, respectively, the quantum and the classical Liouville equation, cf. Sect. 2.2.

To pass from the canonical ensemble distribution to the *canonical probability distribution* we have only to normalize the total probability to 1. We define the *canonical partition function* Z to be the total measure of the ensemble, that is,

$$Z(\beta, V) \equiv e^{-\beta F(\beta, V)} \equiv \begin{cases} \mathrm{Tr}\, e^{-\beta H} & \text{quantum} \\ \int e^{-\beta H} \mu/[(2\pi\hbar)^m N!] & \text{classical.} \end{cases} \tag{2.139}$$

Then the (normalized) *canonical probability distributions* are

$$p_{\mathrm{can}} = \begin{cases} \varrho/Z & \text{quantum} \\ \rho/Z & \text{classical,} \end{cases} \tag{2.140}$$

and the canonical thermal averages at temperature β^{-1} are

$$\langle O \rangle_\beta = \frac{1}{Z} \mathrm{Tr}\big[O\, e^{-\beta H}\big] \qquad\qquad \text{quantum} \tag{2.141}$$

$$\langle O \rangle_\beta = \frac{1}{Z} \int_{\mathfrak{W}} O(p, q)\, e^{-\beta H(q,p)} \frac{\mathrm{d}^m p\, \mathrm{d}^m q}{(2\pi\hbar)^m N!} \qquad \text{classical} \tag{2.142}$$

where O is an operator acting on the Hilbert space in the quantum case, and a function on the phase space $\mathfrak{W}$ in the classical set-up.

For future reference we note some useful relations (we use units where $k = 1$)

$$\frac{\partial}{\partial\beta}\log Z = -\frac{\partial}{\partial\beta}(\beta F) = -\frac{\mathrm{Tr}[H\,\mathrm{e}^{-\beta H}]}{Z} \equiv \langle H \rangle \equiv -U \qquad (2.143)$$

$$(1 - \beta\frac{\partial}{\partial\beta})\log Z = \left(\beta\frac{\partial}{\partial\beta} - 1\right)(\beta F) = S \qquad (2.144)$$

$$C_V = \beta^2 \frac{\partial^2}{\partial\beta^2}\log Z \qquad (2.145)$$

$$P = \frac{1}{\beta}\frac{\partial\log Z}{\partial V}. \qquad (2.146)$$

Entropy In the quantum case the von Newmann/Shannon entropy is

$$S = -\mathrm{Tr}\left[p_{\mathrm{can}}\log p_{\mathrm{can}}\right] = \frac{1}{Z}\mathrm{Tr}\left[\mathrm{e}^{-\beta H}(\log Z + \beta H)\right] =$$
$$= \log Z + \beta\langle H \rangle = \beta(-F + U) = \frac{U - F}{T} \qquad (2.147)$$

which reproduces the well-known expression of the thermodynamical entropy in terms of the free energy F, the internal energy U, and the temperature T.

The rigorous argument in Eqs. (2.129)–(2.132) yields

Theorem 2.4 *Assume that either one of the following holds:*

TL1 the thermodynamic limit of the microcanonical partition function exists and $\log \Omega(U)$ is a **concave** and stable function of $U \in \mathbb{R}_{>0}$

TL2 the thermodynamic limit of the canonical partition function exists and $\log Z(\beta)$ is **convex** and stable function of $\beta \in (\mathbb{R}_{>0})^\vee$.

*where "stable" is in the sense of dual convex cones. Then the thermodynamic limit of the other ensemble also exists and its thermodynamical potential satisfies the required convexity and stability conditions. Moreover the microcanonical and the canonical ensembles are equivalent **in the thermodynamic limit**.*

We recall that concavity (convexity) in temperature is equivalent to $C_V > 0$: heating up a body its temperature *increases* instead of decreasing.

Example 2.1 (A Contrario) Consider the model (2.79) whose microcanonical ensemble describes both positive and "negative" temperature regimes. The canonical partition function is

$$Z = \Big(2\cosh(\beta B)\Big)^N \tag{2.148}$$

The canonical ensemble (as we defined it) only describes equilibrium states at positive temperature. For these states $C_V > 0$ so $\log Z$ is convex and the canonical ensemble is well behaved. But $\log Z(\beta)$ is *not* stable in $(\mathbb{R}_{>0})^\vee$ because its derivative is bounded, so the two ensembles are *not* equivalent for all U, but only for the values of U corresponding to positive temperature.[19]

Example 2.2 Suppose we have the large energy asymptotics

$$\Omega(U, V) \approx \exp[a\, U^b / V^{b-1}] \tag{2.149}$$

for some constants $a > 0$, $b > 1$ and $U \gg 0$. Then

$$S = a\,\frac{U^b}{V^{b-1}}, \qquad \frac{1}{T} = ab\left(\frac{U}{V}\right)^{b-1},$$

$$-\frac{1}{C_V T^2} = \frac{\partial^2 S}{\partial U^2} = \frac{ab(b-1)}{V}\left(\frac{U}{V}\right)^{b-2} > 0 \tag{2.150}$$

The microcanonical partition function is well-defined, the temperature is positive, but the thermal capacity C_V is *negative* when $b > 1$. In facts the canonical partition function diverges for all β

$$\int_0^\infty \Omega(U, V)\, e^{-\beta U}\, dU \approx \int_0^\infty \exp\big[a\, U^b / V^{b-1} - \beta U\big]\, dU = \infty\,! \tag{2.151}$$

and is not defined even formally. The black hole in asymptotic flat spacetime has $b = 2$ hence C_V is proportional to $-1/T^2$ [14]. Giving energy to the black hole its temperature will go down, and the thermal capacity will become even more negative! A particularly interesting case is when $b = 1$. In this case the canonical partition function converges up to the temperature $1/a$, that is, the system at equilibrium cannot have a temperature higher than the maximal one $T_{\max} = 1/a$ no matter how much energy we supply to it. This phenomenon happens in the perturbative string where the maximal temperature is called the *Hagedorn temperature* [15].

---

[19] Mathematically one can extend the canonical ensemble to cover the negative temperature region, but the corresponding states are not equilibrium states in the strict physical sense.

Example 2.3 (The Classical Monoatomic Ideal Gas) For a classical monoatomic ideal gas the only degrees of freedom are the positions of the N particles contained in a box $B \subset \mathbb{R}^3$ of volume V, so

$$Z = \frac{1}{N!} \int_{B^N \otimes \mathbb{R}^{3N}} \exp\left[-\frac{\beta}{2m} \sum_{i=1}^{N} \sum_{a=1}^{3} p_{i,a}^2\right] \frac{d^{3N}q \; d^{3N}p}{(2\pi\hbar)^{3N}} \tag{2.152}$$

The integral over the positions yields a factor V^N while the integral in the momenta $p_{i,a}$ is Gaussian and can be performed explicitly

$$Z = \frac{1}{N!} V^N \left(\frac{m}{2\pi\hbar^2\beta}\right)^{3N/2}. \tag{2.153}$$

The equation of state is

$$P \equiv \frac{1}{\beta} \frac{\partial \log Z}{\partial V} = \frac{kNT}{V} \tag{2.154}$$

in agreement with the result from the microcanonical ensemble. On the other hand

$$U = -\frac{\partial}{\partial \beta} \log Z = \frac{3N}{2} \frac{1}{\beta} = \frac{3NkT}{2} \tag{2.155}$$

confirming that $C_V = \frac{3}{2}Nk$ for a monoatomic ideal gas. The free energy F is

$$-\frac{1}{\beta} \log Z = -\frac{3N}{2} kT \log(kT) - NkT \log(V/N) + NkT \left(\frac{3}{2} \log \frac{2\pi\hbar^2}{m} - 1\right) \tag{2.156}$$

Information-Theoretic Reinterpretation of the Canonical Ensemble

Comparing Eq. (2.136) with (2.96) we get the information-theoretic reinterpretation of the canonical ensemble:

Corollary 2.2 *The canonical ensemble is the probability distribution with the maximal Shannon entropy between the ones with a fixed expectation value $U = \langle H \rangle$ of the energy. Then by Corollary 2.1 its partition function is the Legendre transform of the entropy with respect to U (cf. Eq. (2.147)) as required by Physics.*

Remark 2.6 The microcanonical and canonical ensembles are characterized as the distributions which maximize entropy between the ones with, respectively, fixed energy $H = U$ and fixed energy expectation value $\langle H \rangle = U$. Since the standard deviation of the energy distribution in the canonical case is of order $O(1/\sqrt{N})$ (N being the number of Gibbs' copies), we conclude again that the two ensembles become equivalent in the thermodynamic limit $N \to \infty$.

2.7 Canonical Thermal Expectation Values

For simplicity we write the expressions in the quantum case, leaving to the reader the obvious modifications for the classical and discrete set-up. While we write these relations for the canonical ensemble, they apply *mutatis mutandis* in a wider context, being part of the general formalism of mathematical statistics.

Equation (2.143) is just the identity

$$-\frac{\partial}{\partial \beta} \log Z = -\frac{1}{Z} \frac{\partial}{\partial \beta} \mathrm{Tr}[e^{-\beta H}] = \frac{1}{Z} \mathrm{Tr}[H\, e^{-\beta H}] = \langle H \rangle, \tag{2.157}$$

which states the obvious physical fact that the internal energy U is the *mean value* of the Hamiltonian operator H. Likewise, the quadratic *variance* (the square of the standard deviation σ) of the Hamiltonian around its mean $\langle H \rangle$ is

$$\frac{\partial^2}{\partial \beta^2} \log Z = -\frac{\partial}{\partial \beta} \frac{\mathrm{Tr}[H\, e^{-\beta H}]}{Z} = \frac{\mathrm{Tr}[H^2\, e^{-\beta H}]}{Z} + \frac{\mathrm{Tr}[H\, e^{-\beta H}]}{Z} \frac{\partial}{\partial \beta} \log Z$$

$$= \langle H^2 \rangle - \langle H \rangle^2 \equiv \langle (H - \langle H \rangle)^2 \rangle. \tag{2.158}$$

Comparing with the thermodynamic formula (2.145), in units where $k = 1$,

$$C_V = \frac{1}{T^2} \langle (H - \langle H \rangle)^2 \rangle \geq 0, \tag{2.159}$$

with equality iff $H - \langle H \rangle$ is the zero Hermitian operator, i.e. iff the Hamiltonian is a multiple of the identity operator. The corresponding result in the classical case is that $C_V \geq 0$ with equality iff H is constant in phase space.

The relation (2.159) explains the microscopic origin of the convexity properties of the thermodynamic potentials: the second derivatives are statistical variances which cannot be negative by definition. In other words, they are expectation values on the equilibrium probability distribution of operators/functions which are *positive-definite*. Therefore, as long as the probability distribution is well-defined, the convexity properties are automatically valid. These remarks apply *mutatis mutandis* to all ensembles, not just to the canonical one.

Remark 2.7 The heat capacity C_V measures how the energy U changes in *response* to a variation of the external conditions in the form of an infinitesimal variation δT of the temperature. From Eq. (2.159) we see that this response is proportional to the thermal quadratic fluctuation of the energy. This is a special instance of a general result (the *dissipation-fluctuation theorem*) which says that all responses to variations of the external conditions are proportional to the thermal fluctuation of the appropriate quantity. The general theorem will be discussed in Chap. 6.

Remark 2.8 For systems which obey the Third Law (Sect. 1.8) $C_V \to 0$ as $T \to 0$ more rapidly than $1/\log T$, hence the variance of energy should be $o(T^2/\log T)$ as $T \to 0$.

Remark 2.9 Applying Eq. (2.159) to N copies of our system, we learn (again) that the fluctuations of the energy around its mean is $O(1/\sqrt{N})$, cf. Remark 2.6.

We may go on computing higher derivatives of $\log Z$. The third derivative is

$$\frac{\partial^3}{\partial \beta^3} \log Z = (-1)\Big(\langle H^3 \rangle - 3\langle H^2 \rangle \langle H \rangle + 2\langle H \rangle^3\Big) \tag{2.160}$$

More generally for all $k \in \mathbb{N}$

$$\frac{\partial^k}{\partial \beta^k} \log Z = (-1)^k \, \langle H^k \rangle_{\text{conn}} \tag{2.161}$$

where $\langle \, \cdot \, \rangle_{\text{conn}}$ is the *connected average* over the canonical probability distribution in the usual sense of statistics (to be reviewed in Sect. 5.5 below).

General Correlations for Classical Systems
Suppose we wish to compute a thermal correlation at fixed temperature β^{-1} of the general form

$$\langle O_1 \cdots O_s \rangle_\beta. \tag{2.162}$$

The story in the quantum case is a bit tricky because the operator insertions O_i do not commute: that story will be spelled out in Chap. 6 in the context of the fluctuation-dissipation theorem. Here we limit ourselves to the classical case.

The best method to define the correlations is to modify the Hamiltonian by adding appropriate *source terms*

$$H(\boldsymbol{p}, \boldsymbol{q}) \rightsquigarrow H(\boldsymbol{p}, \boldsymbol{q}; \lambda_i) \equiv H(\boldsymbol{p}, \boldsymbol{q}) + \sum_i \lambda_i \, O_i(\boldsymbol{p}, \boldsymbol{q}), \tag{2.163}$$

where the parameters λ_i are seen as new coupling constants in the modified Hamiltonian $H(\boldsymbol{p}, \boldsymbol{q}; \lambda_i)$. Let

$$Z(\lambda_i) \equiv \mathcal{N} \int e^{-\beta H(\boldsymbol{p}, \boldsymbol{q}; \lambda_i)} \, \mu \tag{2.164}$$

be the partition function for the modified Hamiltonian. We have

$$\langle O_1 \cdots, O_s \rangle_\beta = (-1)^s \beta^{-s} \frac{1}{Z(\lambda_i)} \frac{\partial^s}{\partial \lambda_1 \cdots \partial \lambda_s} Z(\lambda_i) \bigg|_{\lambda_i = 0} \tag{2.165}$$

In particular $\log Z(\lambda_i)$ is a convex function of its arguments; this property corresponds to the fact that the matrix

$$\Delta_{ij} = \langle O_i O_j \rangle - \langle O_i \rangle \langle O_j \rangle = \frac{1}{\beta} \frac{\partial^2}{\partial \lambda_i \, \partial \lambda_j} \log Z(\lambda) \bigg|_{\lambda_i = 0} \tag{2.166}$$

is non-negative. As in the special case of Eq. (2.159), these inequalities are the microscopic properties which underlie the convexity of the thermodynamic potentials. Remark 2.9 applies to all observables: their standard deviation scales with the number N of copies as $1/\sqrt{N}$.

2.8   Classical Canonical Partition Functions: General Results

For *classical continuous* systems the canonical partition function has the form

$$Z = \mathcal{N} \int_{\mathfrak{W}} e^{-\beta H(p,q)} \, d^m q \, d^m p \tag{2.167}$$

where, as before, $\mathcal{N}$ is a normalization constant which is independent of β and V, but may depend on the "number of particles" (when this notion makes sense). We usually choose a $\mathcal{N}$ that makes the thermodynamic limit regular.

Bohr-Van Leeuwen Theorem
The first general result for classical continuous systems of the kind (2.167) is

Theorem 2.5 (Bohr-Van Leeuwen) *In a classical continuous system contained in a finite box K, whose canonical partition function has the form* (2.167), ***switching on arbitrary magnetic fields***, *i.e. replacing*

$$H(q^i, p_j) \rightsquigarrow H(q^i, p_j - e A_j(q)) \qquad \textit{with } A_j(q) \, dq^j \in \Omega^1(Q), \tag{2.168}$$

will leave the partition function Z unchanged. *In particular the magnetization*

$$\langle M \rangle = -\frac{\partial}{\partial B} \log Z \tag{2.169}$$

$(B \equiv$ intensity of the magnetic field$)$ *is always **zero** in the classical set-up. That is: classical physics forbids the existence of all magnetic thermal effects: ferromagnetism, diamagnetism, paramagnetism, ferrimagnetism, etc.*

Remark 2.10 (History) This result was formulated by Bohr in 1911 and Van Leeuwen in 1919 in their respective thesis. Since ferromagnetism is well known to exist in nature, this result convinced Bohr that Classical Physics was (at best) incomplete, and prompted him to lay down the foundations of Quantum Mechanics where all the magnetic properties observed in real solids get elegant explanations.

Remark 2.11 While the *proof* of Theorem 2.5 is elementary, its *meaning* is extremely subtle. It says that at thermal equilibrium the magnetic field induces some new unexpected effects, in addition to the obvious ones, and that the new effects produce a contribution to the magnetization which in the limit $\hbar \to 0$ *exactly cancels* the one produced by the obvious effects. See [16].

Proof In the RHS of (2.167) make the redefinition of the integration variables

$$(q^i, p_j) \rightsquigarrow (q^i, p_j - eA_j(q)) \tag{2.170}$$

whose Jacobian is

$$\det \left(\begin{array}{c|c} 1 & -e\,\partial_i A_j \\ \hline 0 & 1 \end{array} \right) = 1. \tag{2.171}$$

Thus

$$\int_{\mathfrak{W}} e^{-\beta H(q^i,\, p_j - eA_j(q))}\, \mathrm{d}^m q\, \mathrm{d}^m p = \int_{\mathfrak{W}} e^{-\beta H(q^i,\, p_j)}\, \mathrm{d}^m q\, \mathrm{d}^m p. \tag{2.172}$$

$\square$

Classical Partition Functions. Heat Capacities
As a consequence of the Bohr-Van Leeuwen theorem, in classical statistical mechanics (of continuous systems) we can assume the Hamiltonian to be of the form

$$H = \frac{1}{2}m^{ij}(q)\, p_i\, p_j + V(q) \tag{2.173}$$

without loss of generality. Here $m^{ij}(q)$ is the *inverse* of the Jacobi metric[20] $m_{ij}(q)$ [1] on the configuration space Q of our classical system which obeys the monolateral constraint[21] of being contained in the finite box B [1]. When our system consists of N "particles" closed in a box B we have $Q = B^N$.

---

[20] Recall that the Jacobi metric $m_{ij}(q)$ in a classical non-relativistic system is given by the coefficients entering in the kinetic terms in the Lagrangian: $T = \frac{1}{2}m_{ij}(q)\,\dot{q}^i \dot{q}^j$ [1].

[21] As explained in [1] this is most conveniently realized by taking $Q = \mathbb{R}^m$ with a potential V which is $+\infty$ when not all q's are inside the box B.

The integral over the p_i's is Gaussian and can be performed in general

$$\int e^{-\beta H(p,q)} \, d^m p \, d^m q = \left(\frac{2\pi}{\beta}\right)^{m/2} \int_Q e^{-\beta V(q)} \sqrt{\det m_{ij}} \, d^m q =$$
$$= \left(\frac{2\pi}{\beta}\right)^{m/2} \int_Q e^{-\beta V(q)} \, d\mathsf{vol} \tag{2.174}$$

where $d\mathsf{vol}$ is the canonical volume form of the Riemannian manifold $(Q, m_{ij}(q))$.

Corollary 2.3 *For a free system with Hamiltonian*

$$H = \frac{1}{2} m^{ij}(q) \, p_i \, p_j \tag{2.175}$$

where $m^{ij}(q)$ is the inverse of the Riemannian metric $m_{ij}(q)$ on the configuration manifold (with boundary) Q of dimension m, the canonical partition function is

$$Z = \mathcal{N} \left(2\pi/\beta\right)^{m/2} \mathsf{Vol}(Q), \tag{2.176}$$

where $\mathsf{Vol}(Q)$ is the volume of Q as a Riemannian manifold endowed with the Jacobi metric $m_{ij}(q)$. Then from Eq. (2.145) (reinserting the constant k)

$$C_V = \frac{mk}{2} \quad and \quad U = C_V T. \tag{2.177}$$

Example 2.4 (Ideal Gases: Monoatomic, Biatomic, Triatomic) By definition an ideal gas consists of $N = O(10^{24})$ molecules moving freely inside a containing box $B \subset \mathbb{R}^3$. In the monoatomic case the molecules are classically modeled by material points in $\mathbb{R}^3$, so there are $3N$ degrees of freedom and Q is the Riemannian manifold with boundary B^N. In the biatomic case the molecules are classically modeled by two material points at fixed distance. The degrees of freedom of each molecule are the positions of the center of mass of the molecule and the orientation of the line connecting the two atoms, that is, $Q = (B \times S^2)^N$. If the molecule contains three or more atoms, we model it as a (generic) rigid body so that $(B \times SO(3))^N$ [1]. Equation (2.177) gives

$$C_V = \begin{cases} \frac{3}{2} nR & \text{monoatomic} \\ \frac{5}{2} nR & \text{biatomic} \\ 3 nR & \text{triatomic} \end{cases} \tag{2.178}$$

where $n \equiv N/N_A$ is the number of moles.

For a general classical system we have (cf. Eq. (2.145))

$$C_V = \frac{mk}{2} + \frac{1}{k^2 T^2} \frac{\partial^2}{\partial \beta^2} \log \int_Q e^{-\beta V} \, d\text{vol} \tag{2.179}$$

where the second term vanishes[22] for large T as $O(1/T^2)$. We conclude that *the high temperature limit of the heat capacities is universal.*

Virial and Equipartition Theorems
As in [1] we use the short-hand notation

$$(w^i, \ldots, w^{2m}) \equiv (q^i, \ldots, q^m, p_1, \ldots, p_m) \tag{2.180}$$

for the Darboux coordinates of the phase space $\mathfrak{W}$.

Theorem 2.6 (Virial Theorem of Classical Statistical Mechanics) *For all pairs $i, j = 1, \ldots, 2m$ such that $H \to +\infty$ as $|w^j| \to \infty$ we have*

$$\left\langle w^i \frac{\partial H}{\partial w^j} \right\rangle_T = \delta_{ij} \, kT. \tag{2.181}$$

Proof The proof uses integration by parts:

$$Z \left\langle w^i \frac{\partial H}{\partial w^j} \right\rangle \equiv \int e^{-\beta H(w)} \, w^i \frac{\partial H}{\partial w^j} \, d^{2m} w =$$
$$= \frac{-1}{\beta} \int w^i \frac{\partial}{\partial w^j} e^{-\beta H(w)} \, d^{2m} w = \frac{\delta_{ij}}{\beta} \int e^{-\beta H(w)} \, d^{2m} w = \delta_{ij} \, kT \, Z, \tag{2.182}$$

where we used the condition on the behavior of H for $w^j \to \infty$ to set to zero the boundary term in the integration by parts. $\square$

Applications: The Classical Equipartition Theorem
We recover Corollary 2.3 as a special case of the virial theorem. For a free system (2.175) the Hamiltonian satisfies the identity

$$p^i \frac{\partial H}{\partial p_i} = 2 H \tag{2.183}$$

---

[22] Before taking the thermodynamic limit Q is compact and V a non-negative smooth function, so

$$\frac{\partial^2}{\partial \beta^2} \log \int_Q e^{-\beta V} \, d\text{vol} = \langle V^2 \rangle - \langle V \rangle^2 \le \max_Q V^2$$

which gives a bound independent of β. Hence for large but finite N the second terms is $O(1/T^2)$. If the thermodynamic limit exists this remains true as $N \to \infty$.

so that

$$U(T) = \langle H \rangle_T = \frac{1}{2} \left\langle p^i \frac{\partial H}{\partial p^i} \right\rangle = \frac{m}{2} kT. \tag{2.184}$$

Next consider the case where the Jacobi metric m_{ij} is constant and the potential is quasi-homogeneous, i.e. there are non-negative rational numbers d_i such that

$$V(\lambda^{d_i} q^i) = \lambda\, V(q^i) \quad \forall\, \lambda \tag{2.185}$$

so that

$$H = \frac{1}{2} \sum_i p^i \frac{\partial H}{\partial p^i} + \sum_i{}^* d_i\, q^i \frac{\partial H}{\partial q^i} \tag{2.186}$$

(the symbol $\sum^*$ means that in the sum we omit the q^i's which do not appear in the potential i.e. such that $\partial V/\partial q^i \equiv 0$). Hence

$$U = \langle H \rangle = \left(\frac{m}{2} + \sum_i d_i \right) kT \quad \Rightarrow \quad C_V = \left(\frac{m}{2} + \sum_i d_i \right) k. \tag{2.187}$$

A special case is when all d_i's are equal to $1/2$, that is, a coordinate q^i is either cyclic or contributes to the potential V a quadratic term of the form $\frac{1}{2} K_i (q^i)^2$: the system is a collection of ℓ harmonic oscillators and $m - \ell$ free (cyclic) degrees of freedom

$$H = \frac{1}{2} \sum_{i=1}^{\ell} \left(\frac{p_i^2}{m_i} + m_i\, \omega_i^2 (q^i)^2 \right) + \sum_{i=\ell+1}^{m} \frac{p_i^2}{2m_i}, \tag{2.188}$$

where the parameters m_i and ω_i are arbitrary. The energy is then

$$U = \frac{1}{2} (m + 2\ell) kT. \tag{2.189}$$

This result is known as the *equipartition theorem of classical statistical mechanics.*

Corollary 2.4 (Classical Equipartition Theorem) *In a classical system where the Hamiltonian is a quadratic form in the canonical variables, i.e.*[23]

$$H = \frac{1}{2} \sum_i \lambda_i (w^i)^2, \quad \lambda_i \in \mathbb{R}_{>0}, \tag{2.190}$$

---

[23] We can write the Hamiltonian in this form by diagonalizing the quadratic form $H = A_{ij} w^i w^j$.

the total energy U of the system is equipartite *between all the quadratic terms in the Hamiltonian: each such term contributes kT/2 to U independently of the values of the parameters* λ_i *(as long as they are non-zero!).*

The Classical Ultraviolet Catastrophe Just as the Bohr-Van Leeuwen theorem made Bohr to conclude that classical physics cannot be the full story, Lorentz pointed out in 1903 that the equipartition theorem implies that classical physics is wrong since it predicts an *ultraviolet catastrophe* which is not present in the real world.

Vicky Weisskopf formulated this fact by words which are carved in my mind since I heard them some fifty years ago:

Since fire exists, Classical Physics is *obviously* wrong!

Why Fire Is Impossible in a Classical World? We write the most general electromagnetic field inside a cubic box of size L bounded by walls which satisfy the totally reflecting boundary conditions. We use the old-fashioned *radiation gauge*

$$A_0 = \mathbf{\nabla} \cdot \mathbf{A} = 0. \tag{2.191}$$

The Fourier expansions of the 3 components of the gauge fields are

$$A_1(x, y, z) = \sum_{\mathbf{k} \in \mathbb{N}^3} A_1(\mathbf{k}) \cos(2\pi k_1 x/L) \sin(2\pi k_2 y/L) \sin(2\pi k_3 z/L)$$

$$A_2(x, y, z) = \sum_{\mathbf{k} \in \mathbb{N}^3} A_2(\mathbf{k}) \sin(2\pi k_1 x/L) \cos(2\pi k_2 y/L) \sin(2\pi k_3 z/L)$$

$$A_3(x, y, z) = \sum_{\mathbf{k} \in \mathbb{N}^3} A_3(\mathbf{k}) \sin(2\pi k_1 x/L) \sin(2\pi k_2 y/L) \cos(2\pi k_3 z/L)$$

$$\text{with} \quad k_1 A_1(\mathbf{k}) + k_2 A_2(\mathbf{k}) + k_3 A_3(\mathbf{k}) = 0$$

$$\tag{2.192}$$

The details of these expressions are not important. They are written just to inform the reader that we can write them and study the problem in every minute detail *if so needed.* But we don't need. The only point of interest is that for all triple of positive[24] integers $(k_1, k_2, k_3) \equiv \mathbf{k} \in \mathbb{N}^3$ the electromagnetic field has *two* real degrees of freedom; we may choose them to be the Fourier coefficients $A_1(\mathbf{k})$ and $A_2(\mathbf{k})$.

Plugging in the expression for $\mathbf{A}(\mathbf{x})$ written in terms of the (chosen) independent degrees of freedom, $A_1(\mathbf{k})$ and $A_2(\mathbf{k})$, in the Maxwell Lagrangian we get the Lagrangian governing the dynamics of these independent physical variables

$$L = \sum_{i=1,2} \sum_{\mathbf{k} \in \mathbb{N}} \left(\frac{1}{2} \dot{A}_i(\mathbf{k})^2 - \frac{1}{2} \left(\frac{2\pi \mathbf{k}}{L} \right)^2 A_i(\mathbf{k})^2 \right) \tag{2.193}$$

from which we see that our "electromagnetic empty space" bounded by totally reflecting walls is equivalent to a system of non-interacting harmonic oscillators consisting of *two copies*, $A_1(\mathbf{k})$ and $A_2(\mathbf{k})$, of the harmonic oscillator with frequency

$$\nu_\mathbf{k} = |\mathbf{k}|/L \tag{2.194}$$

per wave-number vector $\mathbf{k} \in \mathbb{N}^3$. Now we apply Corollary 2.4, the equipartition theorem: at equilibrium at temperature T, each harmonic oscillator in the system, independently of its frequency, contributes an energy kT to U. Consequently the thermal radiation energy emitted (in electromagnetic waves) at equilibrium at temperature T is *classically*[25]

$$2k_B \sum_{\mathbf{k} \in \mathbb{N}^3} T = \infty \ !! \tag{2.195}$$

In other words, any *classical fire*, which has a positive temperature, would emit an infinite amount of heat and light. Therefore a classical fire is impossible. Every child who sees a fireplace, should immediately conclude that the real world follows the rules of Quantum Physics.

Remark 2.12 (History) Recall from Sect. 1.12 that Boltzmann in 1884 (20 years before Lorentz made its observation!) computed the internal energy of the radiation field from thermodynamic arguments getting the quantum correct result that U is proportional to T^4 not to T (with a divergent coefficient!) as predicted by classical physics. Boltzmann's extraordinary success shows the almighty power of the Second Law.

---

[24] To keep things simple, we are cavalier with the zero modes. They play no role in the argument, and are irrelevant in the computation anyhow.

[25] We write the Boltzmann constant as k_B instead of k not avoid a possible confusion with the norm of the wave-vector $\mathbf{k}$.

2.9 Quantum Systems

What Is *Heat?*

In Chap. 1 we introduced *heat* the way Clausius did in the nineteenth century: if
the conservation of energy should hold in full generality (the First Law) we need
an additional form of energy besides the mechanical, electrical, chemical, etc. ones.
The *only* purpose of this extra form is to enter in the energy conservation equation
to represent the fraction of energy not accounted for by anyone of its "known"
forms. So defined, *heat* looks a rather mysterious quantity, even esoteric, just an ad
hoc book-keeping device. This is highly unsatisfactory, and we need to understand
physically what heat really is.

Quantum Mechanics has a good explanation. Consider a system with discrete
energy levels E_n. The internal energy at temperature β^{-1} is

$$U \equiv \langle H \rangle = \sum_n p_n \, E_n, \qquad p_n = \frac{\mathrm{e}^{-\beta E_n}}{Z}. \tag{2.196}$$

There are two ways to increase U. We can increase the energy levels E_n (by
modifying the Hamiltonian H), or we can modify the probability distribution $\{p_n\}$
increasing the probability of the higher energy levels (at the expense of the lower
ones) while keeping constant each individual energy level E_n. We claim that the first
way to increase U is *work* done on the system, and the second one is *heat* supplied
to the system. To substantiate the claim we have to show two things:

H1   in an adiabatic process (no exchange of heat) the probabilities $\{p_n\}$ do not
change;

H2   in an isovolumetric process (no work) the $\{E_n\}$ remain constant.

The idea is to show **H1, H2** in a simple system. The simplest thermal system is
a classical monoatomic ideal gas. But the proposed physical interpretation of heat
is inherently quantum, so cannot be checked in a classical system. We solve the
conundrum by noticing that an ideal quantum gas behaves as a classical ideal gas in
the limit of zero density,[26] i.e. when there are very few molecules in our macroscopic
volume V. To identify a classical ideal gas with a quantum system we take the most
extreme low-density situation: just one molecule inside our box of volume V with
periodic boundary conditions. The box is a cube of side L, so $L^3 = V$; to simplify
the formulae we set the mass $m = 1/2$, so that the Hamiltonian is simply $H = \mathbf{p}^2$.
The spectrum of each component of the momentum operator is $\{2\pi n/L : n \in \mathbb{Z}\}$.
The one-molecule gas is then a quantum system with energy levels

$$E_{\boldsymbol{n}}(V) = \frac{(2\pi)^2}{V^{2/3}} \, \boldsymbol{n}^2, \qquad \boldsymbol{n} \in \mathbb{Z}^3, \tag{2.197}$$

---

[26] This assertion will be proven in Chap. 3.

which depend only on V, while the canonical probabilities at temperature β^{-1}

$$p_n \equiv \frac{e^{-\beta E_n}}{\sum_m e^{-\beta E_m}} = p_n(kTV^{2/3}) \tag{2.198}$$

depend only on n and the combination $kTV^{2/3} = PV^{5/3}$. Hence for all n the probabilities p_n do not change in any process along which $PV^{5/3} = $ const.: this is precisely the equation of an adiabatic transformation for a monoatomic ideal gas. On the other hand, in a process where $V = $ cost. the energy levels $E_n(V)$ do not change. **H1** and **H2** are thus verified.

Remark 2.13 A time-dependent process is called *adiabatic*:[27]

A1   In Classical Mechanics iff the deformation of the Hamiltonian with time is infinitely slow so that the adiabatic invariants remain constant (see [1]);

A2   In Quantum Mechanics iff the deformation of the Hamiltonian $H(t)$ with time is infinitely slow so that a state which at time t_0 is an eigenstate of $H(t_0)$ with energy $E_n(t_0)$ (the n-th energy level in increasing order) will evolve in such a way that at time t it is an eigenstate of $H(t)$ with energy $E_n(t)$ (same n);

A3   In Thermodynamics iff no heat is exchanged during the process.

The above argument shows that the three definitions agree.

Quantum Harmonic Oscillator

The energy levels of a harmonic oscillator of frequency ω are

$$E_n = n\hbar\omega, \qquad n = 0, 1, 2, \cdots \tag{2.199}$$

where we subtracted the ground state energy $\frac{1}{2}\hbar\omega$ from E exploiting the fact that energy is defined up to an additive constant. Then the partition function[28] is

$$Z = \sum_{n=0}^{\infty} e^{-\beta\hbar\omega n} = \frac{1}{1 - e^{-\beta\hbar\omega}} \tag{2.200}$$

and the energy

$$U \equiv -\frac{\partial}{\partial\beta} \log Z = \frac{\hbar\omega \, e^{-\beta\hbar\omega}}{1 - e^{-\beta\hbar\omega}} \tag{2.201}$$

---

[27] For simplicity of the exposition we formulate the definitions in the context of systems with one degree of freedom.

[28] Equation (2.200) is the partition function *per copy* in a Gibbs' bath of ∞-many copies of the oscillator.

now does depend on ω contrary to the classical situation. However as $\hbar \to 0$ we recover the classical, frequency independent, answer, cf. Eq. (2.189). The same happens asymptotically as $\beta \to 0$ i.e. at high temperatures.

The Quantum Symmetric Rotator

The classical symmetric free rotator is the free system with Hamiltonian

$$H = m^{ij}(q)\, p_i p_j / 2 \tag{2.202}$$

and configuration space $Q = SO(3)/SO(2) = S^2$ (when the inertia momenta are $I_1 = I_2 \neq 0$ and $I_3 = 0$) or $Q = SO(3)$ (when $I_1 = I_2 = I_3 \neq 0$) with Jacobi metric $m_{ij}(q)$ proportional to the round metric on the 2-sphere or, respectively, on the $SO(3)$ group manifold.

Up to the overall normalization (which we may absorb in the definition of β), the quantum Hamiltonian is $H = -\Delta$ where Δ is the Laplacian on S^2 or respectively $SO(3)$. The eigenvalues of H are $\ell(\ell+1)$ ($\ell = 0, 1, \dots$) with multiplicity $2\ell + 1$. Hence in the S^2 case

$$Z = \sum_{\ell=0}^{\infty} (2\ell + 1)\, e^{-\beta \ell(\ell+1)}. \tag{2.203}$$

Let us now focus on the case that Q is the group manifold $SO(3)$. More generally we can consider a quantum system whose configuration space is an arbitrary *compact* Lie group G endowed with the unique (up to overall normalization) bi-invariant metric (cf. [1]), so that, after a suitable rescaling, the Hamiltonian is $-\Delta$ on the group manifold. By the Peter-Weyl theorem [17, theorem 1.12] the Hilbert space is

$$L^2(G) = \bigoplus_R R^{\oplus \dim R} \tag{2.204}$$

where the sum is over all isoclasses of irreducible unitary representations R of G (which are finite-dimensional for G compact). The eigenvalues of the Hamiltonian are the quadratic Casimir invariants $C_2(R)$ of the irreducible representations, and the partition function is

$$Z = \sum_R (\dim R)^2 e^{-\beta C_2(R)}. \tag{2.205}$$

2.10 Generalized Canonical Ensembles

We observed that the microcanonical and canonical thermal distributions $\delta(H - U)$ and $e^{-\beta H}$ are *stationary*—hence in equilibrium—because H is a conserved quantity. These distributions also behave well under the union of systems (in particular multiple copies of the *same* system): the distributions for the compound system read

$$\delta(H_1 + H_2 - U) = \delta(H_1 - U) * \delta(H_2 - U),$$
$$e^{-\beta(H_1+H_2)} = e^{-\beta H_1} \otimes e^{-\beta H_2} \tag{2.206}$$

where $*$ is the convolution product of (operator-valued) distributions, while the Hamiltonians H_1, H_2 act on the different Hilbert spaces $\mathcal{H}_1$ and $\mathcal{H}_2$ of the two subsystems. These good properties of the (micro)canonical ensemble(s) arise from the fact that H is an *additive conserved quantity*. Our microscopic system may have *several* additive conserved quantities in addition to the Hamiltonian H. Suppose the set of additive conserved quantities $H, J_1, \ldots, J_r$ *commute* (as operators in the quantum case, in the Poisson sense in the classical one), so that it make sense to speak of their simultaneous (eigen)values. Then the generalized *microcanonical* and *canonical* distributions

$$\varrho_{\text{micro}} = \delta(H - U) \prod_{i=1}^{r} \delta(J_i - a_i) \tag{2.207}$$

$$\varrho_{\text{can}} = e^{-\beta(H - \sum_{i=1}^{r} \mu_i J_i)} \tag{2.208}$$

are perfectly well defined. The coefficients μ_i are called *chemical potentials* extending to the general case the terminology of the prototypical example, the Grand canonical ensemble to be introduced in Sect. 2.12. Comparing with Eq. (2.100) we get:

> ϱ_{can} is the distribution with the *largest Shannon entropy* between the ones with fixed expectation values $\langle U \rangle$ and $\langle J_i \rangle$ $(i = 1, \ldots, r)$

The partition functions are (say, in the classical case)

$$\Omega(U, V, a_1, \ldots, a_r) = \int_{\mathfrak{W}} \delta(H - U) \prod_{i=1}^{r} \delta(J_i - a_i) \frac{\mu}{(2\pi\hbar)^m N!} \tag{2.209}$$

$$Z(\beta, V, a_1, \ldots, \mu_r) = \int_{\mathfrak{W}} e^{-\beta(H - \sum_{i=1}^{r} \mu_i J_i)} \frac{\mu}{(2\pi\hbar)^m N!}, \tag{2.210}$$

and they are related by the integral transform

$$Z(\beta, V, a_1, \ldots, \mu_r) = \int \Omega(U, V, a_1, \ldots, a_r) e^{-\beta U + \beta \mu_i a^i} dU \prod_i da_i \tag{2.211}$$

There are two situations: either a_i takes only positive values (this happens, say, when $a_i = N_i$ is the number of particles of the i-th species) or the additive conserved quantity a_i takes both positive and negative values (e.g. a_i are components of the magnetization of the system). In the first case convergence usually requires μ_i to be negative, and Eq. (2.211) reduces to a Laplace transform, while in the second case the a_i's typically should be analytically continued to complex values, and (2.211) takes the form of a Fourier transform. Analyticity conditions in the analytically continued parameters is important: it is the starting point of *linear response theory* and associated results (see Chap. 6).

An important special situation is when the conserved quantities J_i are quantized in integral units. The Hilbert space decomposes into simultaneous eigenspaces of the commuting operators J_i

$$\mathcal{H} = \bigoplus_{(n_1,\ldots,n_r)\in\mathbb{Z}^r} \mathcal{H}_{(n_1,\ldots,n_r)}, \quad J_i|_{\mathcal{H}_{(n_1,\ldots,n_r)}} = n_i, \tag{2.212}$$

and the partition function takes the form

$$Z(\beta, V, \mu_i) = \mathrm{Tr}[e^{-\beta H + \beta \sum_i J_i}] = \sum_{(n_1,\ldots,n_r)\in\mathbb{Z}^r} \zeta_i^{n_i}\, \mathrm{Tr}_{\mathbb{H}_{(n_1,\ldots,n_r)}}\left[e^{-\beta H}\right] \tag{2.213}$$

where the quantity

$$\zeta_i \equiv e^{\beta \mu_i} \tag{2.214}$$

is called the *fugacity* of the integral additive conserved quantity J_i.

2.11 The Pressure Ensemble

The *pressure ensemble* is less popular than its cousins but sometimes may be useful (see below for an application). It is an ensemble at fixed T and P, so its natural thermodynamical potential is

$$\mathcal{G}(\beta, P) = \mathcal{F} - V \frac{\partial \mathcal{F}}{\partial V} = \beta\, G(\beta^{-1}, P), \tag{2.215}$$

see Eq. (1.103) (we are using $k = 1$). We have

$$\frac{\partial^2 \mathcal{F}}{\partial V^2} = -\frac{1}{T}\left(\frac{\partial P}{\partial V}\right)_T = \frac{1}{T\, \kappa_T\, V} > 0 \tag{2.216}$$

so that the Legendre relation between the two potentials $\mathcal{F}$ and $\mathcal{G}$ is

$$\mathcal{G}(\beta, P) = \min_{V} \left(\mathcal{F}(\beta, V) + PV \right). \tag{2.217}$$

Applying Lemma 2.2 to the *concave* function of the volume $-\mathcal{F}(\beta, V)$, we get

$$\mathcal{G}(\beta, P) = \left[-\log \int_{\mathbb{R}_{>0}} dV\, Z(\beta, V)\, e^{-\beta P V} \right]_{\substack{\text{thermodynamic} \\ \text{limit}}} \tag{2.218}$$

where $Z(\beta, V)$ is the canonical partition function. In other words: *the constant pressure partition function $\Pi(\beta, \beta P)$ is the Laplace transform of the canonical partition with respect to the volume*

$$e^{-\mathcal{G}(\beta, P)} = \Pi(\beta, \beta P) = \int_0^\infty e^{-\beta P V}\, Z(\beta, V)\, dV. \tag{2.219}$$

By construction the pressure ensemble is equivalent to the canonical partition function *in the thermodynamical limit* provided $\kappa_T > 0$ and $\mathcal{G}$ is stable. Written in this way the formula holds for classical as well quantum systems.

Example 2.5 (Classical Ideal Gas) For the monoatomic ideal gas we have

$$\begin{aligned}
e^{-\mathcal{G}(\beta, P)} &= \int_0^\infty dV\, e^{-\beta P V}\, Z(T, V) = \\
&= \int_0^\infty dV\, e^{-\beta P V}\, \frac{V^N}{(2\pi\hbar)^{3N}\, N!} \left(\frac{2m\pi}{\beta} \right)^{3N/2}
\end{aligned} \tag{2.220}$$

so, in the thermodynamical limit, we get

$$-\frac{1}{kT} G(T, P) = \frac{3}{2} N \log\left(\frac{2m\pi kT}{(2\pi\hbar)^2} \right) + N \log[kT/P]. \tag{2.221}$$

Remark 2.14 In view of Eq. (2.100) we may characterize the pressure ensemble as the probability distribution which maximizes entropy under the constraints $\langle H \rangle = U$ and $\langle \mathsf{Vol} \rangle = V$.

The Feynman Gas

An elegant example of a system which is simpler to study using the *pressure ensemble* is the so-called *Feynman gas*. This is an one-dimensional classical gas of N impenetrable particles moving on a finite box ($\equiv$ segment) of length L see Fig. 2.4. *Impenetrable* means that the particles cannot cross each other, so the order of their coordinates x_i along the line is preserved

$$0 \leq x_1 \leq x_2 \leq x_3 \leq \cdots \leq x_N \leq L. \tag{2.222}$$

Fig. 2.4 The Feynman one-dimensional gas. The first and the last particles are fixed at the endpoints of the segment $[0, L]$ as a choice of boundary condition

The interactions are *short ranged:* each particle interacts only with its nearest neighbors. The Hamiltonian reads

$$H_N = \sum_{i=1}^{N} \frac{p_i^2}{2m} + \sum_{i=2}^{N} V(x_i - x_{i-1}) \tag{2.223}$$

for some potential function $V(x)$. The thermodynamic limit is $N \to \infty$ with L/N fixed. As boundary condition on the finite box $[0, L]$ we add two extra particles whose positions are frozen at the two endpoints $x_0 = 0$ and respectively $x_{N+1} = L$ of the segment. The canonical partition function is

$$Z_N = \left(\frac{2\pi m}{\beta}\right)^{N/2} \int \psi(L - x_N)\, \psi(x_N - x_{N-1}) \cdots \psi(x_2 - x_1)\, \psi(x_1)\, d^N x \tag{2.224}$$

where the pre-factor arises from the Gaussian integral over the momenta and

$$\psi(x) \stackrel{\text{def}}{=} e^{-\beta V(x)}\, \Theta(x) \tag{2.225}$$

with $\Theta(x)$ the step function (2.35). The integral in (2.224) is the N-th fold convolution power of $\psi(x)$ evaluated at L:

$$\psi^{*N}(L) \equiv \overbrace{\psi * \psi * \cdots * \psi}^{N} \text{ factors}(L) \tag{2.226}$$

Therefore the Laplace transform of the canonical partition function with respect to the volume L is (up to the pre-factor) the N-th power of the Laplace transform of $\psi(x)$. But the Laplace transform of the canonical partition function with respect to the volume L is (in the thermodynamic limit) the pressure-ensemble partition function, cf. Eq. (2.219). Then the Gibbs potential $G(\beta, P)$ of the Feynman gas is

$$e^{-\beta G(\beta, P)} = \left(\frac{2\pi m}{\beta}\right)^{N/2} \left(\int_0^\infty e^{-\beta P x - \beta V(x)}\, dx\right)^N \tag{2.227}$$

that is, the chemical potential $\mu \equiv G/N$ satisfies

$$e^{-\beta \mu(\beta, P)} = \sqrt{\frac{2\pi m}{\beta}} \int_0^\infty e^{-\beta P x - \beta V(x)}\, dx. \tag{2.228}$$

2.12   The Grand Canonical Ensemble

The *Grand canonical ensemble* describes a thermal system that can exchange both energy and "particles" with its surroundings which form a thermal bath at fixed temperature and chemical potential μ ($\equiv$ Gibbs free energy per particle). We consider first quantum systems: the classical case is then easily obtained by replacing traces over the Hilbert space $\mathcal{H}$ with integrals on the phase space $\mathfrak{W}$ with the properly normalized Liouville measure. We write N for the Hermitian operator acting on the quantum Hilbert space $\mathcal{H}$ whose eigenvalue is the particle number N assumed to be non-negative. Let $\mathcal{H}_N \subset \mathcal{H}$ be the subspace of states containing N particles; then

$$\mathcal{H} = \bigoplus_{N \geq 0} \mathcal{H}_N \quad \text{where} \quad |\psi\rangle \in \mathcal{H}_N \ \Leftrightarrow \ N|\psi\rangle = N|\psi\rangle. \tag{2.229}$$

The quantum *Grand partition function* is

$$\Xi(T, V, \mu) \overset{\text{def}}{=} \text{Tr}\big[e^{-\beta(H-\mu N)}\big] = \sum_{N \geq 0} e^{\beta \mu N} Z(T, V, N) \tag{2.230}$$

where $Z(T, V, N)$ is the canonical partition function of the system with a *fixed number N of particles*

$$Z(T, V, N) = \text{Tr}_{\mathcal{H}_N}\big[e^{-\beta H}\big]. \tag{2.231}$$

The classical Grand partition function is given by the same formula (2.230) with $Z(T, V, N)$ replaced by the classical canonical partition function at fixed number of particles. Equation (2.230) leads to the *Grand canonical probability distribution*

$$\varrho = \frac{e^{-\beta(H-\mu N)}}{\Xi(T, V, \mu)}. \tag{2.232}$$

The Grand canonical expectation values are

$$\langle O \rangle_{\text{Grand}} = \text{Tr}\big[O\varrho\big] = \frac{\text{Tr}\big[O\,e^{-\beta(H-\mu N)}\big]}{\Xi(T, V, \mu)}, \tag{2.233}$$

and the corresponding obvious formulae for the classical limit and for discrete systems. In particular the mean number of particles in the system is

$$\langle N \rangle = \frac{1}{\beta}\frac{\partial}{\partial \mu} \log \Xi(T, V, \mu). \tag{2.234}$$

The energy is

$$\langle H \rangle = -\frac{\partial}{\partial \beta} \log \Xi(T, V, \mu) + \mu \langle N \rangle, \tag{2.235}$$

while the entropy and pressure are

$$S = \frac{\partial}{\partial T}(T \log \Xi), \tag{2.236}$$

$$P = \frac{1}{\beta}\frac{\partial}{\partial V} \log \Xi. \tag{2.237}$$

We define the thermodynamical *Grand potential* $\Phi(T, V, \mu)$ by the equation

$$\Xi(T, V, \mu) = e^{-\beta \Phi(T, V, \mu)}, \tag{2.238}$$

from which we get the differential relation

$$d\Phi = -S\,dT - P\,dV - N\,d\mu \tag{2.239}$$

in agreement with Chap. 1. Recall from Eq. (1.139) that *in value* the Grand potential Φ is just $-PV$.

Example 2.6 (Classical Ideal Gas (Grand Canonical Approach)) We have

$$\Xi(V, T, \mu) = \sum_{N=0}^{\infty} e^{\beta \mu N} Z(T, V, N) = \sum_{N=0}^{\infty} e^{\beta \mu N} \frac{1}{N!}\left(\frac{(2m\pi)^{3/2} V}{(2\pi \hbar)^3 \beta^{3/2}}\right)^N =$$

$$= \exp\left[\frac{(2\pi m)^{3/2} V e^{\beta \mu}}{(2\pi \hbar)^3 \beta^{3/2}}\right]. \tag{2.240}$$

The Grand potential of the classical ideal gas is then

$$\Phi = -PV = -C\,\beta^{-5/2} V e^{\beta \mu}, \quad \text{where} \quad C = \frac{m^{3/2}}{(\sqrt{2\pi}\,\hbar)^3}. \tag{2.241}$$

The particle number is

$$N = -\left(\frac{\partial \Phi}{\partial \mu}\right)_{T,V} = C\,\beta^{-3/2} V e^{\beta \mu}, \tag{2.242}$$

(continued)

Example 2.6 (continued)
and the pressure

$$P = -\left(\frac{\partial \Phi}{\partial V}\right)_{T,\mu} = C\beta^{-5/2}e^{\beta\mu}. \tag{2.243}$$

Hence the equation of state is

$$PV = \frac{N}{\beta} = kNT, \tag{2.244}$$

as expected. Inverting (2.242) we get

$$\mu = kT \log(N\beta^{3/2}/CV). \tag{2.245}$$

The free energy is

$$F = N\mu - PV = kTN \log(N\beta^{3/2}/CV) - kNT \tag{2.246}$$

in agreement with the canonical result (2.156), confirming ensemble universality.

Problems

2.1 Consider a classical system of monoatomic molecules with *hard cores* moving in a finite box $B \subset \mathbb{R}^3$ of volume V. The Hamiltonian is

$$H = \sum_{i=1}^{N} \frac{\mathbf{p}_i^2}{2m} + \sum_{1 \leq i < j \leq N} V(|\mathbf{r}_i - \mathbf{r}_j|)$$

where (for $0 < \delta \ll V^{1/3}/N_A$)

$$V(x) = \begin{cases} +\infty & \text{for } x \leq \delta \\ 0 & \text{otherwise} \end{cases}$$

and N is of the order of the Avogadro number N_A. Use the microcanonical ensemble to compute the relevant thermodynamical potential in the thermodynamic limit. Find the pressure and the internal energy as a function of T and V.

2.2 Consider a one-dimensional periodic lattice whose vertices are labelled by the elements of $i \in \mathbb{Z}/N\mathbb{Z}$ (the integer i is defined mod N) and links connecting nearest neighbor vertices i and $i+1$, see figure

In the thermodynamic limit one takes $N \to \infty$. On the i-th vertex there is a classical degree of freedom ξ_i taking value in $\mathbb{R}$ with a state measure given by the usual Lebesgue measure $\mathrm{d}\xi_i$. The Hamiltonian is ($\xi_{N+1} \equiv \xi_1$)

$$H(\xi_1, \dots, \xi_N) = 2 \sum_{i=1}^{N} \xi_i^2 - \sum_{i=1}^{N} \xi_i \, \xi_{i+1}.$$

Compute the microcanonical partition function and take the thermodynamic limit.

2.3 Prove the Stirling formula (2.72) (leading terms only) by an asymptotic evaluation of the integral defining the Gamma function (Euler's integral of second kind).

2.4 Consider the Bessel function

$$K_\nu(z) = \frac{1}{2} \left(\frac{z}{2}\right)^\nu \int_0^\infty \exp\left(-t - \frac{z^2}{4t}\right) \frac{\mathrm{d}t}{t^{\nu+1}} \quad \nu \in \mathbb{R}$$

Determine the asymptotic behavior as $z \to \infty$.

2.5 Consider an ideal gas of relativistic particles of mass m in a finite 3-dimensional box of volume V. Determine the canonical partition function and the equation of state. As Hamiltonian take ($c = 1$)

$$H = \sum_{i=1}^{N} \left(\sqrt{\mathbf{p}_i^2 + m^2} - m\right)$$

2.6 Consider a long tube of length L placed vertically on a constant gravitational field of strength g directed in the negative vertical direction. The tube contains a classical perfect gas at equilibrium at the temperature T. The pressure of the gas on the walls is a non-trivial function of the vertical coordinate z. Find this function using the canonical ensemble. Compare with the statics of fluids.

2.7 A quantum system has configuration space the unit 3-sphere $S^3 \subset \mathbb{R}^4$ with the natural metric invariant under $SO(4)$ rotation of the ambient $\mathbb{R}^4$. There is no potential $V \equiv 0$. Compute the canonical partition function.

2.8 Feynman Gas 1

As in Fig. 2.4 we have a one-dimensional classical gas consisting of N impenetrable particles (i.e. the particles cannot cross each other) which move in the segment $[0, V] \subset \mathbb{R}$. The particles screen each other, so that each particle interacts only with the particles next to it along the line. The Hamiltonian is

$$H = \sum_{i=1}^{N} \frac{p_i^2}{2m} - \lambda \sum_{i=1}^{N-1} \log |x_{i+1} - x_i|.$$

Show that in the thermodynamic limit ($N \to \infty$ with V/N fixed) the equation of state is

$$\frac{PV}{N} = kT + \lambda.$$

2.9 Feynman Gas 2

The same set-up of impenetrable particles moving on the segment $[0, L]$ as in Exercise 2.8 but with Hamiltonian

$$H_N = \sum_{i=1}^{N} \frac{p_i^2}{2m} + \sum_{i=2}^{N} \sum_{i=2}^{N} \frac{g}{|x_i - x_{i-1}|}.$$

Write the equation of state (in the thermodynamic limit). HINT: cf. Exercise 2.4.

2.10 Consider a classical system consisting of N particles moving in a circle S^1 of angular coordinate θ (periodic mod 2π) subjected to a pairwise potential of the special form described below. The Lagrangian is

$$L = \frac{1}{2} \sum_{i=1}^{N} m \, \dot{\theta}_i^2 + \lambda \sum_{1 \le i < j \le N} \log\left[2 \sin^2\left(\frac{\theta_i - \theta_j}{2} \right) \right] - \lambda N \log N$$

with $\lambda > 0$ (the additive constant is chosen for convenience). Compute the canonical partition function. HINT: use the Dyson-Wilson β-integral, eqs. (1.10)–(1.12) of [18]

References

1. S. Cecotti, *Analytic Mechanics. A Concise Textbook* (Springer, 2024)
2. P.A.M. Dirac, *Principles of Quantum Mechanics*, 3rd edn. (Oxford University Press, 1947)
3. J.B. Conway, *A Course in Functional Analysis* (Springer, 1990)
4. L.D. Landau, E.M. Lifsitz, *Quantum Mechanics. Non relativistic Theory*, 3rd edn. (Pergamon, 1976)

5. G.E. Andrews, R. Askey, R. Roy, *Special Functions*. Encyclopedia of Mathematics and its Applications, vol. 71 (Cambridge University Press, 2009)
6. D. Ruelle, *Statistical Mechanics. Rigorous Results* (Benjamin, 1969)
7. N.F. Ramsey, Thermodynamics and statistical mechanics at negative absolute temperatures. Phys. Rev. **103**, 20–28 (1956)
8. E.M. Purcell, R.V. Pound, A nuclear spin system at negative temperature. Phys. Rev. **81**, 279 (1951)
9. D.J.C. MacKay, *Information Theory, Inference, and Learning Algorithms* (Cambridge University Press, 2003)
10. E.T. Jaynes, *ProbabilityTheory: The Logic of Science* (Cambridge University Press, 2003)
11. J. Bricmont, *Making Sense of Statistical Mechanics*. Undergraduate Lecture Notes in Physics (Springer, 2022)
12. S. Coleman, *Quantum Field Theory: Lectures of Sidney Coleman* (World Scientific, 2019)
13. K. Iwaki, T. Nakanishi, Exact WKB analysis and cluster algebras. J. Phys. A Math. Theor. **47**, 474009 (2014), `arXiv:1401.7094`
14. D. Grumiller, M.M. Sheikh-Jabbari, *Black Hole Physics. From Collapse to Evaporation*. Graduate Texts in Physics (Springer, 2022)
15. J.J. Atick, E. Witten, The Hagedorn transition and the number of degrees of freedom of string theory. Nucl. Phys. **B310**, 291 (1988)
16. R. Roth, Plasma stability and the Bohr-Van Leeuwen theorem. NASA (1967)
17. A. Knapp, *Representation Theory of Semisimple Groups* (Princeton University Press, 1986)
18. P.J. Forrester, S.O. Warnaar, The importance of the Selberg integral. Bull. Amer. Math. Soc. (N.S.) **45**, 489–534 (2008), `arXiv:0710.3981`

Chapter 3
Non-interacting Quantum Systems

In this chapter we study the Statistical Mechanics of *ideal* quantum gases composed by $N \ggg 1$ non-interacting fundamental *quantum* particles. There are two new ingredients with respect to the classical (monoatomic) ideal gas. First: in addition to the position degrees of freedom $\boldsymbol{x} \in \mathbb{R}^3$, the quantum particles carry a discrete degree of freedom σ, called *spin,* which takes $2S + 1$ values

$$\sigma = -S, -S + 1, \cdots, S - 1, S, \tag{3.1}$$

where $S \geq 0$ is either *integral* or *half-integral,* depending on the particle species: we say that the particle *has spin S* to mean that the spectrum of its operator σ is given by (3.1). The single-particle wave-functions is then a $(2S + 1)$-component vector of square-summable functions

$$\psi(\boldsymbol{x})_\sigma \equiv \langle \boldsymbol{x}, \sigma | \psi \rangle \in L^2(B), \quad B \subset \mathbb{R}^3, \tag{3.2}$$

indexed by the spin quantum number σ which takes the $2S + 1$ values in Eq. (3.1). In Eq. (3.2) $B \subset \mathbb{R}^3$ is the finite-volume "box" in which our system is contained. The N-particle wave-function then has the general form

$$\psi(\boldsymbol{x}_1, \ldots, \boldsymbol{x}_N)_{\sigma_1 \ldots \sigma_N} \qquad \boldsymbol{x}_i \in B. \tag{3.3}$$

The second new feature is that we have to replace the classical statistics with the *quantum statistics* for *indistinguishable particles.* There are two distinct quantum statistics: the *Bose-Einstein* and the *Fermi-Dirac* ones, that we now discuss.

© The Author(s), under exclusive license to Springer Nature Switzerland AG 2024
S. Cecotti, *Statistical Mechanics*, UNITEXT for Physics,
https://doi.org/10.1007/978-3-031-67874-5_3

3.1   Quantum Statistics: Bosons vs. Fermions

Spin and Statistics

Contrary to the classical case, in the quantum set-up fundamental particles of the same kind are *strictly indistinguishable:* the ultimate physical explanation of this fact lays in relativistic quantum physics, that is, in Quantum Field Theory (QFT).[1] The *Spin and Statistics Theorem* of QFT states in full generality that [1]:[2]

Bos When the spin is *integral,* $S \in \mathbb{N}$, the N-particle Hilbert space $\mathcal{H}_N$ is the N-fold *symmetric* tensor power[3] of the one-particle Hilbert space $\mathcal{H}_1$

$$\mathcal{H}_N = \overbrace{\mathcal{H}_1 \odot \mathcal{H}_1 \odot \cdots \odot \mathcal{H}_1}^{N \text{ factors}}, \tag{3.4}$$

that is, the N-particle wave-function (3.3) is invariant under arbitrary permutations of the particles: for all permutation $\pi \in \mathfrak{S}_N$ of the N particles we have

$$\psi(\boldsymbol{x}_{\pi(1)}, \ldots, \boldsymbol{x}_{\pi(N)})_{\pi(\sigma_1)\ldots\pi(\sigma_N)} = \psi(\boldsymbol{x}_1, \ldots, \boldsymbol{x}_N)_{\sigma_1\ldots\sigma_N}. \tag{3.5}$$

In other words: *for S integral the wave-function* (3.3) *transforms in the trivial representation of* $\mathfrak{S}_N$. We express this fact by saying that particles with integral spin *obey the Bose-Einstein statistics,* or simply that integral spin particles are *bosons*;

Fer when $S \in \frac{1}{2}+\mathbb{N}$ the N-particle Hilbert space $\mathcal{H}_N$ is the N-fold *antisymmetric* tensor power of the one-particle Hilbert space $\mathcal{H}_1$

$$\mathcal{H}_N = \overbrace{\mathcal{H}_1 \wedge \mathcal{H}_1 \wedge \cdots \wedge \mathcal{H}_1}^{N \text{ factors}}, \tag{3.6}$$

that is, the N-particle wave-function belongs to the non-trivial one-dimensional representation of $\mathfrak{S}_N$, namely the sign one

$$\psi(\boldsymbol{x}_{\pi(1)}, \ldots, \boldsymbol{x}_{\pi(N)})_{\pi(\sigma_1)\ldots\pi(\sigma_N)} = \mathsf{sign}(\pi)\, \psi(\boldsymbol{x}_1, \ldots, \boldsymbol{x}_N)_{\sigma_1\ldots\sigma_N}. \tag{3.7}$$

We express this fact by saying that particles with half-integral spin *obey the Fermi-Dirac statistics* or that half-integral spin particles are *fermions*.

---

[1] For the second quantization point of view about indistinguishability see Sect. 3.7.

[2] In this form the Spin and Statistics theorem holds iff the dimension of spacetime is at least 4. Through this chapter we assume that space has three dimensions (so spacetime is four dimensional).

[3] In this book the symbol $\odot$ stands for the symmetric tensor product, $\wedge$ for the antisymmetric one.

Non-interacting Quantum Systems

We focus on the case where our indistinguishable particles are *non-interacting*—their Hamiltonian is just the sum of N identical copies of the one-particle Hamiltonian

$$H = \sum_{i=1}^{N} H^{(i)}, \tag{3.8}$$

where the i-th copy $H^{(i)}$ of the 1-particle Hamilton operator H acts on the i-th 1-particle factor Hilbert space $\mathcal{H}_1^{(i)}$: say for bosons

$$H^{(i)}\left(|\psi^{(1)}\rangle \odot \cdots \odot |\psi^{(i)}\rangle \odot \cdots \odot |\psi^{(N)}\rangle\right) \equiv$$
$$\equiv |\psi^{(1)}\rangle \odot \cdots \odot \left(H|\psi^{(i)}\rangle\right) \odot \cdots \odot |\psi^{(N)}\rangle. \tag{3.9}$$

We can choose a basis of $\mathcal{H}_N$ whose elements have the property that each particle is in some definite one-particle state $|a\rangle$ with wave-function

$$\psi_a(\boldsymbol{x})_\sigma \equiv \langle \boldsymbol{x}, \sigma | a \rangle \tag{3.10}$$

where the quantum number a labels the elements of a basis $\{|a\rangle\}$ of one-particle states in $\mathcal{H}_1$. The wave-function of the full N-particle system is then simply the product (symmetric or antisymmetric) of the one-particle wave-functions. Hence the wave-function of N non-interacting indistinguishable *bosons* which occupy the one-particle states with wave-functions $\psi_1(\boldsymbol{x})_\sigma, \psi_2(\boldsymbol{x})_\sigma, \ldots, \psi_N(\boldsymbol{x})_\sigma$ is

$$\psi(\boldsymbol{x}_1, \boldsymbol{x}_2 \ldots, \boldsymbol{x}_N)_{\sigma_1 \ldots \sigma_N} =$$
$$= \frac{1}{N!} \sum_{\pi \in \mathfrak{S}_N} \psi_1(\boldsymbol{x}_{\pi(1)})_{\sigma_{\pi(1)}} \psi_2(\boldsymbol{x}_{\pi(2)})_{\sigma_{\pi(2)}} \cdots \psi_N(\boldsymbol{x}_{\pi(N)})_{\sigma_{\pi(N)}}, \tag{3.11}$$

while for *fermions*

$$\psi(\boldsymbol{x}_1, \boldsymbol{x}_2 \ldots, \boldsymbol{x}_N)_{\sigma_1 \ldots \sigma_N} =$$
$$= \frac{1}{N!} \sum_{\pi \in \mathfrak{S}_N} \text{sign}(\pi)\, \psi_1(\boldsymbol{x}_{\pi(1)})_{\sigma_{\pi(1)}} \cdots \psi_N(\boldsymbol{x}_{\pi(N)})_{\sigma_{\pi(N)}}. \tag{3.12}$$

From equation (3.12) we deduce the

Pauli Exclusion Principle *Two identical fermions cannot be in the same one-particle state.*

Indeed, if two fermions are in the same state, their wave-function $\psi(\boldsymbol{x}_1, \boldsymbol{x}_2)_{\sigma_1 \sigma_2}$ vanishes identically, and the zero function is not a quantum state. Hence the N

fermions must be in N distinct states and Eq. (3.12) can be written in the form

$$\psi(\boldsymbol{x}_1, \boldsymbol{x}_2 \ldots, \boldsymbol{x}_N)_{\sigma_1 \ldots \sigma_N} = C \det_{ij}\big[\psi_i(\boldsymbol{x}_j)_{\sigma_j}\big], \tag{3.13}$$

where C is the obvious normalization constant.

Definition 3.1 Let $\{|a\rangle\}$ be an orthonormal basis of the 1-particle Hilbert space $\mathcal{H}_1$

$$\langle a'|a\rangle = \delta_{a,a'}. \tag{3.14}$$

The *occupation number N_a of the state $|a\rangle$* is the Hermitian operator acting on $\mathcal{H}_N \subset \mathcal{H}^{\otimes N}$

$$N_a \stackrel{\text{def}}{=} n_a^{(1)} + n_a^{(2)} + \cdots + n_a^{(N)}, \tag{3.15}$$

where $n_a^{(i)}$ acts on the i-th factor one-particle Hilbert space, $\mathcal{H}_1^{(i)}$, of the N-particle Hilbert space as

$$n_a^{(i)}|a'\rangle^{(i)} = \delta_{a,a'} \qquad |a'\rangle^{(i)} \in \mathcal{H}_1^{(i)}. \tag{3.16}$$

In other words, N_a acting on a N-particle wave-function of the form

$$\frac{1}{N!}\big(\psi_{a_1} \cdots \psi_{a_N} \pm \text{permutations}\big) \tag{3.17}$$

returns the number of quantum labels a_i equal to a, i.e. the number of particles in the one-particle state $|a\rangle$.

For bosons $N_a = 0, 1, 2, \cdots$, while for fermions $N_a = 0, 1$ by the Pauli exclusion principle. In the Fermi case we say that the a-th one-particle state is *filled* (resp. *empty*) iff $N_a = 1$ (resp. $N_a = 0$).

Character Ring Identities

Let V be a vector space and $A: V \to V$ a linear operator (when $\dim V$ is finite, A is just a $\dim V \times \dim V$ square matrix). We write $V^{\odot k}$ (resp. $V^{\wedge k}$) for the k-fold symmetric (resp. antisymmetric) tensor power of V. The operator A induces operators

$$A^{\odot k}: V^{\odot k} \to V^{\odot k} \quad \text{resp.} \quad A^{\wedge k}: V^{\wedge k} \to V^{\wedge k} \tag{3.18}$$

in the obvious way—say for bosons:

$$A^{\odot k}\big(v_1 \odot v_2 \odot \cdots \odot v_k\big) = (Av_1) \odot (Av_2) \odot \cdots \odot (Av_k). \tag{3.19}$$

We want to compute the generating function (in an indeterminate t) of the traces $\operatorname{Tr} A^{\odot k}$ and $\operatorname{Tr} A^{\wedge k}$ computed, respectively, in the vector spaces $V^{\odot k}$ and $V^{\wedge k}$. These generating functions are given by well-known identities in the ring of characters of a compact group [2] and are essentially equivalent to Newton's theorems on symmetric polynomials.

Lemma 3.1 *We have the identities*

$$\sum_{k=0}^{\infty} t^k \operatorname{Tr} A^{\odot k} = \frac{1}{\det[1 - tA]} = \exp\left[\sum_{k=0}^{\infty} \frac{t^k}{k} \operatorname{Tr}(A^k)\right], \tag{3.20}$$

$$\sum_{k=0}^{\infty} t^k \operatorname{Tr} A^{\wedge k} = \det[1 + tA] = \exp\left[-\sum_{k=0}^{\infty} \frac{(-t)^k}{k} \operatorname{Tr}(A^k)\right]. \tag{3.21}$$

Note that the sum in the second line is a polynomial in t of degree dim V. This is a consequence of the Pauli exclusion principle.

Proof The second equality in each equation follows from the equality

$$\det M = \exp \operatorname{tr} \log M, \tag{3.22}$$

which is valid for all invertible matrix M, together with the power series

$$\log(1 + x) = \sum_{n \geq 1} \frac{(-1)^{n-1}}{n} x^n. \tag{3.23}$$

We turn to the first equalities. Let $x_1, \ldots, x_m$ be the eigenvalues of A:

$$\operatorname{Tr} A^{\wedge k} \equiv e_k(x_1, \ldots, x_m) \overset{\text{def}}{=} \sum_{i_1 < \cdots < i_k} x_{i_1} \cdots x_{i_k} \tag{3.24}$$

$$\operatorname{Tr} A^{\odot k} \equiv h_k(x_1, \ldots, x_m) \overset{\text{def}}{=} \sum_{i_1 \leq \cdots \leq i_k} x_{i_1} \cdots x_{i_k}, \tag{3.25}$$

here e_k (resp. h_k) is the k-th *elementary symmetric function* in the variables $x_1, \ldots, x_m$ (resp. the k-th *complete symmetric function*), see [3]. From the definitions of the symmetric functions we have the obvious identities [3]

$$\sum_{k \geq 0} e_k(x_i) \, t^k = \prod_i (1 + x_i \, t) = \det(1 + At) \tag{3.26}$$

$$\sum_{k \geq 0} h_k(x_i) \, t^k = \prod_i (1 - x_i \, t)^{-1} = \det(1 - At)^{-1} \tag{3.27}$$

which yield the equalities in the **Lemma**.   □

Setting $A = 1$ and $m \equiv \dim V$ in Lemma 3.1 we get

$$\sum_{k=0}^{\infty} \dim V^{\odot k}\, t^k = (1-t)^{-m} = \sum_{k=0}^{\infty} \binom{-m}{k}(-1)^k t^k \equiv \sum_{k=0}^{\infty} \binom{m+k-1}{k} t^k \tag{3.28}$$

$$\sum_{k=0}^{\infty} \dim V^{\wedge k}\, t^k = (1+t)^m = \sum_{k=0}^{m} \binom{m}{k} t^k. \tag{3.29}$$

These formulae yield the total number of states $N(k, m)$ for a system of k bosons (resp. fermions) which may be in m distinct one-particle states

$$N(k, m) = \begin{cases} \binom{m+k-1}{k} & \text{bosons} \\ \binom{m}{k} & \text{fermions.} \end{cases} \tag{3.30}$$

Quantum Grand Partition Functions

To get the quantum Grand partition function $\Xi(\beta, \mu)$ for a *non-interacting* system of particles with one-particle Hilbert space $\mathcal{H}_1$ and one-particle Hamiltonian H, we have simply to replace in Lemma 3.1

$$V \rightsquigarrow \mathcal{H}_1, \qquad A \rightsquigarrow \exp[-\beta H], \qquad t \rightsquigarrow \zeta \equiv \exp[\beta \mu], \tag{3.31}$$

where β is the inverse temperature and μ the chemical potential. We get

$$\Xi(\beta, \mu) = \begin{cases} \det[1 - e^{\beta(\mu-H)}]^{-1} \equiv \displaystyle\prod_{n\geq 0}(1 - e^{\beta(\mu-E_n)})^{-1} & \text{bosons} \\[2mm] \det[1 + e^{\beta(\mu-H)}] \;\;\equiv \displaystyle\prod_{n\geq 0}(1 + e^{\beta(\mu-E_n)}) & \text{fermions} \end{cases} \tag{3.32}$$

where $E_0, E_1, E_2, \cdots$ are the one-particle energy levels (i.e. the eigenvalues of H acting on $\mathcal{H}_1$) listed in non-decreasing order. Notice that computing the Grand canonical partition function for *all* N's is much easier than computing the canonical one for fixed but large N. The deep reason of this "surprising" fact will be elucidated in the last two sections of this chapter: it is a consequence of the quantum particle-field duality. The *Grand potential* is then

$$\Phi(T, V, \mu) = -kT \log \Xi(T, V, \mu) = \pm kT \sum_{n \geq 0} \log\left(1 \mp e^{-\beta(E_n - \mu)}\right)$$

$$(3.33)$$

where the upper (resp. lower) sign is for bosons (resp. fermions)

The mean number of particles is

$$\langle N \rangle = -\left(\frac{\partial \Phi}{\partial \mu}\right)_{T,V} = \sum_{n \geq 0} \frac{1}{e^{\beta(E_n - \mu)} \mp 1},$$

$$(3.34)$$

which obviously can be written as

$$\langle N \rangle = \sum_{n \geq 0} \langle N_n \rangle,$$

$$(3.35)$$

where N_n is the occupation number in the n-th one-particle state of energy E_n. Comparing with (3.34) we get

$$\langle N_n \rangle = \frac{1}{e^{\beta(E_n - \mu)} \mp 1}.$$

$$(3.36)$$

To simplify the notation we shall write simply N and N_n for $\langle N \rangle$ and, respectively, $\langle N_n \rangle$ when there is no danger of confusion. The *internal energy* U is

$$U = \Phi + TS + \mu N = \Phi - T\left(\frac{\partial \Phi}{\partial T}\right)_{V,\mu} + \mu N =$$

$$= \Phi + \beta\left(\frac{\partial \Phi}{\partial \beta}\right)_{V,\mu} + \mu N = -\frac{\partial}{\partial \beta} \log \Xi + \mu N =$$

$$= \sum_n \frac{E_n - \mu}{e^{\beta(E_n - \mu)} \mp 1} + \frac{\mu}{e^{\beta(E_n - \mu)} \mp 1} = \sum_n \frac{E_n}{e^{\beta(E_n - \mu)} \mp 1} \equiv \sum_n E_n \langle N_n \rangle$$

$$(3.37)$$

a result which has an obvious physical interpretation.

Remark 3.1 (For the *Cognoscenti*) The identity

$$\Xi(T, V, \mu)_{\text{bos}} \, \Xi(T, V, \mu + i\pi kT)_{\text{ferm}} = 1$$

$$(3.38)$$

expresses the fact that a system of bosons and fermions with the same energy levels E_n and the same degeneracies is *supersymmetric*. See Eq. (3.254) below.

Fermi Systems at Zero Temperature. Fermi Energy

The number of fermions in the n-th one-particle state of energy E_n as the temperature goes to zero, $\beta \to \infty$, is

$$\lim_{\beta \to \infty} \langle N_n \rangle = \lim_{\beta \to \infty} \frac{1}{e^{\beta(E_n - \mu)} + 1} = \begin{cases} 1 & \text{for } E_n < \mu \\ 0 & \text{for } E_n > \mu, \end{cases} \tag{3.39}$$

so at zero temperature all energy levels below $\mu(T)|_{T=0}$ are *filled* while the ones above $\mu(T)|_{T=0}$ are *empty*. The energy $\mu(T)|_{T=0}$ is called the *Fermi energy*

$$E_F \overset{\text{def}}{=} \mu(T)|_{T=0}. \tag{3.40}$$

Definition 3.2 The *Fermi energy* E_F is the energy such that the number of one-particle states with energy $E \leq E_F$ is equal to the number N of fermions present in the system.

Bose Condensation at Zero Temperature

The lowest energy state of a system of N non-interacting bosons is the state where *all* particles are in the one-particle ground state[4] of energy E_0. This is the state we get in the thermodynamical limit as $T \to 0$ provided the zero-temperature chemical potential $\mu(T)|_{T=0}$ has the appropriate value. When the number N_0 of particles in the ground state is a sizeable portion of all particles in the thermodynamical limit $N \to \infty$, that is,

$$\alpha \equiv \frac{N_0}{N} = O(1) \quad \text{as } N \to \infty, \tag{3.41}$$

we say that the bosonic system is in a quantum phase called *Bose condensation*. In this case

$$N_0 = \frac{1}{e^{\beta(E_0 - \mu)} - 1} \approx \alpha N \ggg 1, \tag{3.42}$$

which requires $\beta(E_0 - \mu) \approx 1/(\alpha N)$ i.e.

$$E_0 - \mu \approx \frac{kT}{\alpha N} \tag{3.43}$$

or $\mu \to E_0^-$ in the limit $N \to \infty$.

---

[4] In this chapter we assume that the one-particle ground state is non-degenerate (i.e. E_0 is an eigenvalue of the one-particle Hamiltonian H with multiplicity 1).

3.2   Quantum Ideal Gases

We consider an ideal gas composed by $N \ggg 1$ (fundamental[5]) quantum particles of spin S moving in a box B of volume V. In addition to the position degrees of freedom[6] $x \in B \subset \mathbb{R}^3$—each particle has a discrete degree of freedom, the spin σ, which takes $2S + 1$ values, cf. Eq. (3.1).

In absence of electromagnetic fields, the Hamiltonian of a (non-relativistic) gas of N identical fundamental particles of mass m is simply

$$H = \sum_{i=1}^{N} \frac{\mathbf{p}_i^2}{2m}. \tag{3.44}$$

In particular H is independent of the spin quantum number σ. The spectrum of the momentum operator

$$\mathbf{p}_i = -i\hbar \frac{\partial}{\partial x^i} \tag{3.45}$$

in the finite box B depends on the boundary conditions on the walls of the box. In the thermodynamic limit

$$N, V \to \infty \quad \text{with} \quad V/N = \text{const.} \tag{3.46}$$

the walls of B are pushed at infinite distance, so that, *if the thermodynamic limit is unique*, the details of the boundary conditions are immaterial, and we may choose them in any convenient way. We shall see later in this book that there are important systems whose thermodynamic limit is *not* unique, and the choice of boundary conditions matters *dramatically* even in the thermodynamic limit. The non-uniqueness issue does not apply to the present situation since the system consists of $N \to \infty$ *non-interacting* copies of a system with a finite number of degrees of freedom and a unique ground state.

We choose the box $B \subset \mathbb{R}^3$ to be cubic with all sides of length L ($V = L^3$) and use periodic boundary conditions, i.e. we identify the opposite walls of the cubic box see Fig. 3.1. Mathematically speaking, B is a flat 3-torus $(S^1)^3$ and the one-particle wave-functions $\psi(x_1, x_2, x_3)_\sigma$ satisfy the periodicity conditions

$$\psi(x_1 + L, x_2, x_3)_\sigma = \psi(x_1, x_2 + L, x_3)_\sigma =$$
$$= \psi(x_1, x_2, x_3 + L)_\sigma = \psi(x_1, x_2, x_3)_\sigma \tag{3.47}$$

---

[5] For the purpose of this chapter "fundamental" means that the particle has no other degrees of freedom besides its position $x \in \mathbb{R}^3$ and its discrete spin σ.

[6] As stressed in the previous footnote, we neglect rotational degrees of freedom, so we are considering the quantum version of the monoatomic ideal gas. Adding the rotational degrees of freedom is straightforward, see Sect. 2.9.

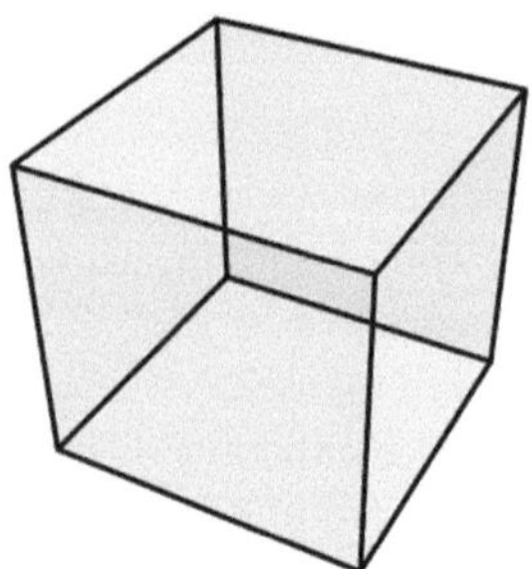

Fig. 3.1 A cubic box B in $\mathbb{R}^3$ with sides of length L. The opposite faces are periodically identified

Therefore the one-particle wave-function $\psi(x)_\sigma$ has a Fourier series expansion, and we have a basis of states $|k, \sigma\rangle$ with $k \in \mathbb{Z}^3$ and $\sigma = -S, -S+1, \ldots, S$ whose wave-functions have the form

$$\langle x, \sigma' | k, \sigma \rangle = \exp\left[\frac{2\pi i}{L} k \cdot x\right] \delta_{\sigma\sigma'}, \tag{3.48}$$

with momentum $\mathbf{p} = 2\pi \hbar k / L$ and energy

$$E_{k,\sigma} = \frac{2\pi^2 \hbar^2}{mL^2} k^2. \tag{3.49}$$

Plugging the energy levels (3.49) in the Grand potential formula (3.33) we get

$$\Phi = \pm kT \sum_{k \in \mathbb{Z}^3} (2S + 1) \log\left[1 \mp \exp\left(-\beta \frac{2\pi^2 \hbar^2}{mL^2} k^2 + \beta\mu\right)\right] \tag{3.50}$$

upper (lower) sign for bosons (resp. fermions). From Eq. (3.34)

$$\langle N \rangle = (2S + 1) \sum_{k \in \mathbb{Z}^3} \frac{1}{\exp\left(\beta \frac{2\pi^2 \hbar^2}{mL^2} k^2 - \beta\mu\right) \mp 1}. \tag{3.51}$$

The Canonical Ensemble

While it is convenient to work with the Grand canonical ensemble, there is a formula which is most easily deduced using the canonical ensemble with partition function

$$Z_N = \mathrm{Tr}_{\mathcal{H}_N} e^{-\beta H} \equiv \sum_{k \in \mathbb{Z}^3} \exp\left(-\beta \frac{2\pi^2 \hbar^2}{mL^2} k^2\right) \tag{3.52}$$

which depends only on the combination $\beta L^{-2} = \beta V^{-2/3}$ so that

$$\beta \frac{\partial}{\partial \beta} \log Z_N = -\frac{3}{2} V \frac{\partial}{\partial V} \log Z_N, \tag{3.53}$$

that is,

$$PV = \frac{2}{3} U, \tag{3.54}$$

a relation which is *true for all non-relativistic ideal gases,* classical, bosonic, or fermionic, as long as the particles are really "elementary", that is, they have no other degrees of freedom except the position and the spin, and space is three-dimensional. Classical gases of pluri-atomic molecules have in addition rotational and vibrational degrees of freedom and the numerical coefficient in front of U is different.

Remark 3.2 The same scaling argument shows, more generally, that when the energy is a homogeneous function of the momentum of degree α, $E \propto |\boldsymbol{p}|^{\alpha}$, and the space is d-dimensional, we have

$$PV = \frac{\alpha}{d} U. \tag{3.55}$$

MATH INTERLUDE: Euler-Maclaurin Summation
To simplify the formulae (3.50) and (3.51) it is convenient to replace the sums over $\boldsymbol{k} \in \mathbb{Z}^3$ with integrals. We have to make this replacement in a careful and controlled way: any discrepancy between sums and integrals which survives in the thermodynamic limit represents an observable physical phenomenon. The precise math procedure is to evaluate the sums (3.50), (3.51) with (a baby version of) the *Euler-Maclaurin summation formula,* cf. [4] §. 4.2.

Lemma 3.2 (Euler-Maclaurin Formula [4]) *If $f(x)$ is of class C^1 for $x \geq a$, $a \in \mathbb{Z}$, and $f(\infty) = 0$, we have*

$$\sum_{n>a} f(n) = \int_a^\infty \Big(f(x) + \psi(x)\, f'(x) \Big) dx - \frac{1}{2} f(a) \tag{3.56}$$

where[7]

$$\psi(x) \overset{\text{def}}{=} x - [x] - \tfrac{1}{2}, \qquad |\psi(x)| \leq \tfrac{1}{2}. \tag{3.57}$$

We leave the easy proof to the reader.

---

[7] When $x \in \mathbb{R}$, the symbol $[x]$ stands for the *floor function* i.e. largest integer such that $[x] \leq x$ (also known as the *integral part* of the real number x).

The idea of the Euler-Maclaurin formula is that when $f'(x)$ is small, omitting the second term in the integrand we get a good approximation to the sum. If f' is not small but f'' is small, we may evaluate the second term by using the same formula twice, and so on, recursively, until we get a small higher ℓ-th derivative $f^{(\ell)}$ which we can safely disregard. For more details and generalizations, see [4].

3.3   MATH INTERLUDE: Fermi-Dirac and Bose-Einstein Integrals

Before going to the physics of quantum ideal gases, we replenish our box of math tools. In this section we introduce and study a class of special functions [5] which play a significant role in the analysis of quantum systems and are central in many other math contexts.

Definition 3.3 The *Fermi-Dirac integral* $\mathsf{F}_s(z)$ and the *Bose-Einstein integral* $\mathsf{G}_s(z)$ *of index s* are the special functions[8] of the "fugacity" z

$$\mathsf{F}_s(z) = \frac{1}{\Gamma(s)} \int_0^\infty \frac{t^{s-1}\,dt}{e^t z^{-1} + 1} \tag{3.58}$$

$$\mathsf{G}_s(z) = \frac{1}{\Gamma(s)} \int_0^\infty \frac{t^{s-1}\,dt}{e^t z^{-1} - 1}. \tag{3.59}$$

When $|z| < 1$ one has

$$\mathsf{F}_s(z) = \frac{1}{\Gamma(s)} \int_0^\infty \frac{e^{-t} z\, t^{s-1}\,dt}{1 + e^{-t} z} = \frac{1}{\Gamma(s)} \sum_{n \geq 0} (-1)^n z^{n+1} \int_0^\infty e^{-(n+1)t}\, t^{s-1} dt$$

$$= \sum_{n \geq 0} (-1)^n \frac{z^{n+1}}{(n+1)^s} = -\mathsf{Li}_s(-z),$$

$$\tag{3.60}$$

and analogously

$$\mathsf{G}_s(z) = \mathsf{Li}_s(z), \tag{3.61}$$

---

[8] See e.g. [5]. The relation between our functions $\mathsf{F}_s(z)$ and their functions $F_s(x)$ is $\mathsf{F}_s(z) \equiv F_{s+1}(\log z)$.

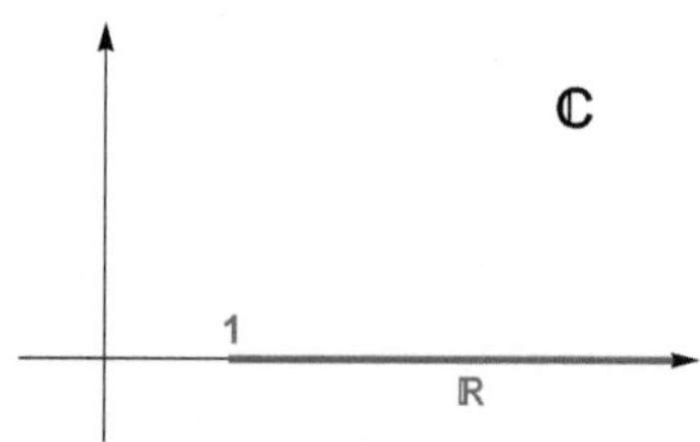

Fig. 3.2 The domain where the principal polylog is holomorphic. There is a branch cut along the positive real axis starting at 1 and going to ∞ (red line)

where the *polylogarithm function of index* s, $\mathrm{Li}_s(z)$, is defined by

$$\mathrm{Li}_s(z) = \sum_{n \geq 1} \frac{z^n}{n^s} \quad \text{when } |z| < 1 \tag{3.62}$$

and by analytic continuation (as a *multivalued* holomorphic function) elsewhere.[9]

Principal Determination The *principal determination* of the polylogarithm with $s > 0$ is the branch of the multivalued function $\mathrm{Li}_s(z)$ which is holomorphic in $\mathbb{C}$ minus a cut along the real axis from 1 to $+\infty$ and agrees with the sum (3.62) for $|z| < 1$, see Fig. 3.2. Comparing with our previous discussion, we conclude that: *the Fermi-Dirac and Bose-Einstein integrals are given by the principal determination of the polylog with the same index.* Indeed the function $\mathbf{G}_s(z)$ is holomorphic everywhere but for a cut along the real axis from 1 to ∞ with discontinuity (here $z \in \mathbb{R}$)

$$\mathbf{G}_s(z + i\epsilon) - \mathbf{G}_s(z - i\epsilon) = \frac{z}{\Gamma(s)} \int_0^\infty t^{s-1} dt \left(\frac{1}{e^t - z - i\epsilon} - \frac{1}{e^t - z + i\epsilon} \right) =$$

$$= \frac{2\pi i\, z}{\Gamma(s)} \int_0^\infty t^{s-1}\, dt\, e^{-t}\, \delta(t - \log z) = \begin{cases} \frac{2\pi i}{\Gamma(s)} (\log z)^{s-1} & z \geq 1 \\ 0 & z < 1 \end{cases}$$

$$\tag{3.63}$$

where we used the fundamental identity (here $\delta(x)$ is Dirac's delta-function):

$$\frac{1}{x - i\epsilon} - \frac{1}{x + i\epsilon} = 2\pi i\, \delta(x). \tag{3.64}$$

We are particularly interested in the behavior of these functions in a neighborhood of the singularity at $z = 1$ of the principal polylogarithm $\mathrm{Li}_s(z)$. We have

$$\frac{d}{dw}\mathrm{Li}_s(e^w) = \mathrm{Li}_{s-1}(e^w) \quad \text{and} \quad \mathrm{Li}_s(1) = \zeta(s), \tag{3.65}$$

---

[9] $\mathrm{Li}_s(z)$ extends as a univalued holomorphic function on the universal Abelian cover of the 3-punctured Riemann sphere $\mathbb{P}^1 \setminus \{0, 1, \infty\}$.

where $\zeta(s)$ is the *Riemann zeta-function* i.e. the meromorphic function in $\mathbb{C}$ which for $\operatorname{Re} s > 1$ has the Dirichlet sum representation [4]

$$\zeta(s) = \sum_{n=0}^{\infty} \frac{1}{n^s}. \tag{3.66}$$

From Eqs. (3.65) and (3.66) we get the power series expansion of $\mathsf{Li}_s(e^w)$ around $w = 0$ which is valid when $s \neq 1, 2, 3, \ldots$ and converges for $|w| \leq 2\pi$

$$\mathsf{Li}_s(e^w) = \Gamma(1 - s)(-w)^{s-1} + \sum_{k=0}^{\infty} \zeta(s - k) \frac{w^k}{k!}, \tag{3.67}$$

which, as advertised, shows a root branch cut from $w = 0$ $(z = 1)$.[10] As $w \to 0^-$

$$\mathsf{Li}_{5/2}(e^w) = \zeta(5/2) + \zeta(3/2)w + \frac{4\sqrt{\pi}}{3}(-w)^{3/2} + O(w^2) \tag{3.68}$$

$$\mathsf{Li}_{3/2}(e^w) = \zeta(3/2) - 2\sqrt{\pi}(-w)^{1/2} + O(w) \tag{3.69}$$

$$\mathsf{Li}_{1/2}(e^w) = \sqrt{\pi}(-w)^{-1/2} + O(1). \tag{3.70}$$

Since the fugacity $\zeta = e^{\beta\mu}$ is non-negative[11] we conclude:

Proposition 3.1

(1) The Fermi-Dirac integrals $\mathsf{F}_s(\zeta)$ with $s > 0$ are holomorphic (analytic) functions in the physical region $\zeta \geq 0$.

(2) The Bose-Einstein integrals $\mathsf{G}_s(\zeta)$ with $s > 0$ are analytic in the half-open interval $0 \leq \zeta < 1$, but they have a branch root singularity at $\zeta = 1$. If $s \neq 1, 2, 3, \ldots$ the derivatives

$$\frac{d^k}{dw^k} \mathsf{G}_s(e^w) = \mathsf{Li}_{s-k}(e^w) \tag{3.71}$$

are continuous as $w \to 0$ iff $k < s - 1$.

This result implies that in the Bose case the physical region in the fugacity space is $0 \leq \zeta \leq 1$ with peculiar phenomena at the endpoint $\zeta = 1$. In Sect. 3.5 we shall understand the physical meaning of this math statement.

Remark 3.3 As we will discuss in Chap. 4, a non-analyticity of the thermodynamic potential at some physically allowed value of a natural variable represents a

---

[10] To connect this formula with Eq. (3.63), use the functional equation of the Γ-function: $\left(e^{-i\pi(s-1)} - e^{i\pi(s-1)}\right)\Gamma(1 - s) = 2\pi i / \Gamma(s)$.

[11] Before the thermodynamic limit $\zeta > 0$; after we can only say that $\zeta \geq 0$.

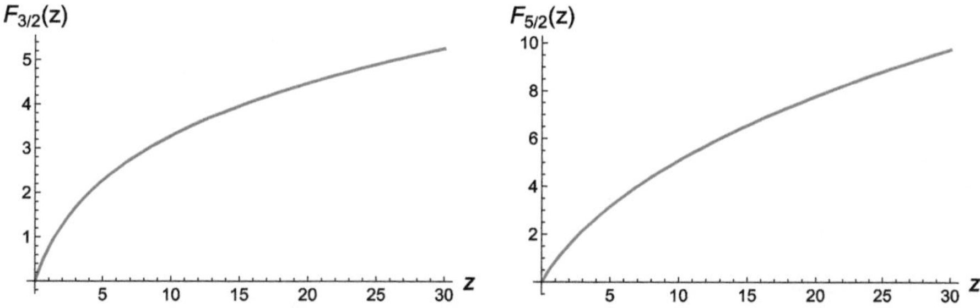

Fig. 3.3 (**a**) Fermi-Dirac integral $F_{3/2}(z)$. (**b**) Fermi-Dirac integral $F_{5/2}(z)$

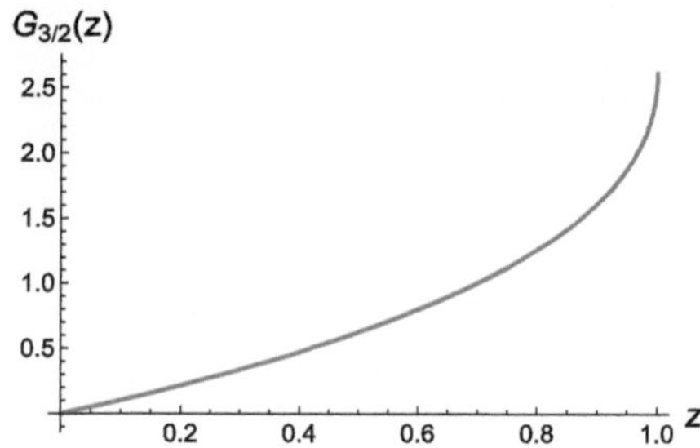

Fig. 3.4 Plot of the Bose-Einstein integral $G_{3/2}(z)$ for $0 \leq z \leq 1$. Notice that its derivative diverges as $z \to 1$

phase transition between two phases with *different physics*. We conclude from Proposition 3.1 that the ideal bosonic gas has *two* distinct phases. In Sect. 3.5 we shall confirm this prediction and discuss the physics of the two phases and the nature of the transition.

The Fermi-Dirac functions $F_{3/2}(z)$, $F_{5/2}(z)$ are plotted in Fig. 3.3a, b while the Bose-Einstein function $G_{3/2}(z)$ is plotted in Fig. 3.4. From the equation

$$\frac{dF_s(e^w)}{dw} = F_{s-1}(e^w) > 0 \quad \text{for } s > 1, \tag{3.72}$$

we see that the function $F_s(e^w)$ with $s > 1$ is a *strictly increasing monotonic function of w* whose image is the full positive real axis $\mathbb{R}_{>0}$. The same argument shows that the Bose-Einstein integrals $G_s(z)$ with $s > 1$ are *monotonic increasing* for $0 \leq z \leq 1$ with image the interval $[0, \zeta(s)] \subset \mathbb{R}$.

The small z expansions of the Fermi-Dirac and Bose-Einstein integrals can be read from the series expansion Eq. (3.62) (convergent for $|z| < 1$).

Asymptotic Expansion While the physical region of the fugacity for a Bose gas is bounded to the interval $0 \leq \zeta \leq 1$, for a Fermi gas it extends to the full positive real axis $\zeta \geq 0$. Therefore in the Fermi case we need the large $\zeta \gg 1$ asymptotics given

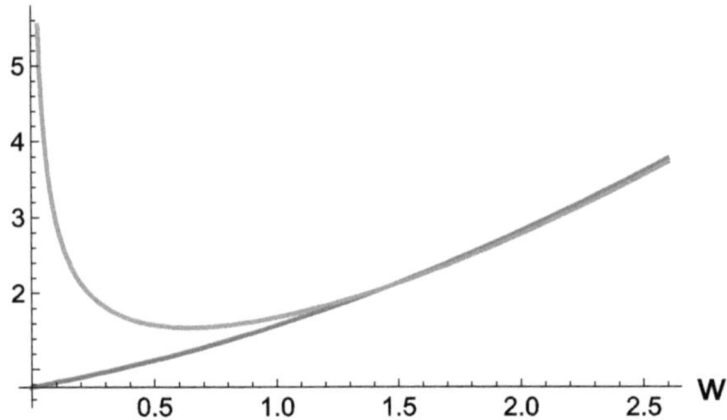

Fig. 3.5 The blue curve represents the function $F_{3/2}(e^w)$, while the orange curve is its asymptotic expression truncated to the second term $\frac{w^{3/2}}{\Gamma(5/2)} + \frac{\pi^2}{6}\frac{w^{-1/2}}{\Gamma(1/2)}$ We see that the asymptotic approximation is already accurate for w as small as 1.5

by the expansion (cf. eq. (11.1) in [6])[12]

$$
F_s(e^w) = 2\sum_{k=0}^{\infty}(1 - 2^{1-2k})\,\zeta(2k)\,\frac{w^{s-2k}}{\Gamma(s+1-2k)} =
$$

$$
= \frac{w^s}{\Gamma(s+1)} + \frac{\pi^2}{6}\frac{w^{s-2}}{\Gamma(s-1)} + O(w^{s-4}),
\tag{3.73}
$$

see Fig. 3.5. In the second line of (3.73) we used

$$
\zeta(0) = -\frac{1}{2}, \qquad \zeta(2) = \frac{\pi^2}{6}.
\tag{3.74}
$$

More generally

$$
\zeta(2k) = \frac{1}{2}(2\pi)^{2k}\frac{|B_{2k}|}{(2k)!} \quad \text{for } k = 1, 2, 3, \ldots,
\tag{3.75}
$$

where B_n are *Bernoulli numbers* (sequences **OEIS:A027641**, **OEIS:A027642** [7], see also [5]). Other useful formulae are

$$
\Gamma(n + 1/2) = \sqrt{\pi}\,\frac{1\cdot 3\cdot 5\cdot 7\cdots(2n-1)}{2^n} = \sqrt{\pi}\,\frac{(2n)!}{2^{2n}n!}
\tag{3.76}
$$

$$
\frac{1}{\Gamma(1/2-n)} = \frac{(-1)^n}{\pi}\,\Gamma(n+1/2) = \frac{(-1)^n}{\sqrt{\pi}}\,\frac{(2n)!}{2^{2n}n!}
\tag{3.77}
$$

From these expressions we see that the ratio of the $(k+1)$-th term to the k-th term in the expansion (3.73) is for $k \gg 1$ equal to $4k^2/w^2$, so that the asymptotic series should be truncated after $k \sim w/2 \gg 1$ terms to get an optimal approximation.

---

[12] Equation (3.73) is a mere asymptotic expansion which is not convergent. To get an optimal approximation the sum should be truncated at the $k \sim \frac{1}{2}w$ term.

3.4   Ideal Fermi Gas

We apply the Euler-Maclaurin formula to the sum (3.51) for fermions (lower sign)

$$\langle N \rangle = (2S+1) \sum_{n_1 \in \mathbb{Z}} \sum_{n_2 \in \mathbb{Z}} \sum_{n_3 \in \mathbb{Z}} F\left(\frac{n_1^2 + n_2^2 + n_3^2}{V^{2/3}}\right), \tag{3.78}$$

where

$$F(z) \overset{\text{def}}{=} \frac{1}{\exp\left(\beta \frac{2\pi^2 \hbar^2}{m} z - \beta\mu\right) + 1} \tag{3.79}$$

We first compute the sum over n_3 using (3.56): this is legitimate since the function $F(z)$ is analytic for all $z \in \mathbb{R}$, so a fortiori of class C^1. The first derivative of the summand with respect to n_3 is

$$\frac{\partial}{\partial n_3} F\left(\frac{n_1^2 + n_2^2 + n_3^2}{V^{2/3}}\right) = \frac{2n_3}{V^{2/3}} F'. \tag{3.80}$$

We are interested in the thermodynamic limit $V \to \infty$. In this limit $\langle N \rangle$ is of order $O(V)$. From Eq. (3.80) we see that the second term in the integrand of the Euler-Maclaurin formula (3.56) is of order $O(V^{-2/3})$ relative to the first one. Moreover each term in the triple sum (3.78) is bounded by $(2S+1)$ so the boundary terms in the Euler-Maclaurin formula are $O(V^{-1/3})$ relative to the bulk integral, and may be neglected in the thermodynamic limit. Then, in this limit,

$$\langle N \rangle = (2S+1) \int \frac{\mathrm{d}^3 x}{\exp\left(\beta \frac{2\pi^2 \hbar^2}{m L^2} x^2 - \beta\mu\right) + 1}. \tag{3.81}$$

Writing $y = (2\pi\hbar\sqrt{\beta})x/\sqrt{2mL^2}$ we get

$$\begin{aligned}
\langle N \rangle &= V \frac{(2mkT)^{3/2}}{(2\pi\hbar)^3}(2S+1) \int \frac{\mathrm{d}^3 y}{e^{y^2 - \beta\mu} + 1} + O(V^{1/3}) = \\
&= V \frac{4\pi(2mkT)^{3/2}}{(2\pi\hbar)^3}(2S+1) \int \frac{y^2 \, \mathrm{d}y}{e^{y^2 - \beta\mu} + 1} + O(V^{1/3})
\end{aligned} \tag{3.82}$$

which confirms our claim that $\langle N \rangle / V$ is finite in the thermodynamic limit. In a similar fashion the internal energy is

$$
U = (2S+1) \int \frac{4\pi^2 \hbar^2 x^2}{2mL^2} \frac{\mathrm{d}^3 x}{\exp\left(\beta \frac{2\pi^2 \hbar^2}{mL^2} x^2 - \beta\mu\right) + 1} =
$$

$$
= V \frac{4\pi (2m)^{3/2}}{(2\pi\hbar)^3} (kT)^{5/2} (2S+1) \int \frac{y^4 \, \mathrm{d}y}{e^{y^2 - \beta\mu} + 1}. \tag{3.83}
$$

Thermodynamical Quantities

To simplify the expressions we introduce the *thermal de Broglie wavelength*

$$
\lambda \stackrel{\text{def}}{=} \frac{2\pi\hbar}{\sqrt{2m\pi kT}}. \tag{3.84}
$$

Equations (3.82) and (3.83) may be rewritten in terms of Fermi-Dirac integrals $F_s(\zeta)$:

$$
N = \frac{V}{\lambda^3} (2S+1) \, F_{3/2}(e^{\beta\mu}) \tag{3.85}
$$

$$
\frac{U}{kT} = \frac{3}{2} \frac{V}{\lambda^3} (2S+1) \, F_{5/2}(e^{\beta\mu}) \tag{3.86}
$$

The Grand potential is $\Phi \equiv -PV$. Then from Eq. (3.54)

$$
\Phi \equiv -\frac{2}{3} U = -\frac{V}{\lambda^3} kT \, (2S+1) \, F_{5/2}(e^{\beta\mu}) \tag{3.87}
$$

To get the equation of state, we have first to invert Eq. (3.85) and write

$$
\mu = \mu(T, V, N). \tag{3.88}
$$

From the structure of Eq. (3.85), we see that the fugacity $\zeta \equiv \exp(\beta\mu)$ is a function of the *single* a-dimensional quantity

$$
\xi \stackrel{\text{def}}{=} \frac{N\lambda^3}{(2S+1)V} = \frac{1}{2S+1} \frac{N}{V} \left(\frac{4\pi^2 \hbar^2}{2m\pi kT}\right)^{3/2} \geq 0. \tag{3.89}
$$

The quantity ξ has the following properties:

$\xi 1$ is linear in the density N/V (so it is small for a *dilute* gas);
$\xi 2$ scales with $\hbar$ as $\hbar^3$ (so it goes to zero in the *classical regime*);
$\xi 3$ scales with T as $T^{-3/2}$ (so it vanishes at *high temperature*).

Equation (3.85) takes the form

$$F_{3/2}(e^{\beta \mu}) = \xi \tag{3.90}$$

The relation $F_{3/2}(\zeta) = \xi$ is globally invertible since $F_{3/2} \colon \mathbb{R}_{\geq 0} \to \mathbb{R}_{\geq 0}$ is strictly monotonic increasing (cf. Eq. (3.72)): hence

$$\beta \mu = \log f(\xi), \tag{3.91}$$

where $f \colon \mathbb{R}_{\geq 0} \to \mathbb{R}_{\geq 0}$ is the unique function such that

$$F_{3/2}\big(f(x)\big) = x. \tag{3.92}$$

From Eq. (3.87) and $\Phi = -PV$ we see that the equation of state is

$$PV = NkT \, \xi^{-1} \, F_{5/2}(f(\xi)). \tag{3.93}$$

We study the limit forms of this equation of state in various physical regimes.

High Temperature For T large, ξ is small, hence $\zeta \equiv e^{\beta \mu} \ll 1$,

$$F_s(\zeta) = -\mathrm{Li}_s(-\zeta) = \zeta - \frac{\zeta^2}{2^s} + O(\zeta^3), \tag{3.94}$$

and Eq. (3.90) reduces to

$$\zeta - \frac{\zeta^2}{2^{3/2}} + O(\zeta^3) = \xi. \tag{3.95}$$

The inverse function is

$$\zeta(\xi) = \xi \left(1 + \frac{\xi}{2^{3/2}} + O(\xi^2) \right), \tag{3.96}$$

while

$$-\frac{\Phi}{N\,kT} = \frac{F_{5/2}(\zeta)}{\xi} = \xi^{-1} \left(\zeta - \frac{\zeta^2}{2^{5/2}} + O(\zeta^3) \right) = 1 + \frac{\xi}{2^{5/2}} + O(\xi^2). \tag{3.97}$$

More generally, inverting the power series (3.60) we get the Grand potential written as a (convergent) power series in the a-dimensional quantity ξ. ξ is proportional to the density N/V. In statistical mechanics the expansion in powers of the density of a gas is known as the "virial expansion": thus the recursive solution to the above equations in power series of ξ is the quantum virial expansion. The equation of state at high temperature is then

$$\frac{PV}{NkT} = 1 + \frac{\xi}{2^{5/2}} + O(\xi^2) \tag{3.98}$$

If we neglect the small correction $O(\xi)$ we get back the classical equation of state for a perfect gas $PV = NkT$. This shows that at high temperature the quantum effects are suppressed, and we recover the classical behavior. Indeed we have seen above that the expansion in powers of ξ may be interpreted in three different ways:

(a)   low density ("virial") expansion;
(b)   a semiclassical expansion in powers of $\hbar^3$;
(c)   a large T expansion in powers of $T^{-3/2}$.

Low-density, high temperature, and classical regimes are all physically equivalent.

In classical *non-ideal* gases, we also have a virial expansion in powers of the density. The first correction to the ideal equation of states is positive (resp. negative)—that is, the pressure is larger (resp. smaller) than the ideal gas one— iff the interactions between the molecules are *repulsive* (resp. *attractive*). Since in the ideal Fermi gas the first "virial" correction is positive, we see that the quantum effects act effectively as a repulsion: this is a manifestation of the Pauli exclusion principle. Two fermions with the same spin have a wave function

$$\psi(x_1, x_2) = -\psi(x_2, x_1) = O(x_1 - x_2) \quad \text{for } x_1 \approx x_2, \tag{3.99}$$

and the probability of them approaching each other is small (see also Sect. 3.7 below).

Low Temperature   As $T \to 0$ we use the asymptotic series (3.73):

$$\frac{1}{2S+1} \frac{N}{V} \left(\frac{4\pi^2 \hbar^2 \beta}{2m\pi} \right)^{3/2} \equiv \xi = F_{3/2}(e^{\beta\mu}) =$$
$$= \frac{4(\beta\mu)^{3/2}}{3\sqrt{\pi}} \left(1 + \frac{\pi^2}{8}(\beta\mu)^{-2} + O((\beta\mu)^{-4}) \right), \tag{3.100}$$

and

$$\frac{1}{2S+1} \frac{N}{V} \left(\frac{4\pi^2 \hbar^2 \beta}{2m\pi} \right)^{3/2} \frac{\beta U}{N} = \frac{3}{2} F_{5/2}(e^{\beta\mu}) =$$
$$= \frac{4(\beta\mu)^{5/2}}{5\sqrt{\pi}} \left(1 + \frac{5\pi^2}{8}(\beta\mu)^{-2} + \dots \right), \tag{3.101}$$

that is,

$$N = \frac{(2S+1)}{4\pi^2} V \left(\frac{2m}{\hbar^2}\right)^{3/2} \left(\frac{2}{3}\mu^{3/2}\right) \left(1 + \frac{\pi^2}{8}\frac{(kT)^2}{\mu^2} + \cdots \right) \tag{3.102}$$

$$U = \frac{(2S+1)}{4\pi^2} V \left(\frac{2m}{\hbar^2}\right)^{3/2} \left(\frac{2}{5}\mu^{5/2}\right) \left(1 + \frac{5\pi^2}{8}\frac{(kT)^2}{\mu^2} + \cdots \right) \tag{3.103}$$

Recall from Sect. 3.1 that the *Fermi energy* is defined as $E_F = \mu(T)|_{T=0}$; setting $T = 0$ in (3.102) we get

$$N = \frac{(2S+1)}{4\pi^2} V \left(\frac{2m}{\hbar^2}\right)^{3/2} \frac{2}{3} E_F^{3/2}. \tag{3.104}$$

Hence we can rewrite Eq. (3.102) as

$$E_F = \mu \left(1 + \frac{\pi^2}{8}\frac{(kT)^2}{\mu^2} + \cdots \right)^{2/3} = \mu + \frac{\pi^2}{12}\frac{(kT)^2}{\mu} + \cdots, \tag{3.105}$$

which yields

$$\mu = E_F - \frac{\pi^2}{12}\frac{(kT)^2}{E_F} + \cdots \tag{3.106}$$

Note that μ decreases as the temperature increases. We write

$$\begin{aligned}
U &= U(T)|_{T=0} \left(\frac{\mu}{E_F}\right)^{5/2} \left(1 + \frac{5\pi^2}{8}\frac{(kT)^2}{\mu^2} + \cdots \right) = \\
&= U(T)|_{T=0} \left(1 - \frac{5\pi^2}{24}\frac{(kT)^2}{E_F^2} + \cdots \right)\left(1 + \frac{5\pi^2}{8}\frac{(kT)^2}{E_F^2} + \cdots \right) = \\
&= U(T)|_{T=0} \left(1 + \frac{5\pi^2}{12}\frac{(kT)^2}{\mu^2} + \cdots \right)
\end{aligned} \tag{3.107}$$

By definition,

$$\begin{aligned}
U(T)|_{T=0} &= (2S+1)V \frac{4\pi}{(2\pi\hbar)^3} \int_0^{\sqrt{2mE_F}} \frac{p^4}{2m}\, dp = \\
&= \frac{2}{5}(2S+1)\frac{V}{4\pi^2}\left(\frac{2m}{\hbar^2}\right)^{3/2} E_F^{5/2} = \frac{3}{5}NE_F,
\end{aligned} \tag{3.108}$$

where in the last equality we used (3.104). The energy at low temperature is then

$$U = \frac{3}{5} N E_F \left(1 + \frac{5\pi^2}{12} \frac{(kT)^2}{\mu^2} + \dots \right) \tag{3.109}$$

The higher order corrections (in powers of T^2) can be read from the asymptotic expansion (3.73). The *heat capacity* at constant volume for T small is then

$$C_V = \left(\frac{\partial U}{\partial T} \right)_{V,N} = Nk \frac{\pi^2}{2} \frac{kT}{E_F} + O(T^3). \tag{3.110}$$

3.5   Ideal Bosonic Gas

The Two Phases In a *finite* Grand canonical ensemble—before the thermodynamic limit—the number of particles in the ground state of energy $E_0 = 0$ is

$$N_0 = \frac{2S + 1}{e^{-\beta\mu} - 1}. \tag{3.111}$$

In the physically realizable situations $N_0 \leq N$ is positive and finite, so $\zeta \equiv e^{\beta\mu} < 1$ i.e. $\mu < 0$. In the thermodynamic limit this bound becomes

$$\zeta \leq 1 \quad \text{i.e.} \quad \mu \leq 0. \tag{3.112}$$

This fact explains physically the mathematical observation in Sect. 3.3 about the holomorphic domain in the fugacity ζ-plane for the Bose case.

There are two possible physical regimes:

BG1   in the thermodynamic limit $N, V \to \infty$, with V/N finite, the (negative) chemical potential remains bounded away from zero

$$\mu < -\epsilon < 0. \tag{3.113}$$

In this situation the sums (3.33), (3.34), (3.37), (3.51) (with the upper sign) satisfy the assumption of the Euler-Maclaurin summation formula and we can replace the sums by integrals. In this regime the formulae deduced before for fermions, Eqs. (3.85)–(3.87) apply to the ideal bosonic gas with the obvious replacement

$$\mathsf{F}_s(\zeta) \rightsquigarrow \mathsf{G}_s(\zeta). \tag{3.114}$$

BG2   as $N, V \to \infty$, with V/N finite, $\mu \to 0^-$ with

$$N\beta\mu \to -K < 0 \quad \text{finite.} \tag{3.115}$$

In this second situation the number N_0 of particles in the ground state for $N \ggg 1$ is

$$\langle N_0 \rangle = \frac{2S+1}{e^{-\beta\mu} - 1} \approx -\frac{2S+1}{\beta\mu} = \frac{2S+1}{K} N, \tag{3.116}$$

so in the thermodynamic limit a *sizeable portion* of all particles are frozen in the ground state

$$\lim_{N\to\infty} \frac{\langle N_0 \rangle}{N} = \frac{2S+1}{K}. \tag{3.117}$$

In this case the assumptions of Lemma 3.2 are *not* satisfied and we cannot replace the sum by an integral *á la* Euler-Mclaurin. However, if we rewrite Eq. (3.51) in the form

$$\langle N \rangle = \langle N_0 \rangle + (2S+1) \sum_{\substack{k\in\mathbb{Z}^3 \\ k\neq 0}} \frac{1}{\exp\left(\beta\frac{4\pi^2\hbar^2}{2mL^2}k^2 - \beta\mu\right) - 1} \tag{3.118}$$

singling out the contribution of the one-particle ground state from the contributions of the excited states, we easily check that in the thermodynamic limit we are allowed to replace the sum over all excited states by an integral. In this regime the bosonic counterpart to the fermionic equation (3.85) becomes

$$N = N_0 + \frac{V}{\lambda^3} (2S+1) \, \mathsf{G}_{3/2}(e^{\beta\mu}), \tag{3.119}$$

that is,

$$1 = \frac{N_0}{N} + \xi^{-1} \, \mathsf{G}_{3/2}(\zeta), \tag{3.120}$$

where the first term in the RHS may be non-zero only if $\zeta \equiv e^{\beta\mu} = 1$.

As mentioned in Remark 3.3 the point $\zeta = 1$ is a *critical point* ($\equiv$ a point where the thermodynamic potential is non-analytic). At a critical point we have a transition between different phases. Now we see that the two phases are the "ordinary" one in which $N_0/N = 0$, a phase which is smoothly connected at high temperature to the classical gas phase with equation of state

$$PV = NkT, \tag{3.121}$$

and the *deep quantum phase* where a sizeable portion of all particles are in the ground state $E_0 = 0$. This phase is known as *Bose condensation* (or *Bose condensate*).

The limit $\xi \to 0$ is both the large temperature regime and the classical limit. Hence at high temperature the Bose gas is in its "ordinary" (classical-like) phase. On the other hand, at zero temperature we expect all bosons to be in the ground state. Hence there must exist a *critical temperature* T_c such that for $T > T_c$ the bosonic gas is in the ordinary phase while for $T < T_c$ it is in the Bose condensate one.

We compute the critical temperature T_c at which the Bose transition happens. By definition T_c is the infimum of the temperatures such that $N_0/N = 0$. T_c corresponds to a critical value $\xi_c \equiv \xi(T_c)$ which is the supremum of the ξ's for which $N_0/N = 0$. In this "ordinary" phase

$$\xi = \mathsf{G}_{3/2}(\zeta) \quad \Rightarrow \quad \xi_c = \sup_{0 \le \zeta \le 1} \mathsf{G}_{3/2}(\zeta) = \mathsf{G}_{3/2}(1) = \zeta\left(\tfrac{3}{2}\right). \tag{3.122}$$

Comparing with the definition of ξ, Eq. (3.89), we get the critical temperature

$$k\,T_c(v) = \frac{1}{\zeta\left(\tfrac{3}{2}\right)^{2/3}} \frac{2\pi\hbar^2}{m} \left(\frac{v}{2S+1}\right)^{2/3}, \tag{3.123}$$

where

$$v \equiv \frac{V}{N} \tag{3.124}$$

is the inverse of the *density* σ (i.e. v is the volume per particle). Since

$$\xi = \xi_c \left(\frac{T_c(v)}{T}\right)^{3/2} = \zeta\left(\tfrac{3}{2}\right)\left(\frac{T_c(v)}{T}\right)^{3/2}, \tag{3.125}$$

Eq. (3.120) for $T \le T_c$ may be written as

$$\frac{N_0}{N} = 1 - \left(\frac{T}{T_c(v)}\right)^{3/2}, \tag{3.126}$$

so that

$$\frac{N_0}{N} = \begin{cases} 1 - (T/T_c(v))^{3/2} & T \le T_c \\ 0 & T \ge T_c, \end{cases} \tag{3.127}$$

showing the phase transition associated to the *Bose condensation* at $T = T_c$.

High Temperature Regime

For $T > T_c$, Eq. (3.97) holds with the obvious modification (3.114)

$$\frac{PV}{NkT} = \frac{\mathsf{G}_{5/2}(\zeta)}{\xi} \equiv \frac{\mathsf{G}_{5/2}(\zeta)}{\zeta(3/2)} \left(\frac{T}{T_c(v)}\right)^{3/2},\tag{3.128}$$

that is,

$$P = (2S+1)\,\frac{kT}{\lambda^3}\,\mathsf{G}_{5/2}(\zeta)\tag{3.129}$$

where we used the expression (3.89) for ξ and λ is de Broglie thermal length.

We study the asymptotic behaviour as $T \to \infty$. We already know that this regime corresponds to the classical limit and also to the dilute gas. For T large, ξ is small, hence $\zeta \equiv e^{\beta\mu} \ll 1$, and

$$\mathsf{G}_s(\zeta) = \mathsf{Li}_s(\zeta) = \zeta + \frac{\zeta^2}{2^s} + O(\zeta^3).\tag{3.130}$$

The equation of state differs from the fermionic one (3.98) just for a few signs

$$\frac{PV}{NkT} = 1 - \frac{\xi}{2^{3/2}} + O(\xi^2)\tag{3.131}$$

In this regime the pressure is *smaller* than for a classical gas. This corresponds to the fact that the Bose statistics gives an effective attractive interaction (the opposite effect of the Fermi statistics).

Low Temperature Phase

At low temperature $T < T_c$, Eqs. (3.128), (3.129) are still valid because the ground state, having zero energy, does not contribute to U hence to $PV = \frac{2}{3}U$. But now ζ is frozen to the value 1. Thus for $T < T_c$

$$P = (2S+1)\,\zeta\!\left(\tfrac{5}{2}\right)\frac{kT}{\lambda^3} = (2S+1)\,\zeta\!\left(\tfrac{5}{2}\right)\left(\frac{m}{2\pi\hbar^2}\right)^{3/2}(kT)^{5/2},\tag{3.132}$$

and *the pressure is independent of the volume V*. The isothermal compressibility κ_T is then *infinite* in this deep quantum regime.

Comparing Eqs. (3.131) and (3.132) we see that the equations of state of the Bose gas in the two phases have quite different analytic forms.

Heat Capacity

We compute the heat capacity at fixed volume and number of bosons

$$C_V = \left(\frac{\partial U}{\partial T}\right)_{V,N}. \tag{3.133}$$

Since $U = \frac{3}{2}PV$, for $T < T_c$ Eq. (3.128) yields

$$C_V\Big|_{<T_c} = \frac{3}{2}Nk\,\frac{\zeta(5/2)}{\zeta(3/2)}\frac{\partial}{\partial T}\left(\frac{T^{5/2}}{T_c(v)^{3/2}}\right) = \frac{15}{4}\frac{\zeta(5/2)}{\zeta(3/2)}Nk\left(\frac{T}{T_c(v)}\right)^{3/2} \tag{3.134}$$

In particular the heat capacity grows monotonically from 0 at $T = 0$ to

$$\frac{15}{4}\frac{\zeta(5/2)}{\zeta(3/2)}Nk \approx 1.92567\,Nk \quad \text{at } T_c(v). \tag{3.135}$$

When $T > T_c$ we set $w \equiv \log\zeta$

$$\left(\frac{\partial U}{\partial T}\right)_{V,N} = \left(\frac{\partial U}{\partial T}\right)_{V,w} + \left(\frac{\partial U}{\partial w}\right)_{V,T}\left(\frac{\partial w}{\partial T}\right)_{V,N} =$$

$$= \frac{3}{2}\frac{Nk\,\mathsf{G}_{5/2}(e^w)}{\zeta(3/2)}\frac{\partial}{\partial T}\left(\frac{T^{5/2}}{T_c(v)^{3/2}}\right) + \frac{3}{2}\frac{NkT}{\zeta(3/2)}\left(\frac{T}{T_c(v)}\right)^{3/2}\mathsf{G}_{3/2}(e^w)\left(\frac{\partial w}{\partial T}\right)_{V,N} =$$

$$= \frac{15}{4}\frac{Nk}{\zeta(3/2)}\left(\frac{T}{T_c(v)}\right)^{3/2}\mathsf{G}_{5/2}(e^w) + \frac{3}{2}Nk\,\mathsf{G}_{3/2}(e^w)\frac{T}{\xi}\left(\frac{\partial w}{\partial T}\right)_{V,N} \tag{3.136}$$

where in the last term we used Eq. (3.125). From the equation $\xi = \mathsf{G}_{3/2}(e^w)$ and the fact that ξ is a homogeneous function of T of degree $-3/2$ (cf. Eq. (3.89)) we get

$$-\frac{3}{2}\frac{\xi}{T} \equiv \frac{\partial\xi}{\partial T} = \frac{\partial\mathsf{G}_{3/2}(e^w)}{\partial w}\frac{\partial w}{\partial T} = \mathsf{G}_{1/2}(e^w)\frac{\partial w}{\partial T}, \tag{3.137}$$

that is,

$$\frac{T}{\xi}\frac{\partial w}{\partial T} = -\frac{3}{2}\frac{1}{\mathsf{G}_{1/2}(e^w)} \tag{3.138}$$

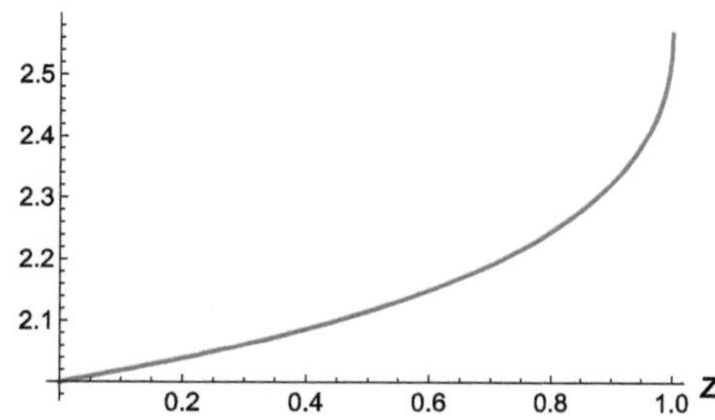

Fig. 3.6 The function $5\frac{G_{5/2}(z)}{G_{3/2}(z)} - 3\frac{G_{3/2}(z)}{G_{1/2}(z)}$ for $0 \le z \le 1$

and

$$
C_V\Big|_{>T_c} = \frac{15}{4}\frac{Nk}{\zeta(3/2)}\left(\frac{T}{T_c(v)}\right)^{3/2} G_{5/2}(e^w) - \frac{9}{4}Nk\frac{G_{3/2}(e^w)}{G_{1/2}(e^w)} \equiv
$$
$$
\equiv \frac{3}{4}Nk\left(5\frac{G_{5/2}(e^w)}{G_{3/2}(e^w)} - 3\frac{G_{3/2}(e^w)}{G_{1/2}(e^w)}\right),
$$
(3.139)

As a check, note that as $\hbar \to 0$, that is, as $e^w \to 0$ the ratios of Bose integrals inside the large parenthesis go to 1 and we get back the classical answer $C_V = \frac{3}{2}Nk$. The function in the large parenthesis is monotonically increasing, see Fig. 3.6. Since e^w is a monotonically decreasing function of T, in the region $T > T_c$ the heat capacity C_V is monotonically decreasing with T from its maximal value (3.135) at $T = T_c$ to $\frac{3}{2}Nk$ at $T = \infty$.

Near T_c (i.e. for w approaching zero from the left) Eqs. (3.68)–(3.70) yield

$$
C_V\Big|_{>T_c} - C_V\Big|_{<T_c} \approx -\frac{9}{4}\frac{\zeta(3/2)}{\sqrt{\pi}}(-w)^{1/2} + O(w)
$$
(3.140)

which goes to zero as $w \to 0^-$. Therefore *the heat capacity at constant volume C_V is* continuous *at the Bose condensation phase transition.* Now

$$
\frac{\partial}{\partial T}\left(C_V\Big|_{>T_c} - C_V\Big|_{<T_c}\right)\Big|_{w\to 0^-} = \frac{9}{8}Nk\frac{\zeta(3/2)}{\sqrt{\pi}}\frac{1}{(-w)^{1/2}}\frac{\partial w}{\partial T}\Big|_{w\to 0^-} =
$$
$$
= -\frac{27}{16}\frac{\zeta(3/2)}{\sqrt{\pi}}Nk\frac{\xi_c}{T_c}\frac{1}{(-w)^{1/2}G_{1/2}(e^w)}\Big|_{w\to 0^-} =
$$
$$
= -\frac{27}{16\pi}\zeta(3/2)^2\frac{Nk}{T_c} \approx -3.66577\frac{Nk}{T_c}
$$
(3.141)

so that the *derivative* of the heat capacity with respect to the temperature T is discontinuous at T_c. The dependence of C_V from T is represented in Fig. 3.7.

Fig. 3.7 The dependence of C_V from T in an ideal bosonic gas (schematically)

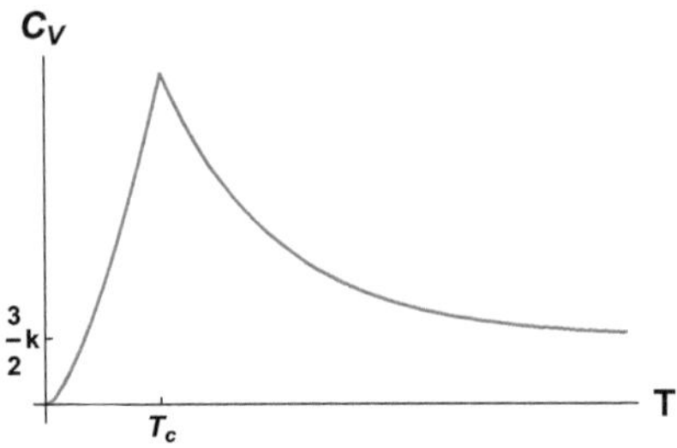

3.6   Ultra-Relativistic Gases. Black Body Radiation

Ultra-Relativistic Gases

In the previous sections we considered quantum ideal gases which was *non-relativistic*, that is, the single-particle Hamiltonian was

$$H = \frac{\mathbf{p}^2}{2m}. \tag{3.142}$$

The non-relativistic treatment is justified as long as the thermal energy per degree of freedom is much less that the mass of the particle

$$kT \ll kT_c \equiv mc^2. \tag{3.143}$$

Otherwise the formulae should be modified to take care of the relativistic effects. For instance, in the relativistic case the expression of the Grand canonical number of particles gets modified to

$$N = \frac{4\pi V}{(2\pi\hbar)^3} \int_0^\infty \frac{p^2\,\mathrm{d}p}{\zeta^{-1}\exp(\beta\sqrt{p^2+m^2}) \mp 1} \tag{3.144}$$

where we used units with $c = 1$. For massive particles T_c is a pretty high temperature. One eV of energy is equivalent to 1.16×10^4 degrees Kelvin, and even for a very light fundamental particle like the electron, which has a mass of 0.5×10^6 eV, the temperature T_c is of the order of $10^{10} K$. However for massless particles (or essentially massless particles like the *neutrini*) the gas is *ultra-relativistic* at *all* temperatures. The basic example is a gas of *photons*, the massless quanta of the electromagnetic field. This system has two dual descriptions:

D1   we can see the system as the electromagnetic field inside a box with reflective walls—this is the original definition of the *black body*—and the thermal physics describes the black body radiation at temperature T, the very problem that led Planck to introduce his celebrated constant $\hbar$. In the classical set-up this system is meaningless because of the *ultraviolet catastrophe* (cf.

Sect. 2.8), but at the quantum level the thermodynamic potentials (hence the equations of state) are perfectly well-behaved;

D2   we can see the system as an ideal[13] quantum gas of ultra-relativistic massless bosonic particles with 2 polarizations per momentum state.

Similar considerations apply to other ultra-relativistic gases, in particular to the neutrino gases which are *fermionic* relativistic gases.

Physical consistency requires the two approaches **D1** and **D2** to be equivalent. In order for the two viewpoints to agree two "miracles" must happen:

M1   the quantization of the Maxwell (resp. Dirac) field theory automatically reproduces the bosonic (resp. fermionic) statistics;

M2   for ultra-relativistic massless particles (bosonic as well as fermionic) the fugacity $\zeta \equiv 1$ automatically.

M1 is just (a special case of) the *Spin and Statistics* theorem of QFT which was the starting point of this chapter. We shall elaborate a little more on it in the next two sections. **M2** is a basic fact from Thermodynamics:

Theorem 3.1 *All ultra-relativistic massless gases (bosonic and fermionic) have identically zero Gibbs free energy $G(T, V) \equiv 0$, hence the particle fugacity $\zeta \equiv 1$ in all conditions.*

Proof In the massless case $E = |\boldsymbol{p}|$. From Remark 3.2 we have

$$PV = \frac{1}{3} U \tag{3.145}$$

By the Boltzmann theorem in Sect. 1.12, this equation entails the Stefan law

$$U = C V T^4 \quad \Rightarrow \quad T = (CV)^{-1/4} U^{1/4}, \tag{3.146}$$

where the integration constant C depends on the particular ultra-relativistic gas (photons, neutrini, gravitons, etc). In particular

$$T \equiv \frac{\partial U}{\partial S} = (U/CV)^{1/4} \quad \Rightarrow \quad dS = (CV)^{1/4} \frac{dU}{U^{1/4}}, \tag{3.147}$$

so that the entropy is

$$S = \frac{4}{3} (CV)^{1/4} U^{3/4}. \tag{3.148}$$

---

[13] Maxwell theory leads to an ideal gas since it is a linear theory, i.e. a linear combination of solutions is also a solution. This is equivalent to saying that there are no interactions and the corresponding gas is ideal.

The Gibbs free energy then is

$$G = U - ST + PV =$$

$$= U - \left(\frac{4}{3}(CV)^{1/4}U^{3/4}\right)\left((CV)^{-1/4}U^{1/4}\right) + \frac{1}{3}U \equiv 0. \tag{3.149}$$

$\square$

As we shall see momentarily, the fact that $\mu \equiv 0$ means that the Grand canonical ensemble of the particle description coincides with the canonical ensemble of the field picture. To make sense, this interchange of the two ensembles requires the equality of the respective thermodynamical potentials:

Corollary 3.1 *For a massless ultra-relativistic gas the Grand potential Φ and the free energy F are equal.*

Indeed, by Eq. (1.135) $\Phi = F - \mu N \equiv F$ when $\mu \equiv 0$.

Black Body Radiation: The Particle Viewpoint
The Grand canonical partition is given by Eq. (3.33) with the upper sign, $\mu = 0$, and $E_p = |\boldsymbol{p}|$. In the thermodynamic limit we can replace sums with integrals. In a cubic box of volume V with periodic boundary conditions we get

$$-\frac{F}{kT} = \log \Xi = -\frac{8\pi\, V}{(2\pi\hbar)^3} \int_0^\infty \log\left(1 - e^{-\beta p}\right) p^2 \, dp, \tag{3.150}$$

where the additional factor 2 arises from the two polarizations of photons. The integral is easily computed by parts

$$\int_0^\infty \log\left(1 - e^{-\beta p}\right) p^2 \, dp = \int_0^\infty dp \, \log\left(1 - e^{-\beta p}\right) \frac{d}{dp} \frac{p^3}{3} =$$

$$= \log\left(1 - e^{-\beta p}\right) \frac{p^3}{3}\bigg|_0^\infty - \frac{\beta}{3} \int_0^\infty \frac{p^3 \, dp}{e^{\beta p} - 1} = -\frac{1}{3\beta^3} \int_0^\infty \frac{t^3 \, dt}{e^t - 1}. \tag{3.151}$$

The reader may notice that the equality LHS = RHS is equivalent to Eq. (3.145) (in view of $\Phi = -PV$). The integral in the RHS is a known Bose-Einstein integral. Using results from Sect. 3.3 we get

$$\int_0^\infty \log\left(1 - e^{-\beta p}\right) p^2 \, dp = -\frac{1}{3\beta^3}\Gamma(4)\,\mathrm{Li}_4(1) \equiv$$

$$\equiv -\frac{1}{3\beta^3} 3!\, \zeta(4) \equiv -\frac{\pi^4}{45}(kT)^3, \tag{3.152}$$

where we used Eq. (3.75) and $B_4 = -1/30$ [5]. Then

$$- PV = \Phi = -kT \, \log \Xi = -\frac{\pi^2 k^4}{45 \, \hbar^3} \, V \, T^4 = -\frac{\pi^2 k^4}{45 \, \hbar^3 \, c^3} \, V \, T^4, \qquad (3.153)$$

where in the RHS we have reinserted the velocity of light c which before we set to 1 by a choice of units. The energy is given in terms of the same Bose-Einstein integral

$$U = \frac{8\pi \, V}{(2\pi \hbar)^3} \int_0^\infty \frac{p^3 \, dp}{e^{\beta p} - 1} \equiv 3 \, PV = \frac{\pi^2 k^4}{15 \, \hbar^3 \, c^3} \, V \, T^4 \qquad (3.154)$$

in full agreement with Eq. (3.145). Note that the energy diverges in the classical limit $\hbar \to 0$ as a consequence of the classical ultraviolet catastrophe.

Remark 3.4 The energy q radiated by a black body at temperature T per unit area and time is related to the energy density $u \equiv U/V$ of the photon gas by the formula

$$q = \frac{c}{4} \, u \equiv \sigma \, T^4, \qquad \sigma \equiv \frac{\pi^2 k^4}{60 \, \hbar^3 \, c^2}. \qquad (3.155)$$

σ is known as the *Stefan constant*.

We write the main thermodynamical quantities for the photon gas

$$F \equiv \Phi = -\frac{4\sigma}{3c} V T^4 \qquad\qquad S = -\left(\frac{\partial F}{\partial T} \right)_V = \frac{16\sigma}{3c} V T^3 \qquad (3.156)$$

$$E = F + TS = \frac{4\sigma}{c} V T^4 \qquad\qquad P = -\left(\frac{\partial F}{\partial V} \right)_T = \frac{4\sigma}{3c} T^4 \qquad (3.157)$$

$$C_V = T \left(\frac{\partial S}{\partial T} \right)_V = \frac{16\sigma}{c} V T^3 \qquad\qquad \kappa_S = \frac{9c}{16\sigma} \left(\frac{V}{S} \right)^{4/3} \qquad (3.158)$$

Note that κ_T is infinite since P is independent of V.

Black Body Radiation: The Field Viewpoint

As in the discussion of the *classical* ultraviolet catastrophe (Sect. 2.8), we can decompose the electromagnetic field inside a cubic box of size L into Fourier components. It is simpler to use the periodic boundary conditions than the reflecting-wall ones, so that we have *two* Fourier modes per $\boldsymbol{n} \in \mathbb{Z}^3$ where the *two* arises from the two possible physical polarizations of the electromagnetic field once we take into account the gauge condition. Each such Fourier mode is a *quantum* harmonic oscillator of frequency

$$\omega_{\boldsymbol{n}} = \frac{2\pi \, |\boldsymbol{n}|}{L}, \qquad \boldsymbol{n} \in \mathbb{Z}^3. \qquad (3.159)$$

The only difference with respect to the classical discussion in Sect. 2.8 is that the field oscillators are now *quantum* not classical. In particular the troublesome *equipartition theorem* does not apply in the quantum set-up. A single harmonic oscillator of frequency ω with Hamiltonian redefined so that the ground state has energy zero

$$H = \frac{1}{2m}p^2 + \frac{m\omega^2}{2}q^2 - \frac{1}{2}\hbar\omega \tag{3.160}$$

has quantum energy levels

$$E_n = n\hbar\omega \quad (n = 0, 1, 2, \ldots) \tag{3.161}$$

so that the single oscillator canonical partition function is

$$Z = \mathrm{Tr}\,\mathrm{e}^{-\beta H} = \sum_{n \geq 0} \mathrm{e}^{-\beta E_n} = \sum_{n \geq 0} \mathrm{e}^{-\beta\hbar\omega n} = \frac{1}{1 - \mathrm{e}^{-\beta\hbar\omega}} \tag{3.162}$$

and the free energy of a single quantum harmonic oscillator is

$$F = -kT \log Z = kT \log(1 - \mathrm{e}^{-\beta\hbar\omega}). \tag{3.163}$$

Putting all field oscillators together, and converting sums over $\boldsymbol{n} \in \mathbb{Z}^3$ into integrals, the canonical partition function of the electromagnetic field in a period box of volume V becomes

$$\begin{aligned}
F = -kT \log Z &= \frac{2V}{(2\pi)^3} kT \int \mathrm{d}^3\boldsymbol{\omega} \, \log(1 - \mathrm{e}^{-\beta\hbar|\boldsymbol{\omega}|}) = \\
&= \frac{8\pi V}{(2\pi\hbar)^3} \int_0^\infty p^2 \, \mathrm{d}p \, \log(1 - \mathrm{e}^{-\beta p})
\end{aligned} \tag{3.164}$$

in perfect agreement with the Grand potential for the Grand ensemble of the particle viewpoint. The energy is

$$U = -\frac{\partial}{\partial\beta} \log Z = \frac{8\pi V\hbar}{(2\pi)^3} \int_0^\infty \frac{\omega^3 \, \mathrm{d}\omega}{1 - \mathrm{e}^{-\beta\hbar\omega}} \tag{3.165}$$

The distribution of black body radiation with frequency

$$\phi(\beta, \omega) = \frac{\hbar c}{(2\pi)^2} \frac{\omega^3 \, \mathrm{d}\omega}{1 - \mathrm{e}^{-\beta\hbar\omega}} \tag{3.166}$$

is called the *Planck distribution*, see Fig. 3.8.

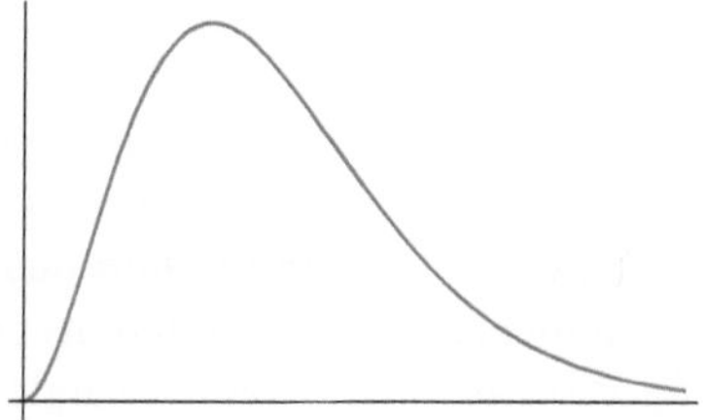

Fig. 3.8 The Planck distribution in frequency of the black body radiation. The horizontal axis is the frequency ω and the vertical axis the frequency energy density

3.7 Second Quantization

The duality between the field and particle interpretations, that we observed for the photon gas as a consequence of the *Spin and Statistics theorem,* extends to all quantum systems, bosonic and fermionic, relativistic or not. One way to realize this duality is to use the *second quantization* formalism: the particle Grand canonical ensemble will coincide with the field canonical partition function where we switch on one fugacity for each conserved charge of the (possibly non-relativistic) field theory.[14]

Bosonic Multi-Particle Systems

Let us start with the bosonic case. We have a single-particle Hilbert space[15] $\mathcal{H}_1 \simeq L^2(Q)$ with energy eigenstates $|E_k\rangle$ ($k \in \mathbb{N}$) whose Schrödinger eigenfunctions are

$$\psi_k(\boldsymbol{x}) \equiv \langle \boldsymbol{x}|E_k\rangle. \tag{3.167}$$

Physical consistency requires the single-particle Hamiltonian H to be bounded below and we may assume $E_k > 0$ for all k with no loss.

Next we consider the Hilbert space $\mathcal{H}$ of the quantum system containing an arbitrary number of such Bose particles

$$\mathcal{H} = \bigoplus_{n=0}^{\infty} \mathcal{H}_n, \qquad \mathcal{H}_n = \mathcal{H}_1^{\odot n} \tag{3.168}$$

where $\mathcal{H}_n \subset \mathcal{H}$ is the subspace of states containing n particles (*n-particle states*). Two states $|a, n\rangle \in \mathcal{H}_n$ and $|b, n'\rangle \in \mathcal{H}_{n'}$ are orthogonal if $n' \neq n$. Thus we can introduce a Hermitian operator N acting on the big Hilbert space $\mathcal{H}$ such that

$$N\big|_{\mathcal{H}_n} = n, \tag{3.169}$$

---

[14] Pure Maxwell theory has no non-trivial conserved charge, so it must have $\mu \equiv 0$ as we found.

[15] Q is the one-particle configuration space of dimension m. In this paragraph $\boldsymbol{x}$ is a point in Q.

which counts the number of particles. N is called the *particle number* operator. In $\mathcal{H}$ we have a unique special state $|0\rangle$, called the *vacuum*, which spans $\mathcal{H}_0 \simeq \mathbb{C}$. $|0\rangle$ is the state where no particle is present, so it has naturally zero energy.

The formalism we are after is called *second quantization* for the following historical reason. In the *first* quantization for single-particle systems, the classical dynamical quantities (functions on the single-particle phase space $\mathfrak{W}_1$) got promoted to *operators* acting on the single-particle Hilbert space $\mathcal{H}_1$, while the states $|\psi\rangle \in \mathcal{H}_1$ got represented by wave-functions $\psi(x) \in L^2(Q)$ which are *classical* functions, i.e. valued in the commutative algebra $\mathbb{C}$.

In the *second* quantization the wave-function $\psi(x)$ gets promoted to a *field operator* $\boldsymbol{\psi}(x)$ which acts on the multi-particle Hilbert space $\mathcal{H}$ in Eq. (3.168). The relation with the first quantization is obtained by restricting to the subspace of one-particle states $\mathcal{H}_1 \subset \mathcal{H}$. Let $|\alpha, 1\rangle \in \mathcal{H}_1$: its Schrödinger wave-function $\psi_\alpha(x)$ in the sense of the first quantization is

$$\psi_\alpha(x) = \langle 0|\boldsymbol{\psi}(x)|\alpha, 1\rangle. \tag{3.170}$$

Taking the time derivative of this equation we see that the field operator $\boldsymbol{\psi}(x)$ satisfies the Schrödinger equation

$$i\hbar \frac{\partial}{\partial t}\boldsymbol{\psi}(x) - H\boldsymbol{\psi}(x) = 0, \tag{3.171}$$

where H is the one-particle Hamiltonian which we see as a linear differential operator acting on wave-functions in $L^2(Q)$. The solutions to Eq. (3.171) are the extrema of the action functional

$$S[\boldsymbol{\psi}(x), \boldsymbol{\psi}(x)^\dagger] = \int dt \int d^m x \left[i\boldsymbol{\psi}(x)^\dagger \partial_t \boldsymbol{\psi}(x) - \boldsymbol{\psi}(x)^\dagger H \boldsymbol{\psi}(x)\right] \tag{3.172}$$

To simplify the computations it is convenient to expand the field $\boldsymbol{\psi}(x)$ into one-particle energy eigenfunctions $\psi_k(x)$

$$\boldsymbol{\psi}(x) = \sum_{k=0}^{\infty} a_k\, \psi_k(x) \tag{3.173}$$

where the a_k's are *quantum operators* acting on the multi-particle Hilbert space $\mathcal{H}$ and the complex coefficients $\psi_k(x)$ are the one-particle energy eigenfunctions of the first quantization formalism which satisfy the differential equation

$$H\psi_k(x) = E_k\, \psi_k(x), \tag{3.174}$$

and are normalized so that

$$\int d^m x \; \psi_k^*(x)\, \psi_h(x) = \delta_{kh}, \qquad k, h = 0, 1, 2, \cdots . \tag{3.175}$$

For one-particle states $|E_k, 1\rangle \in \mathcal{H}$ of energy E_k—with one-particle wave-function $\psi_k(x) \equiv \langle x|E_k\rangle$—Eq. (3.170) reduces to

$$\langle 0|a_k|E_h, 1\rangle = \delta_{kh}. \tag{3.176}$$

Plugging Eq. (3.173) in Eq. (3.172) the action functional S simplifies

$$S = \int dt \sum_{k=0}^{\infty} \left[i a_k^\dagger \dot{a}_k - E_k\, a_k^\dagger a_k \right]. \tag{3.177}$$

From this equation we see that $i a_k^\dagger$ is the *conjugate momentum* to the operator a_k and that the field Hamiltonian of the multi-boson system is

$$H = \sum_k E_k\, a_k^\dagger a_k \equiv \int d^m x\, \psi(x)^\dagger H \psi(x). \tag{3.178}$$

The quantum operator a_k and its conjugate momentum $i a_k^\dagger$ satisfy the canonical commutation relations (written in units where $\hbar = 1$)

$$[q_i, p_j] = i\delta_{ij}, \qquad [q_i, q_j] = [p_i, p_j] = 0, \tag{3.179}$$

where we identify $q \rightsquigarrow a,\ p \rightsquigarrow i a^\dagger$, that is,

$$a_k a_j^\dagger - a_j^\dagger a_k \equiv [a_k, a_j^\dagger] = \delta_{kj}, \tag{3.180}$$

$$a_k a_j - a_j a_k = a_k^\dagger a_j^\dagger - a_j^\dagger a_k^\dagger = 0 \tag{3.181}$$

an algebra which is called the *Bose algebra* (or the *canonical commutation relations*). In term of the field operator $\psi(x)$ the canonical algebra reads

$$[\psi(x), \psi(x')] = [\psi(x)^\dagger, \psi(x')^\dagger] = 0 \tag{3.182}$$

$$[\psi(x), \psi(x')^\dagger] = \sum_{k\geq 0} \psi_k(x)\, \psi_k(x')^* = \delta(x - x'), \tag{3.183}$$

where we used the completeness of one-particle energy eigenfunctions.

The multi-particle Hamiltonian H (3.178) is just a sum of decoupled quantum harmonic oscillators of frequencies $\omega_k \equiv E_k$

$$H = \frac{1}{2} \sum_k \left(p_k^2 + E_k^2 q_k^2 \right) \tag{3.184}$$

where $\{q_k, p_j\}$ are canonical operators satisfying (3.179) related to the operators $\{a_k, a_j^\dagger\}$ by the linear redefinitions

$$a_k = \frac{1}{\sqrt{2\omega_k}}(\omega_k q_k + i p_k), \qquad a_k^\dagger = \frac{1}{\sqrt{2\omega_k}}(\omega_k q_k - i p_k). \tag{3.185}$$

Each oscillator has a $U(1)$ symmetry $a_k \to e^{i\phi_k} a_k$, corresponding to a shift in its angle variable [8, Chapter 10] whose generating charge is the corresponding action variable (which is classically an adiabatic invariant [8])

$$n_k \overset{\text{def}}{=} a_k^\dagger a_k, \quad \text{with} \quad [n_k, n_j] = 0. \tag{3.186}$$

We have

$$[n_k, a_h^\dagger] = a_k^\dagger a_k a_h^\dagger - a_h^\dagger a_k^\dagger a_k = \delta_{kh}\, a_h^\dagger \tag{3.187}$$

It follows from this equality that, if $|\psi\rangle \in \mathcal{H}$ is an eigenstate of n_k with eigenvalue $\ell \in \mathbb{N}$, $a_k^\dagger|\psi\rangle$ is an eigenstate of n_k with eigenvalue $\ell + 1$: indeed (no sum over repeated indices!)

$$n_k\, a_k^\dagger|\psi\rangle = ([n_k, a_k^\dagger] + a_k^\dagger n_k)|\psi\rangle = a^\dagger(1 + n_k)|\psi\rangle = (1 + \ell)a_k^\dagger|\psi\rangle. \tag{3.188}$$

Conversely, $a_k|\psi\rangle$, *if non-zero*, is an eigenstate of n_k with eigenvalue $\ell - 1$. Since $H = \sum_k E_k n_k$, the operator a_k lowers the energy by $E_k > 0$. Since the vacuum has the minimal possible energy (zero) we must have

$$a_k|0\rangle = 0 \quad \text{for all } k, \tag{3.189}$$

while $a_k^\dagger|0\rangle$ is a one-particle state in the k-th energy eigenstate $|E_k, 1\rangle$. Indeed the wave-function of this state is

$$\langle 0|\boldsymbol{\psi}(\boldsymbol{x})\, a_k^\dagger|0\rangle = \sum_j \psi_j(\boldsymbol{x})\,\langle 0|a_j a_k^\dagger|0\rangle = \sum_j \psi_j(\boldsymbol{x})\,\langle 0|[a_j, a_k^\dagger]|0\rangle =$$

$$= \sum_j \psi_j(\boldsymbol{x})\,\delta_{jk} = \psi_k(\boldsymbol{x}). \tag{3.190}$$

By Eq. (3.187), the state $(a_k^\dagger)^\ell |0\rangle$ is an eigenstate of n_k with eigenvalue ℓ. In other words $(a_k^\dagger)^\ell |0\rangle$ is a ℓ-particle state with ℓ particles all in the k-th energy eigenstate. Thus the quantum operator n_k is the *occupation number of the k-th energy eigenstate* and the operator counting the total number of particles is

$$N = \sum_k n_k \equiv \int d^m x \; \psi(x)^\dagger \psi(x). \tag{3.191}$$

All states in the n-particle sector $\mathcal{H}_n \subset \mathcal{H}$ of the Hilbert space are of the form

$$a_{k_1}^\dagger a_{k_2}^\dagger \cdots a_{k_n}^\dagger |0\rangle, \tag{3.192}$$

for some n-tuple $\{k_1, \ldots, k_n\}$ (repetitions allowed). The operator $a_k^\dagger$ adds to the state a particle in the E_k-eigenstate while a_k subtracts a particle in the E_k-eigenstate if it was present and otherwise annihilates the state. For these reasons the $a_k^\dagger$'s are called *creation* operators, and their Hermitian conjugate a_k's *annihilation* operators.

The canonical commutation relations (3.180), (3.181) imply that the state (3.192) is *totally symmetric* in its indices $\{k_1, \ldots, k_n\}$, that is, that the state is an element of $\odot^n \mathcal{H}_1$ whose wave-function is totally symmetric in the n particles. We conclude that the canonical commutation relations (3.182), (3.183) for the field $\psi(x)$ *imply* the Bose-Einstein statistics for the particles. In particular: in the second quantization framework the particles created out of the vacuum $|0\rangle$ by the field $\psi(x)^\dagger$ cannot be distinguished in any way.

The non-relativistic field theory with action (3.172) has two natural conserved quantities which are *extensive* (i.e. given by integrals in $d^m x$ of local functions of $\psi(x)$ and its derivatives) namely the Hamiltonian H (Eq. (3.178)) and the particle number N (Eq. (3.191)). The natural (generalized) canonical ensemble has then partition function

$$Z(\beta, \mu) \equiv \mathrm{Tr}_{\mathcal{H}}\left[e^{-\beta H + \beta \mu N} \right] \tag{3.193}$$

where β is the inverse temperature, μ is the chemical potential, and $\zeta \equiv e^{\beta \mu}$ is fugacity of the particle-number N. The second quantized canonical partition function (3.193) for the field $\psi(x)$ is equal to the Grand canonical partition function (3.32) of the particle picture. Indeed

$$\mathrm{Tr}_{\mathcal{H}}\left[e^{-\beta H + \mu N} \right] = \sum_{n \geq 0} \zeta^n \, \mathrm{Tr}_{\mathcal{H}_n}\left[e^{-\beta H} \right] =$$

$$= \sum_{n \geq 0} \zeta^n \, \mathrm{Tr}_{\mathcal{H}_1^{\odot n}}\left[e^{-\beta H^{\odot n}} \right] = \exp\left[-\mathrm{Tr}_{\mathcal{H}_1} \log\left(1 - \zeta e^{-\beta H} \right) \right] \tag{3.194}$$

Remark 3.5 The particle number N is the Noether charge of the $U(1)$ symmetry associated to the phase of the field $\psi(x)$. The fact that we cannot add a non-trivial chemical potential to the photon gas mirrors the fact that the Maxwell field theory has no symmetry which is associated to the number of photons.

Fermionic Multi-Particle Systems

In the fermionic case the multiparticle wave-function is totally *anti*-symmetric. We expand the field operator $\psi(x)$ in one-particle energy eigenfunctions as before

$$\psi(x) = \sum_k b_k \, \psi_k(x). \tag{3.195}$$

The state with s Fermi particles in the H-eigenstates with energies $E_{k_1}, \ldots, E_{k_s}$ is

$$b_{k_1}^\dagger \cdots b_{k_s}^\dagger |0\rangle. \tag{3.196}$$

The condition that this state is totally antisymmetric yields the algebra of the *fermion creation* (resp. *annilation*) operators $b_k^\dagger$ (resp. b_k)

$$b_i^\dagger b_j^\dagger + b_j^\dagger b_i^\dagger = 0 \quad \Rightarrow \quad b_i b_j + b_j b_i = 0 \tag{3.197}$$

Again $b_j|0\rangle = 0$ for all j. For each k the interchange $b_k \leftrightarrow b_k^\dagger$ is an automorphism of the Fermi-Dirac algebra generated by the b_j, $b_l^\dagger$ since counting particles or *holes* ($\equiv$ absence of particles in a given energy state) is equivalent by the Pauli exclusion principle. The number of particles in the k-th state $n_k = b_k^\dagger b_k$ (not summed over k!) and the number of holes in the k-th state

$$\tilde{n}_k \equiv 1 - n_k = b_k b_k^\dagger \tag{3.198}$$

are Hermitian operators acting on $\mathcal{H}$ whose spectra consist of two eigenvalues 0 and 1, that is, $(n_k)^2 = n_k$ and $(\tilde{n}_k)^2 = \tilde{n}_k$. These equations imply

$$b_j^\dagger b_k + b_k \, b_j^\dagger = \delta_{kj}. \tag{3.199}$$

The operator algebra generated by the b_k's and $b_j^\dagger$, subjected to the three *anti*commutation relations (3.197), (3.199), is called the *Fermi-Dirac algebra* (or canonical *anti*commutation relations). When we have a finite number n of one-particle states the Fermi-Dirac algebra is isomorphic to the *complex Clifford algebra in $2n$ dimensions* [9]. However in the applications we have in mind (such as a fermionic gas in a periodic box $(S^1)^3$) there are countably many one-particle states. As we have seen above, the $b_j^\dagger$ are creation and the b_j annihilation operators. Interchanging particles with holes the role of creators and annihilators gets interchanged. Again we have two natural conserved extensive operators

$$N = \sum_k b_k^\dagger b_k \equiv \int d^m x \; \psi(x)^\dagger \psi(x) \tag{3.200}$$

$$H = \sum_k E_k \, b_k^\dagger b_k \equiv \int d^m x \; \psi(x)^\dagger H \psi(x) \tag{3.201}$$

where now the wave-field $\psi(x)$ is given by Eq. (3.195) so that it is an *odd* (i.e. anticommuting) object

$$\psi(x)\,\psi(x') = -\psi(x')\,\psi(x) \tag{3.202}$$

while in the bosonic case the field operator $\psi(x)$ was commuting, cf. Eq. (3.182). The minus sign in Eq. (3.202) explains why the Fermi statistics has a repulsive effect between indistinguishable particles.

Again, the second quantized (generalized) partition function for the field $\psi(x)$ agrees with the Grand canonical partition of the particle picture

$$\mathrm{Tr}_{\mathcal{H}}\big[e^{-\beta H + \mu N}\big] = \sum_{n \geq 0} \zeta^n \; \mathrm{Tr}_{\mathcal{H}_n}\big[e^{-\beta H}\big] =$$

$$= \sum_{n \geq 0} \zeta^n \; \mathrm{Tr}_{\mathcal{H}_1^{\wedge n}}\big[e^{-\beta H^{\wedge n}}\big] = \exp\Big[\mathrm{Tr}_{\mathcal{H}_1} \log\big(1 + \zeta e^{-\beta H}\big)\Big]. \tag{3.203}$$

3.8 Path Integrals and Partition Functions

Trotter Formula We consider a continuous quantum system with Hamiltonian

$$H = H_0(p_i) + V(q^i), \tag{3.204}$$

where H, H_0 and V are (essentially) self-adjoint operators acting on a Hilbert space $\mathcal{H}$. If there is no danger of confusion we shall leave the indices implicit and write q and p for respectively q^i and p_j. For all positive integer n

$$e^{-\beta H} = (e^{-\beta H/n})^n, \tag{3.205}$$

that is, in terms of integral kernels

$$\langle q^{(n)}|e^{-\beta H}|q^{(0)}\rangle = \int dq^{(n-1)}\cdots dq^{(1)}\langle q^{(n)}|e^{-\beta H/n}|q^{(n-1)}\rangle \times$$

$$\times \langle q^{(n-1)}|e^{-\beta H/n}|q^{(n-2)}\rangle \cdots \langle q^{(2)}|e^{-\beta H/n}|q^{(1)}\rangle\langle q^{(1)}|e^{-\beta H/n}|q^{(0)}\rangle. \tag{3.206}$$

The Baker-Campbell-Hausdorff formula [10] yields for $n \gg 1$

$$e^{-\beta H/n} = e^{-\beta H_0/n}e^{-\beta V/n}\left(1 + O(n^{-2})\right) \tag{3.207}$$

which produces the *Trotter formula* (cf. **Theorem 3.2.2** of [11])

$$e^{-\beta H} = \operatorname*{strong\ lim}_{n\to\infty}\left(e^{-\beta H_0/n}e^{-\beta V/n}\right)^n \tag{3.208}$$

where, for the Hamiltonian (3.204),

$$\langle q^{(k+1)}|e^{-\beta H/n}|q^{(k)}\rangle = \langle q^{(k+1)}|e^{-\beta H_0/n}|q^{(k)}\rangle\, e^{-\beta V(q^{(k)})/n}\left(1 + O(n^{-2})\right). \tag{3.209}$$

First Quantization: The Feymann-Kac Formula

We apply the formula (3.208) to compute the canonical partition function of a first quantized quantum system with Hamiltonian of the form

$$H = \frac{1}{2m}\sum_{i=1}^{\ell} p_i^2 + V(x_1,\ldots,x_\ell), \qquad [x_i, p_j] = i\hbar\,\delta_{ij}. \tag{3.210}$$

We first compute the integral kernel $\langle x'|e^{-\beta H/n}|x\rangle$ of the operator $e^{-\beta H/n}$. We stress that replacing $\beta \rightsquigarrow it/\hbar$ we get the kernel of the time-evolution operator $e^{-itH/\hbar}$, that is, the Green's function of the Schrödinger operator. For $n \gg 1$ Eq. (3.209) yields

$$\langle x'|e^{-\beta H/n}|x\rangle \approx \langle x'|e^{-\beta p^2/(2mn)}|x\rangle e^{-\beta V(x)/n} =$$

$$= \int \frac{d^\ell p}{(2\pi)^\ell}\, e^{ip(x'-x)}e^{-\beta p^2/(2mn)}\, e^{-\beta V(x)/n} = \tag{3.211}$$

$$= \exp\left[-\epsilon\left(\frac{m}{2}\left(\frac{x'-x}{\epsilon}\right)^2 + V(x)\right)\right]$$

where in the last line we set $\epsilon \equiv \beta/n \ll 1$. Writing $x(k\epsilon)$ for $x^{(k)}$ we have

$$\langle x(n\epsilon)|e^{-\beta H}|x(0)\rangle = \int \prod_{k=0}^{n-1} dx(k\epsilon)\langle x((k+1)\epsilon)|e^{-\epsilon H}|x(k\epsilon)\rangle =$$

$$= \int \prod_{k=0}^{n-1} dx(k\epsilon) \exp\left[-\epsilon \sum_k \left(\frac{m}{2}\frac{(x(k\epsilon+\epsilon)-x(k\epsilon))^2}{\epsilon^2} + V(x(k\epsilon))\right)\right] \tag{3.212}$$

As $n \to \infty$, $\epsilon \to 0$ so that for $k\epsilon = t$

$$\frac{x(k\epsilon+\epsilon)-x(k\epsilon)}{\epsilon} \to \dot{x}(t) \tag{3.213}$$

and we get the *path integral*[16] representation of the quantum amplitude

$$\langle x_f|e^{-\beta H}|x_i\rangle = \int [Dx(t)]_i^f \, \exp\left(-\int_0^\beta dt \, L_E(x(t))\right) \tag{3.214}$$

where $[Dx(t)]_i^f$ stands for integration "over all paths" $t \in [0,\beta] \mapsto x_j(t)$ with initial condition $x_j(0) = (x_i)_j$ and final condition $x_j(\beta) = (x_f)_j$. $L_E(x(t))$ is the *Euclidean Lagrangian* (that is, the Lagrangian continued to imaginary times $t = i\tau$)

$$L_E = \frac{m}{2}\sum_{i=1}^{\ell}(\dot{x}_i)^2 + V(x_1,\ldots,x_\ell) \tag{3.215}$$

Equation (3.214) is the analytic continuations to imaginary times of the Feymann-Kac formula for the kernel $\langle x_f|e^{-itH/\hbar}|x_i\rangle$ of the real time unitary evolution operator $U(t) \equiv e^{-itH/\hbar}$. We stress that this path integral formula for the kernel is mathematically well defined (in terms of Wiener measure [11]).

The canonical partition function is then

$$Z = \text{Tr}\,e^{-\beta H} = \int dx \, \langle x|e^{-\beta H}|x\rangle = \int [Dx(t)]_{\text{per}} \, e^{-\int_0^\beta dt \, L_E} \tag{3.216}$$

where in the RHS the functional integral is over all paths periodic in Euclidean ($\equiv$ imaginary) time with period β

$$x_i(t+\beta) = x_i(t). \tag{3.217}$$

---

[16] For an introduction to path integrals in field theory see e.g. [12].

The canonical thermal expectation value have the form

$$\langle O \rangle = \frac{1}{Z} \int dx \, dy \langle x|O|y\rangle \langle y|e^{-\beta H}|x\rangle = \frac{1}{Z} \int [Dx(t)]_{\mathrm{per}} \, e^{-\int_0^\beta dt \, L_E} \, O(0) \tag{3.218}$$

Second Quantization: Grand Partition Function

The real time action for the second quantized theory is given in Eq. (3.177). Its imaginary time version is

$$S[a(t)_k^*, a(t)_k] = \sum_{k=0}^{\infty} \int_0^\beta dt \left[a(t)_k^* \left(\frac{d}{dt} - E_k \right) a(t)_k \right], \tag{3.219}$$

where now we see $a_k(t)$, $a_k(t)^*$ not as operators acting on a Hilbert space but as complex functions we integrate over in the path integral. Since the second quantized theory has a conserved charge N, we are free to add a chemical potential term to the imaginary-time action

$$S \rightsquigarrow S - \beta \mu N = \sum_{k=0}^{\infty} \int_0^\beta dt \left[a(t)_k^* \left(\frac{d}{dt} - E_k - \mu \right) a(t)_k \right] \tag{3.220}$$

Let us consider first the case where the particles are bosons, that is, the integration variables $a_k(t)$, $a_k(t)^*$ are *ordinary* ($\equiv$ commuting) variables. The Grand partition function is then the path integral

$$\mathcal{Z} = \int [Da_k \, Da_k^*]_{\mathrm{per}} \, \exp\left[-\int_0^\beta dt \sum_k a(t)_k^* \left(\frac{d}{dt} - E_k - \mu \right) a(t)_k \right]. \tag{3.221}$$

The integrand is a quadratic functional, so that the integral is Gaussian. Gaussian path integrals will be studied in Chap. 5; here we limit to an ad hoc treatment. Consider first a finite-dimensional Gaussian integral in $\mathbb{C}^m \simeq \mathbb{R}^{2m}$ of the form

$$\int \exp\left(-\bar{z}_k \, A_{kl} \, z_l \right) d^m \bar{z}_k \, d^m z_k \tag{3.222}$$

where the $m \times m$ matrix $A = (A_{kl})$ is Hermitian and an overbar means complex conjugation. Let $\{\lambda_k\}$ be the eigenvalues of A ($\prod_k \lambda_k = \det A$); we have

$$\int e^{-\bar{z}_k \, A_{kl} \, z_l} \, d^m \bar{z}_k \, d^m z_k = \prod_{k=1}^{m} \int d\bar{z} \, dz \, e^{-\lambda_k \bar{z} z} = \pi^m (\det A)^{-1}. \tag{3.223}$$

The path integral (3.221) differs from the finite-dimensional one (3.223) only because the $m \times m$ matrix A acting on $\mathbb{C}^m$ is replaced by a differential operator acting on an infinite-dimensional Hilbert space. Therefore the path integral (3.221) is just

$$\mathcal{N} \prod_k \left[\mathsf{Det}\left(\frac{\mathrm{d}}{\mathrm{d}t} - E_k - \mu \right)_\beta \right]^{-1} \tag{3.224}$$

where $\mathsf{Det}(\mathrm{d}_t - E - \mu)_\beta$ is the *functional determinant* of the differential operator $\mathrm{d}_t - E - \mu$ acting on the Hilbert space of periodic functions of t with period β, an $\mathcal{N}$ is a normalization constant whose value is immaterial.

Since the differential operator has constant coefficients, we can compute its determinant by elementary methods. More advanced techniques, which may be applied to more general situations, will be described in Chap. 5. The eigenfunction equation

$$\frac{\mathrm{d}\psi_\lambda}{\mathrm{d}t} - (E + \mu)\psi_\lambda = \lambda\,\psi_\lambda \tag{3.225}$$

has the solution

$$\psi_\lambda(t) = C \exp[(\lambda + E + \mu)t] \tag{3.226}$$

which is periodic of period β if and only if

$$\lambda + E + \mu = \frac{2\pi i n}{\beta} \quad n \in \mathbb{Z}. \tag{3.227}$$

The eigenvalues are then

$$\lambda_n = \frac{2\pi i}{\beta}n - E - \mu, \qquad n \in \mathbb{Z} \tag{3.228}$$

and therefore

$$\frac{\partial}{\partial\mu} \log \mathsf{Det}(\mathrm{d}_t - E - \mu)_\beta = \frac{\partial}{\partial\mu} \sum_n \log \lambda_n = -\sum_{n \in \mathbb{Z}} \frac{1}{2\pi i n/\beta - E - \mu} =$$

$$= \frac{1}{E + \mu} + 2(E + \mu) \sum_{n \geq 1} \frac{1}{(E + \mu)^2 + 4\pi^2 n^2/\beta^2} \tag{3.229}$$

Now recall the identity

$$\coth z = \frac{1}{z} + 2z \sum_{n \geq 1} \frac{1}{z^2 + \pi^2 n^2}. \tag{3.230}$$

Comparing with Eq. (3.229) we get

$$\frac{\partial}{\partial \mu} \log \mathsf{Det}(\mathrm{d}_t - E - \mu)_\beta = \frac{\beta}{2} \coth\left(\frac{\beta(E + \mu)}{2}\right), \tag{3.231}$$

that is,

$$\mathsf{Det}\left(\frac{\mathrm{d}}{\mathrm{d}t} - E - \mu\right)_\beta = 2 \sinh\left(\tfrac{1}{2}\beta(E + \mu)\right) = \mathrm{e}^{-\beta(E+\mu)/2}\left(1 - \mathrm{e}^{-\beta(E+\mu)}\right) \tag{3.232}$$

The first factor in the RHS arises from the zero-point energy of the harmonic oscillator with frequency $\omega \equiv E + \mu$ (cf. discussion around Eq. (3.184)). Shifting the Hamiltonian so that the ground state is at zero energy, we get

$$-\log \mathcal{Z} = \sum_k \log \mathsf{Det}(\mathrm{d}_t - E_k - \mu)_\beta + \frac{\beta}{2}(E + \mu) = \sum_k \log(1 - \mathrm{e}^{-\beta(E_k+\mu)}) \tag{3.233}$$

in full agreement with Eq. (3.33) from the particle perspective.

Second Quantization: Fermionic Partition Functions

In the fermionic case the fields[17] a_k, a_k^* are *anticommuting*

$$a_k a_l + a_l a_k = a_k^* a_l^* + a_l^* a_k^* = a_k^* a_l + a_l a_k^* = 0, \tag{3.234}$$

and the algebra they generate is a *Grassmann algebra* where all generators are *odd* i.e. they anticommute with all generators including themselves. In particular all generators are nilpotent i.e. $(a_k)^2 = (a_k^*)^2 = 0$ for all k.

Integration over Odd Coordinates To get a path integral representation of the fermionic Grand canonical partition function we have first to define what we mean by integration over *odd* ($\equiv$ anticommuting) fields.

---

[17] To avoid all misunderstandings, we stress that these are "classical" Fermi fields ($\equiv$ integration variables in path integrals) not Fermi operators. They generate a *Grassmann* algebra whereas the quantum Fermi operators generate a *Clifford* algebra.

We start with an algebra generated by finitely-many *odd* indeterminates, $\theta_1, \ldots, \theta_s$. We first define the finite-dimensional integral over the θ_a's and then extend the result to infinite dimension i.e. to path integrals over fermionic fields.

Let $f(\theta)$ be a function of the single odd variable θ. We wish to define its integral

$$\int \mathrm{d}\theta \, f(\theta) \tag{3.235}$$

in such a way that the following three properties hold:

IO1   is not identically zero, i.e. (3.235) is a non-zero complex number for some choice of the function $f(\theta)$;

IO2   the number (3.235) is linear in the integrand, that is,

$$\int \mathrm{d}\theta \left(a_1 f_1(\theta) + a_2 f_2(\theta) \right) = a_1 \int \mathrm{d}\theta \, f_1(\theta) + a_2 \int \mathrm{d}\theta \, f_2(\theta) \tag{3.236}$$

   for all $a_1, a_2 \in \mathbb{C}$;

IO3   The integral is translational invariant, that is,

$$\int \mathrm{d}\theta \, f(\theta + a) = \int \mathrm{d}\theta \, f(\theta) \tag{3.237}$$

   for all (odd) constant a.

Since $\theta^2 = 0$, the most general function $f(\theta)$ of a single *odd* variable θ is just a polynomial of degree 1: $f(\theta) = f_0 + \theta f_1$ for some constants f_0, f_1. Then by **IO3**

$$\int \mathrm{d}\theta \, a = \int \mathrm{d}\theta \, (\theta + a) - \int \mathrm{d}\theta \, \theta = 0, \tag{3.238}$$

which means that the integral of any constant vanishes. Then the integral of the most general function of a single Grassmann odd variable θ is

$$\int \mathrm{d}\theta \left(f_0 + \theta f_1 \right) = f_1 \int \mathrm{d}\theta \, \theta \tag{3.239}$$

and the last integral must be non-zero by **IO1**. Without loss we normalize it to 1 so that

$$\int \mathrm{d}\theta \left(f_0 + \theta f_1 \right) \stackrel{\mathrm{def}}{=} f_1. \tag{3.240}$$

Note that $\int \mathrm{d}\theta \cdot 1$ must be zero because the integral is a number ($\equiv$ Grassmann *even*) while the differential $\mathrm{d}\theta$, as θ itself, is an anticommuting quantity (*odd*).

The integral in several anticommuting variables is then constructed by induction on their number (paying attention to the order of variables and differentials since they anticommute). Suppose we make a linear change of odd variables

$$\xi_i = A_{ij}\theta_i \tag{3.241}$$

where A_{ij} is an invertible matrix of complex numbers. We have (notice orders!)

$$
\begin{aligned}
1 &= \int d\xi_1 d\xi_2 \cdots d\xi_s \, (\xi_s \cdots \xi_2 \xi_1) = \\
&= A_{s j_s} \cdots A_{2, j_2} A_{1, j_1} \int d\xi_1 d\xi_2 \cdots d\xi_s \, (\theta_{j_s} \cdots \theta_{j_2}\theta_{j_1}) = \\
&= \sum_{\pi \in \mathfrak{S}_s} (-1)^{\mathsf{sign}(\pi)} A_{s\pi(s)} \cdots A_{2,\pi(2)} A_{1,\pi(1)} \int d\xi_1 \cdots d\xi_s \, (\theta_s \cdots \theta_2 \theta_1) \\
&= \det A \int d\xi_1 d\xi_2 \cdots d\xi_s \, (\theta_s \cdots \theta_2 \theta_1)
\end{aligned}
\tag{3.242}
$$

that is,

$$
d\xi_1 \, d\xi_2 \cdots d\xi_s = \frac{1}{\det A} \, d\theta_1 \, d\theta_2 \cdots d\theta_s = \left(\det \frac{\partial \xi_i}{\partial \theta_j}\right)^{-1} d\theta_1 \, d\theta_2 \cdots d\theta_s
\tag{3.243}
$$

Fact 3.2 *The Jacobian for a change of odd variables $\xi_i \rightsquigarrow \theta_i$ the Grassmannian volume form $d\xi_1 \cdots d\xi_s$ is the **inverse** of the determinant of the matrix*

$$\frac{\partial \xi_i}{\partial \theta_j}. \tag{3.244}$$

This result is natural since the integral over anticommuting variables is essentially the same thing as the derivative

$$\int d\theta \, (f_0 + \theta f_1) = f_1 = \frac{\partial}{\partial \theta}(f_0 + \theta f_1). \tag{3.245}$$

This also entails that the Grassmannian δ-function ($\equiv$ the reproducing kernel) defined by the property

$$\int d\theta \, \delta(\theta - \theta') \, f(\theta) = f(\theta') \tag{3.246}$$

is the identity function

$$\delta(\theta - \theta') = \theta - \theta'. \tag{3.247}$$

Indeed,

$$\int d\theta \left(\theta - \theta'\right)\left(f_0 + \theta f_1\right) = \int d\theta \left(\theta f_0 - \theta' f_0 + \theta \theta' f_1\right) = f_0 + \theta' f_1. \qquad (3.248)$$

An important Grassmann integral is the Gaussian one

$$\int \exp\left(-\bar{\xi}_i A_{ij}\xi_j\right) \prod_k d\bar{\xi}_k \, d\xi_k, \qquad (3.249)$$

where ξ_i and $\bar{\xi}_j$ are independent Grassmann variables[18] and A_{ij} is a matrix of complex numbers (say Hermitian). We can compute the integral (3.249) in three ways: either we expand the exponential in power series, getting a finite polynomial since the ξ_k, $\bar{\xi}_k$ are nilpotent, or we make the change of variables $\theta_i = A_{ij}\xi_j$ (without touching the $\bar{\xi}_k$) or else we go to a basis where A is diagonal. Either way we get

$$\int \exp\left(-\bar{\xi}_i A_{ij}\xi_j\right) \prod_k d\bar{\xi}_k \, d\xi_k = \det A. \qquad (3.250)$$

Again, this is the inverse of the result for the corresponding integral with ξ_k bosonic (commuting) variables. In that case the integral is the *inverse* of the determinant of A (up to a power of π that we absorb in $\mathcal{N}$).

The Fermionic Grand Partition Function We have the same Gaussian path integral as in Eq. (3.221) except that now the integration fields a_k and a_k^* are fermionic (*anticommuting*). This leads to two modifications:

FGP1 to pass from the kernel of $e^{-\beta H}$ to the Hilbert-space trace we have to reverse the order of fields at the initial and final (imaginary) times, which produces a flip of sign. Hence to get the correct partition function we have to integrate over fields which are *anti*-periodic instead of periodic [13]

$$a_k(t + \beta) = -a_k(t), \qquad a_k(t + \beta)^* = -a_k(t)^* \qquad (3.251)$$

FGP2 the bosonic Gaussian integral is the inverse of the determinant of the operator in the action, while the fermionic Gaussian integral is the determinant itself, see Eq. (3.250).

---

[18] Morally speaking the $\bar{\xi}_k$'s are the "complex conjugates" of the ξ_k.

The effect of flipping periodic $\leftrightarrow$ anti-periodic boundary conditions is to subtract $i\pi/\beta$ from the eigenvalues, see Eqs. (3.226) and (3.227). The new eigenvalues are

$$\lambda_n = \frac{2\pi i}{\beta}n - E - \mu - \frac{i\pi}{\beta}. \tag{3.252}$$

Thus the flip periodic $\leftrightarrow$ anti-periodic is equivalent to adding an imaginary term to the chemical potential

$$\mu \rightsquigarrow \mu + ikT\pi. \tag{3.253}$$

Therefore the fermionic Grand canonical partition function is related to the bosonic one (with the same energy levels $\{E_n\}$) by the functional relation

$$\mathcal{Z}(\mu)_{\text{ferm}} = \frac{1}{\mathcal{Z}(\mu + ikT\pi)_{\text{bos}}} \tag{3.254}$$

which is the result we got in Eqs. (3.33) and (3.38).

Problems

3.1 Compute the entropy and the specific heat of a gas a fermions in the low temperature regime.

3.2 Compute the isothermal compressibility of an ideal bosonic gas just above and below the critical temperature.

3.3 Compute the coefficient of thermal expansion (1.109) for the ideal Bose gas at low temperatures.

3.4 Calculate C_V, C_P, κ_T for the ideal Fermi gases in the limit of extreme dilution up to the order $\hbar^3$.

3.5 Calculate the chemical potential $\mu(T, N/V)$ for a two-dimensional ideal Fermi gas.

3.6 Compute the Grand canonical partition function of a gas of massless spin-$\frac{1}{2}$ particles ("photinos").

References

1. R.F. Streater, A.S. Wightman, *PCT, Spin and Statistics, and All That* (Princeton University Press, 2000)
2. J.P. Serre, *Representations of Finite Groups*. Graduate Texts in Mathematics, vol. 42 (Springer, 1977)
3. B.E. Sagan, *The Symmetric Group. Representations, Combinatorial Algorithms, and Symmetric Functions*. Graduate Texts in Mathematics, vol. 203 (Springer, 2001)
4. H. Iwaniec, E. Kowalski, *Analytic Number Theory*. AMS Colloquium Publications, vol. 53 (AMS, 2004)
5. D.M. Lozier, R.F. Boisvert, C.W. Clark (eds.), *NIST Handbook of Mathematical Functions* (Cambridge University Press, 2010). On-line version: *NIST Digital Library of Mathematical Functions,* available on-line https://dlmf.nist.gov
6. D.C. Wood, The computation of polylogarithms. Technical Report 15-92. University of Kent Computing Laboratory, Canterbury, UK. Retrieved 2005-11-01
7. *The on-line encyclopedia of integer sequences,* available on-line https://oeis.org/A027641 or https://oeis.org/A027642
8. S. Cecotti, *Analytic Mechanics. A Concise Textbook* (Springer, 2024)
9. M. Postnikov, *Leçon de géometrie. Groupes et algèbres de Lie* (Mir, 1982)
10. J. Fuchs, C. Schweigert, *Symmetries, Lie Algebras, and Representations. A Graduate Course for Physicists*. Cambridge Monographs in Mathematical Physics (Cambridge University Press, 1997)
11. J. Glimm, A. Jaffe, *Quantum Physics. A Functional Integral Point of View* (Springer, 1981)
12. R.J. Rivers, *Path Integral Methods in Quantum Field Theory*. Cambridge Monographs in Mathematical Physics (Cambridge University Press, 1987)
13. S. Cecotti, L. Girardello, Functional measure, topology and dynamical supersymmetry breaking. Phys. Lett. B **110**, 39 (1982)

Chapter 4
Phase Transitions and Lattice Systems

In this chapter we start the discussion of phase transitions, with particular focus on (classical) discrete systems which model ferromagnetic materials. The hero of the chapter is the *Ising model*. Apart for the introductory section which presents the physical motivations, the discussion in this chapter is rather rigorous and mathematically oriented. More "physical" approximation schemes to study ferromagnets and phase transitions will be introduced in Chap. 5. Fancier systems with a more intricate geography of phases will be touched upon in Chap. 7.

4.1 Phase Transitions: Generalities

Phenomenology: Water

It is a common experience that *water*, that is, a system composed of a large number of H_2O molecules, exists in several forms. The ones at thermal equilibrium are called *phases*. Water has three phases: *liquid*, *solid* (ice), and *gas* (vapor). The equations of state of water in the three phases are quite different both in their analytic form and in their qualitative behavior. Since the equations of state are obtained from the thermodynamical potentials (written in the respective natural variables), this entails that each thermodynamical potential of a system of H_2O molecules has several distinct analytic forms valid for different ranges of the control parameters (temperature and pressure).

© The Author(s), under exclusive license to Springer Nature Switzerland AG 2024   157
S. Cecotti, *Statistical Mechanics*, UNITEXT for Physics,
https://doi.org/10.1007/978-3-031-67874-5_4

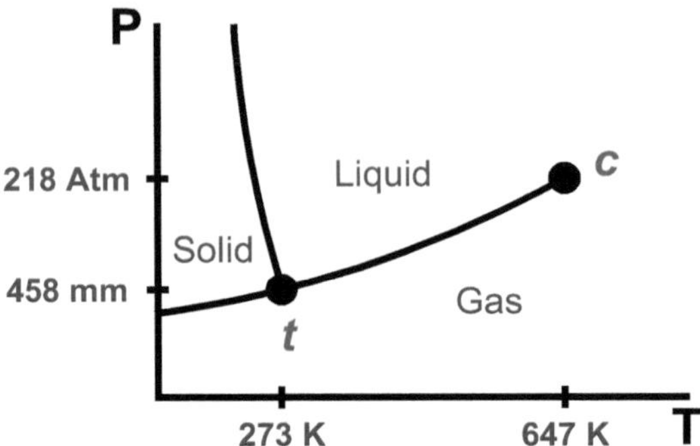

Fig. 4.1 The phase diagram of water in the T-P plane. The curves separate the different phases: along a curve two different phases coexist at the same values of (T, P). The point t is the *triple point* where all three phases (solid, liquid and gas) coexist. c is the *critical point:* at temperatures higher than $T_c = 647$ K we can pass *smoothly* from the liquid to the gas phase

We know from Sect. 1.11 that two distinct phases coexist along the codimension-1 loci in the T-P plane where the respective chemical potentials, ($\equiv$ Gibbs free energy per molecule) agree

$$\begin{aligned}
\text{liquid-gas curve:} \quad & \mu(T, P)_{\text{liq.}} = \mu(T, P)_{\text{gas}} \\
\text{liquid-solid curve:} \quad & \mu(T, P)_{\text{liq.}} = \mu(T, P)_{\text{sol.}} \\
\text{solid-gas curve:} \quad & \mu(T, P)_{\text{sol.}} = \mu(T, P)_{\text{gas}}
\end{aligned} \tag{4.1}$$

Two phases coexist on a codimension-1 space, i.e. on curves in the T-P plane, while all three phases coexist in a codimension-2 locus, indeed at an unique point called the *triple point*, see Fig. 4.1. Crossing a coexistence curve we pass from one phase to a different one: this process is called *phase transition* and is triggered by a suitable change in temperature or/and pressure. However the liquid-gas curve terminates in a point c, called the *critical point*: hence we can pass from liquid to gas (and viceversa) without going through any phase transition, provided we heat-up water to sufficiently high temperature (or exert enough pressure).

From Eq. (4.1) we see that the Gibbs free energy $G(T, P)$ of a system of $N = O(10^{24})$ H_2O molecules is a *continuous function* of its natural variables T, P. However the function $G(T, P)$ cannot be smooth. Say along the liquid-gas curve, we have

$$T\left(\frac{\partial G_{\text{liq.}}}{\partial T}\right)_P - T\left(\frac{\partial G_{\text{gas}}}{\partial T}\right)_P = T(S_{\text{gas}} - S_{\text{liq.}}) = N\,L(T) \tag{4.2}$$

where $L(T) > 0$ is the *latent heat* per particle of the liquid to gas transition along the liquid-gas curve, and

$$\left(\frac{\partial G_{\text{liq.}}}{\partial P}\right)_T - \left(\frac{\partial G_{\text{gas}}}{\partial P}\right)_T = N\left(\frac{1}{\rho_{\text{liq.}}}(T) - \frac{1}{\rho_{\text{gas}}}(T)\right) \tag{4.3}$$

where $\rho_{\text{liq.}}(T) = N/V_{\text{liq.}}(T)$ and $\rho_{\text{gas}}(T) = N/V_{\text{gas}}(T)$ are the densities of the liquid and gas, respectively. The discontinuities of the first derivatives of the Gibbs potential, that is, the latent heat $L(T)$ and the variation in the specific volume

$$\Delta v(T) \equiv \rho_{\text{liq.}}(T)^{-1} - \rho_{\text{gas}}(T)^{-1} \tag{4.4}$$

are decreasing functions of T which vanish at $T = T_c$. For $T > T_c$ the derivatives of $G(T, P)$ are continuous, in facts in this region the Gibbs free energy is an analytic function of its natural variables. We conclude:

Empirical Fact *The thermodynamical potential $G(T, P)$ of water is a real-analytic function in the T-P plane "cut" along the coexistence curves C_a. Moreover $G(T, P)$ extends* continuously *across the cuts.*

This observation suggests that phase transitions happen at loci in the space of thermal variables where the relevant thermodynamical potential is *non-analytic*. We can analytically continue the thermodynamical potential only along a path γ in control variables' space which does not cross any phase transition "cut": on the two sides of the "cut" the physics is qualitatively different, as it is the case of liquid water and solid ice, and hence the equations of state should change dramatically when we cross the "cut".

Remark 4.1 We can go from liquid water to vapor without crossing any phase transition by following a path beyond the critical point. Instead there is *no way* to go from solid (ice) to any other phase without crossing a transition. This reflects the fact that ice is an *ordered* phase where the molecules lay in a regular order in a crystal lattice, whereas the other two phases are *disordered* with molecules in random positions. We cannot get order from the analytic continuation of disorder, and hence the transition from fluid to solid cannot be analytic.

Example: Bose Condensation
As another example of phase transition, consider the *Bose condensation* in an ideal bosonic gas that we studied in Chap. 3. The Grand potential is not analytic at the critical temperature T_c, and we have a phase transition there. However the transition in the ideal Bose gas is much milder than in the water-ice case. In the Bose gas the internal energy at fixed volume, and hence the entropy S, is continuous across the transition, i.e. *there is no latent heat $L \equiv 0$.* Also the derivative of S with respect to

T, the heat capacity C_V at fixed volume,

$$C_V = T \left(\frac{\partial S}{\partial T} \right)_{V,N} \tag{4.5}$$

is continuous across the Bose condensation transition, see Eq. (3.140). However the second derivative of entropy

$$\left(\frac{\partial^2 S}{\partial T^2} \right)_{V,N} \tag{4.6}$$

is *discontinuous* at T_c as we proved in Eq. (3.141). Hence in the Bose gas the thermodynamic potential may be extended over the transition as a C^2 function.

Ehrenfest Classification

Following the above line of thought, Ehrenfest classified the phase transitions according to the non-analytic behavior of the thermodynamic potentials along the critical locus. One says that the phase transition is of *order k* iff the thermodynamic potential extends over the critical locus as a function of class C^{k-1} but not as a C^k function. The transition has *infinite order* iff the extension is of class C^∞ but non-analytic over the critical locus. While this classification of phase transitions according to their order is useful and widely used, it is not fully intrinsic and is quite coarse. The modern theory of phase transitions prefers the finer and fully intrinsic classification in *universality classes* which we shall encounter at the end of this chapter.

Thermodynamic Limit: Pure/Mixed Phases

Phase transition are associated to loci in thermal variable space where the thermo-dynamic potentials are not analytic. For the usual systems, discrete or continuous, the thermodynamic potentials are trivially analytic *before* taking the thermodynamic limit, that is, as long as the system has a huge *yet finite* number of degrees of free-dom.[1] Thus, strictly speaking, the phase transitions arise only in the thermodynamic limit, and are a property of the limit. They are also related to *non-uniqueness* of the thermodynamic limit: at a point (T, P) on the liquid-solid coexistence curve in Fig. 4.1 we have several inequivalent thermodynamic limits: for all $0 \leq \alpha \leq 1$ we have a limit where αN out of the N H_2O molecules are in the liquid phase and $(1 - \alpha)N$ of them are in the solid state. The general equilibrium state is then a *convex linear combination* of the two *pure* states (ice and liquid water) which correspond respectively to $\alpha = 0$ and 1. By "convex linear combination" we mean that the thermodynamic limit probability measure is a convex combination of the

---

[1] This principle will be demonstrated several times in this chapter from different viewpoints: Yang-Lee theory, Frobenius decomposition, etc.

pure phase ones

$$d\mu_\alpha = \alpha \, d\mu_{\textbf{water}} + (1 - \alpha) \, d\mu_{\textbf{ice}}, \quad 0 \le \alpha \le 1. \tag{4.7}$$

so that the thermal expectation value of a quantity O in the general phase α is

$$\langle O \rangle_\alpha = \alpha \, \langle O \rangle_{\textbf{water}} + (1 - \alpha) \, \langle O \rangle_{\textbf{ice}}. \tag{4.8}$$

We shall also refer to a pure phase as a *pure state*; the terms *state* and *phase* are used interchangeably in the present context.[2] We identify a phase α with the corresponding linear functional $\langle \cdots \rangle_\alpha$ on the space of statistical observables.

There are various characterization of the *pure* phases (pure states) some of which will be discussed in this textbook. Non-pure states are called *mixed*.

Non-uniqueness of the Thermodynamic Limit When multiple phases coexist, the thermodynamic limit cannot be unique: this means that the limit crucially depends on the *details of the way the thermodynamic limit*

$$N, V \to \infty \quad \text{with} \quad V/N = v \text{ fixed} \tag{4.9}$$

is defined, in particular on the boundary conditions we use when taking the limit.

Spontaneous Symmetry Breaking
An important instance of coexistence of several equilibrium pure states is provided by the phenomenon of *spontaneous symmetry breaking:* the Hamiltonian H is invariant under the action of the group G on the microscopic degrees of freedom, i.e. the group G (which may be discrete or a non-trivial Lie group) is a *symmetry* of the Hamiltonian. We say that *the symmetry G is broken to a subgroup $I \subset G$ in the pure state* $\langle \cdots \rangle_{\text{pu}}$ if the correlation functions in the pure state are not invariant under the G-action

$$\langle g \cdot O_1 \, g \cdot O_2 \, \cdots \, g \cdot O_s \rangle_{\text{pu}} \neq \langle O_1 \, O_2 \, \cdots \, O_s \rangle_{\text{pu}}, \quad g \in G \tag{4.10}$$

but only under the action of the subgroup I, that is, (4.10) becomes and equality when $g \in I$, the *unbroken* subgroup. G acts on the microscopic degrees of freedom, hence it acts on the states. Since G is a symmetry of the dynamics, the action of $g \in G$ on an equilibrium pure state yields another equilibrium pure state, while the action of $g \in I$ gives back the same state. Hence the G-action produces a family of coexisting pure states parametrized by the coset G/I. Spontaneous breaking of symmetries is a main theme in the rest of the book.

---

[2] Mathematically there is a slight difference between the two notions of no consequence for reasonable systems.

Ferromagnetism

The examples of phase transition which are best understood theoretically are the *ferromagnetic* systems. Many materials,[3] called *paramagnetic,* have the property that in presence of a strong external magnetic field $\mathbf{B}$ their valency electrons align their spins with the field, producing a macroscopic magnetic dipole moment $\langle \mathbf{M} \rangle$ in the direction $-\mathbf{B}$. It is well known that there are also materials—such as iron, nichel, cobalt—with the property that the magnetization $\langle \mathbf{M} \rangle$ persists even after we switch off the external magnetic field: after setting $\mathbf{B}$ to zero, the non-zero vector $\langle \mathbf{M} \rangle$ still remembers the original direction of $\mathbf{B}$ ("memory effect"). Materials with this property are called *ferromagnetic.* A non-zero magnetization $\langle \mathbf{M} \rangle \neq 0$ at zero external field, $\mathbf{B} = 0$, is called *spontaneous.* Spontaneous magnetization is an example of *spontaneous symmetry breaking*: after switching off the external magnetic field, the system's Hamiltonian becomes invariant under $SO(3)$ spatial rotations. A non-zero magnetization $\langle \mathbf{M} \rangle \neq 0$ is a vector pointing in a particular direction and the $SO(3)$ symmetry breaks down to the $SO(2)$ isotropy subgroup of the vector $\langle \mathbf{M} \rangle \in \mathbb{R}^3$. Due to the "memory effect", in a ferromagnetic phase $\langle \mathbf{M} \rangle$ is a function of $\mathbf{B} \equiv B\hat{z}$ which is discontinuous at $B = 0$

$$\lim_{B \to 0^+} \langle \mathbf{M} \rangle \neq \lim_{B \to 0^-} \langle \mathbf{M} \rangle \equiv - \lim_{B \to 0^+} \langle \mathbf{M} \rangle \neq 0. \tag{4.11}$$

However, there is a temperature T_c, called the *Curie temperature,* such that for $T > T_c$ there is no spontaneous magnetization

$$\langle \mathbf{M} \rangle \Big|_{B=0^+} = \begin{cases} \neq 0 & T < T_c \\ = 0 & T > T_c. \end{cases} \tag{4.12}$$

Thus a ferromagnet has two phases separated by a transition with no latent heat ($\equiv$ second or higher order in T). The two phases are distinguished by the value of the expectation value $\langle \mathbf{M} \rangle$ of the magnetization operator $\mathbf{M}$.

The study of ferromagnetic models paved the way to the modern theory of phase transitions. Ferromagnetism is the main topic of this chapter and the next one.

Yang-Lee Theory

The *Yang-Lee theory* of phase transitions [1, 2] aims to prove *rigorously* that the thermodynamical potentials are analytic away from some special loci in their natural variable space, so that we can exclude the existence of phase transitions except, possibly, at these special loci. The Yang-Lee theory applies in a variety of situations:

YL1   for the Grand potential of a classical gas of particles with a hard core of finite diameter a and pairwise interactions of finite range;

YL2   for the Grand potential of lattice gases with finite range interactions;

---

[3] *Paramagnetism, ferromagnetism,* etc. are strictly quantum phenomena, cf. Theorem 2.5.

YL3   for the free energy of (classical) ferromagnetic lattice systems with finite range interactions, possibly with an external magnetic field.

While the physical interpretations of the systems **YL1–YL3** are quite different, mathematically the three set-ups are equivalent and one may translate the results from one system to the others.

Before stating the original Yang-Lee theorem, we explain the general idea beyond it. In most systems[4] the finite-ensemble partition function Z_N is trivially an analytic function of the control parameters *before* taking the thermodynamic limit $N \to \infty$. For instance: for any reasonable continuous quantum system the finite-N canonical partition function

$$Z_N(\beta) = \mathrm{Tr}_N \, e^{-\beta H_N} \tag{4.13}$$

is analytic in the half-plane $\mathrm{Re}\,\beta > 0$ and often it can be extended to a holomorphic function in some larger domain $\Omega \subset \mathbb{C}$ (such as the plane less a cut along the negative real axis). Now suppose that, for all finite N, the finite-ensemble partition function $Z_N(\zeta)$ is a holomorphic function of the control parameter ζ in some domain Ω; assume (as it is typically the case) that the physically allowed values of ζ are *real positive,* and that these allowed values belong to the holomorphy domain i.e. $\mathbb{R}_{>0} \subset \Omega$. This is a very common situation: for systems **YL1–YL3** the partition function Z_N is a polynomial in a suitable variable ζ (that we call loosely "fugacity")

$$Z_N(\zeta) = \sum_{n=0}^{d(N)} A_n \, \zeta^n, \qquad A_n > 0, \tag{4.14}$$

hence $Z_N(\zeta)$ is an *entire function* holomorphic everywhere in the complex plane $\mathbb{C}$. For simplicity we focus on this polynomial case where, typically, the degree $d(N) = O(N)$. The free energy (per particle) at *finite N*

$$F_N(\zeta) \equiv -\frac{1}{N}kT \log Z_N(\zeta) \tag{4.15}$$

is **not** holomorphic in $\mathbb{C}$: $F_N(\zeta)$ has singularities at the $d(N)$ roots $\zeta_a^{(N)} \in \mathbb{C}$ of the finite-N partition function, $Z_N(\zeta_a^{(N)}) = 0$. However, for finite N, $F_N(\zeta)$ is still analytic along the positive real axis, since in the physical region the partition function is a sum of positive contributions (cf. (4.14)), hence non-zero. Fix a point $\zeta_* \in \mathbb{R}_{>0}$. The Taylor series of $F_N(\zeta)$ centered in ζ_* has a positive radius of convergence

$$\rho_{*,N} = \inf_a |\zeta_a^{(N)} - \zeta_*| > 0. \tag{4.16}$$

---

[4] An exception is the ideal bosonic gas Grand partition which has a pole at $\mu = 0$.

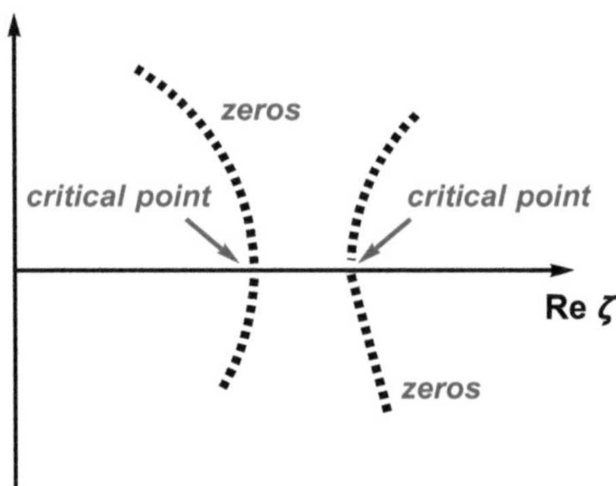

Fig. 4.2 Critical values of the fugacity ζ in Yang-Lee theory

Consider first the case where the infimum

$$\rho_* = \inf_N \rho_{*,N} \tag{4.17}$$

is *positive*. Inside the (open) disk D_* of radius ρ_*, centered at ζ_*, all finite-N free energies $F_N(\zeta)$ are holomorphic. Assume that the thermodynamic limit exists, so that the sequence $\{F_N(\zeta)\}$ converges *point-wise* in the physical subset $\mathbb{R}_{>0} \cap D_*$. Under mild conditions, Vitali-like theorems [3] guarantee that the convergence is uniform in all compacts $\Subset D_*$ and that the limit is a holomorphic function. To this end it suffices, say, that the sequence of functions $\{F_N(\zeta)\}$ is locally bounded [3]. In this case the thermodynamic-limit free energy $\lim_{N\to\infty} F_N(\zeta)$ is analytic in $\mathbb{R}_{>0} \cap D_*$ and we cannot have phase transitions in this region. We draw the

Informal Lesson *Under mild conditions, a point in the physical domain of a control parameter ζ may be a phase transition point only if it is an accumulation point of complex zeros of the finite-N partition functions $Z_N(\zeta)$. See Fig. 4.2.*

Yang and Lee made this idea in a precise technical statement which applies to systems **YL1–YL3**. We review their work following the original literature [1]. We stress, however, that the scenario applies to a wider context: for instance for a continuous quantum system $[\mathrm{Tr}_N \exp(-\beta H_N)]^{1/N}$ is an analytic function of β in the largest simply-connected region $\Omega_N \supset \mathbb{R}_{>0}$ in the right half-plane $\{\mathrm{Re}\,\beta > 0\}$ which is free of zeros. Moreover

$$\left| [\mathrm{Tr}_N \exp(-\beta H_N)]^{1/N} \right| \leq \left[\mathrm{Tr}_N \exp(-\mathrm{Re}\,\beta\, H_N) \right]^{1/N}, \tag{4.18}$$

so when the sequence of function $[\mathrm{Tr}_N \exp(-\beta H_N)]^{1/N}$ converges point-wise for real positive β (i.e. iff the thermodynamic limit *exists*) the sequence is locally bounded in the half-plane $\mathrm{Re}\,\beta > 0$ hence in $\bigcap_N \Omega_N$. By Vitali's theorem phase transitions may arise only for $\beta \notin \bigcap_N (\Omega_N \cap \mathbb{R}_{>0})$ i.e. only at accumulation points of complex zeros of the partition functions $\{Z_N(\beta)\}$ on the positive real axis.

Theorem 4.1 (Yang and Lee [1]) *Let $Q(V, \zeta)$ be the partition function of an ensemble at finite volume V which depends on a control parameter $\zeta = \mathrm{e}^w$ (called "fugacity") having the two properties:*

YLT1   *The thermodynamic limit*

$$\lim_{V \to \infty} \frac{1}{V} \log Q(V, \zeta) = \Xi(\zeta) \qquad (4.19)$$

exists;

YLT2   *At finite volume $Q(V, \zeta)$ is a polynomial in ζ of degree $d \leq c\,V$ for some constant c.*

Assume that there is a region R in the complex ζ-plane such that

R1   *R contains a segment on the positive real axis;*
R2   *R does not contain roots of the polynomial $Q(V, \zeta)$ for all V*

$$Q(V, \zeta) \neq 0 \quad \text{for } \zeta \in R \text{ and all } V > 0. \qquad (4.20)$$

Then the sequence of quantities

$$\frac{\partial^k}{\partial w^k} \frac{1}{V} \log Q(V, \zeta), \qquad k = 0, 1, 2, \cdots \qquad (4.21)$$

*have thermodynamic limits as $V \to \infty$ which are analytic in ζ for $\zeta \in R$. Furthermore the operations $\partial/\partial w$ and $\lim_{V \to \infty}$ commute in R. Therefore there is **no** phase transition in ζ in the region*

$$\zeta \in \mathbb{R}_{>0} \cap R. \qquad (4.22)$$

Theorem 4.1 says that the physical values of the "fugacity" ζ for which we have phase transitions are the points on the positive real axis which are accumulation points of the zeros of $Q(V, \zeta)$ as $V \to \infty$, see Fig. 4.2. Theorem 4.1 is proven in the Appendix following the original paper [1].

Comments on Definitions

Up to now we gave two distinct definitions of "phase transition points":

(i) as the locus where the thermodynamic potentials are not analytic;
(ii) as the locus where some thermal expectation value $\langle M \rangle$ ("order parameter") is discontinuous, cf. Eq. (4.11).

For typical systems the thermodynamic potentials are piece-wise analytic and the two definitions agree, but there is no general proof of their equivalence [2]. The heuristics beyond the identification of the two notions is that the expectation values of order parameters may be written as derivatives of the thermodynamical potentials (see Sect. 2.7) and hence a discontinuity in the order parameter signals a non-analyticity of the thermodynamical potentials. So, formally, the two notions are expected to agree at least for transitions of *finite* order. However for some peculiar system it may happen that the thermodynamic potential is analytic at criticality or—on the contrary—that the thermodynamic potential is non-analytic in an open domain of the parameter space. In all these "unusual" situations, the peculiar analytic properties of the thermodynamic potentials is associated to new interesting physical phenomena.

A "counterexample" to non-analyticity at criticality is the Ising model on a Caley tree:[5] its free energy is analytic as function of T for all $T > 0$, while it has a phase transition at some positive temperature T_c. This is related to a subtlety of the model: if we consider a sequence of finite sub-graphs $\Gamma_k \subset \Gamma_{k+1}$ such that $\cup_k \Gamma_k$ is the Caley tree, the ratio of the number of degrees of freedom in the boundary $\Gamma_{k+1} \setminus \Gamma_k$ and in the bulk of the "box" Γ_{k+1} remains finite as $k \to \infty$, that is, in the thermodynamic limit a finite portion of the degrees of freedom live on the boundary. Hence the thermodynamic limit of the free energy will depend *dramatically* on the chosen boundary condition. There is a prescription which produces an analytic free energy, but also other ones which lead to singular answers. See Problem 4.4.

In Chap. 7 we shall study a subtle class of systems called *spin glasses*. As discussed there, the free energy is non-analytically (presumably not even C^∞) below a certain curve in the T-B plane. Heuristically: the system has infinitely many pure states, so ∞-many phase transitions, and the critical points get dense in the low temperature region, so that the free energy is nowhere analytic in this open domain.

4.2 Magnetic Systems: The Ising Model

We now focus on *ferromagnetic* materials.[6] Let's recall the phenomenology: when we switch on an external magnetic field **B** most materials *magnetize,* i.e. develop a non-zero magnetic dipole moment $\langle \mathbf{M} \rangle \neq 0$ in the direction opposite to the field. The ferromagnetic materials have the peculiarity that when we switch off the external field the magnetization $\langle \mathbf{M} \rangle \neq 0$ *will persist,* provided the temperature T is lower than a certain critical temperature T_c called the *Curie temperature.* We say that these materials undergo *spontaneous magnetization,* that is, their magnetization is not the response to an external magnetic field.[7] Let us have a look at what is happening at the microscopic level: the electrons in the material have spins $\mathbf{s}_i$ and magnetic moments $\boldsymbol{\mu}_i \propto \mathbf{s}_i$; if a sizable proportion of all valency electrons have their spin aligned in the same direction, the material develops a *macroscopic* magnetic

---

[5] See e.g. Chap. 14 of [4].

[6] Some aspects of **anti**ferromagnetic systems will be discussed in Chap. 7.

[7] Responses to external fields will be studied in full generality in Chap. 6.

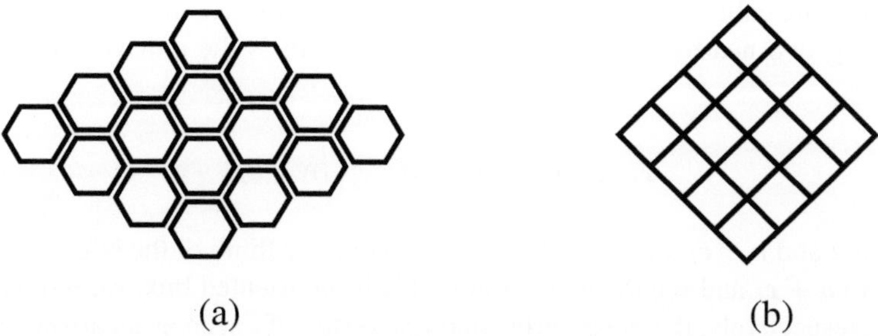

Fig. 4.3 (**a**) Portion of hexagonal lattice in two dimensions. (**b**) Portion of square lattice in two dimensions

moment which produces the overall magnetization (per unit volume)

$$\langle \mathbf{M} \rangle = \frac{1}{V} \sum_i \boldsymbol{\mu}_i. \tag{4.23}$$

Lattices

The simplest and more widely-studied statistic mechanical model of a ferromagnetic system is the *Ising model*.[8]

The d-dimensional Ising model is based on a *lattice* $\Lambda \subset \mathbb{R}^d$ which may have various shapes (regular or not). We stress that here the term *lattice* is **not** used in the Group Theoretical sense. In the general case Λ is a mere *graph* that we think embedded in $\mathbb{R}^d$ for convenience (to "develop physical intuition"). When *regular,* Λ is a lattice in the sense of Crystallography. Listing only the *regular* lattices, in $d = 2$ we have three different lattices: the triangular, the square, and the hexagonal one (also called the *honeycomb* lattice). In Fig. 4.3 we draw small portions of the hexagonal and square lattices in $\mathbb{R}^2$. Informally we think of a regular lattice $\Lambda \subset \mathbb{R}^d$ as a "discretized version" of its ambient manifold $\mathbb{R}^d$ where the distances between nearby points are thought to be of order ϵ. The continuous space $\mathbb{R}^d$ is *formally* recovered from the regular lattice $\Lambda \subset \mathbb{R}^d$ in the limit $\epsilon \to 0$.

While there are discrete statistical systems where the detailed form of the lattice is crucial, for a ferromagnetic model the physics becomes essentially independent of these details in the thermodynamic limit, and we do not lose much by focusing on the simplest *cubic lattice in d dimensions:* we consider a cubic box of size L in $\mathbb{Z}^d$

$$\Lambda = \{\boldsymbol{n} \equiv (n_1, \ldots, n_d) \colon 1 \le n_i \le L\} \subset \mathbb{Z}^d \tag{4.24}$$

---

[8] A basic resource about the Ising model is the book [5] which analyzes it from different viewpoints.

Points in Λ are called *sites* (or *nodes* or *vertices*). An *edge* (or *link*) of Λ is a segment connecting two nearby sites $\boldsymbol{n}$ and $\boldsymbol{n} + \boldsymbol{e}_i$, where $\boldsymbol{e}_i$ is the unit vector in the i-th direction

$$\boldsymbol{e}_i = (0, \ldots, 0, \overset{i}{1}, 0, \ldots, 0) \in \mathbb{Z}^d, \tag{4.25}$$

and both $\boldsymbol{n}$ and $\boldsymbol{n} + \boldsymbol{e}_i$ are in Λ. When convenient, we think of the edges as oriented from $\boldsymbol{n}$ to $\boldsymbol{n} + \boldsymbol{e}_i$ and see them as arrows.[9] If l is an oriented link, we write $h(l)$ and $t(l)$ for, respectively, the *head* vertex and *tail* vertex of l seen as an arrow. A square with the (ordered) vertices

$$\boldsymbol{n}, \ \boldsymbol{n} + \boldsymbol{e}_i, \ \boldsymbol{n} + \boldsymbol{e}_i + \boldsymbol{e}_j, \ \boldsymbol{n} + \boldsymbol{e}_j, \quad i \neq j, \ i, j = 1, \ldots, d, \tag{4.26}$$

all four contained in Λ, will be called an (oriented) *plaquette* at $\boldsymbol{n}$ in the i,j directions. A *k-plaquette* at $\boldsymbol{n}$ in the directions $i_1, \ldots, i_k$ ($i_a \neq i_b$ for $a \neq b$) is a fundamental k-dimensional cube with the 2^k vertices

$$\left\{ \boldsymbol{n} + \sum_{a \in A} \boldsymbol{e}_{i_a} : A \subset \{1, \ldots, k\} \right\} \tag{4.27}$$

0-plaquettes (resp. 1-plaquettes) are sites (resp. edges). The d-plaquettes are called *(fundamental) cells*. Two nearby sites $\boldsymbol{n}, \ \boldsymbol{n}' \in \Lambda$ which are connected by an edge are called *nearest neighbors*. The symbol $\langle \boldsymbol{n}, \boldsymbol{n}' \rangle$ means that the sites $\boldsymbol{n}, \boldsymbol{n}'$ are nearest neighbors; we also use the symbol $\langle \boldsymbol{n}, \boldsymbol{n}' \rangle$ to label the edge with endpoints $\boldsymbol{n}, \boldsymbol{n}'$ (forgetting orientation). The set of all oriented links (resp. k-plaquettes) in Λ will be written Λ_1 (resp. Λ_k). The free Abelian group $\mathbb{Z}\Lambda_k$ over the *oriented* k-plaquettes is the group of *k-chains* on Λ. The *boundary* $\partial\eta$ of a k-chain η is defined in the obvious way and $\partial^2 = 0$. Similar definitions apply to lattices of other shapes in the obvious manner.

The cubic lattice Λ is *periodic* (of period L)[10] iff we have the identification of nodes (and edges connecting them)

$$\boldsymbol{n} \sim \boldsymbol{n} + L\boldsymbol{e}_i \quad \text{for all} \ i = 1, \ldots, d, \tag{4.28}$$

i.e. if Λ is a "discretization" of the d-torus $(S^1)^d$.

---

[9] We stress that the orientation of the edges is a mere book-keeping device, with no intrinsic meaning. In particular all physical quantities are independent of the lattice orientation.

[10] The periods in diverse directions may be different.

Fig. 4.4 The dual of the
2-dimensional square lattice.
The sites of the dual lattice
(which is also square) are the
centers of the plaquettes of
the original lattice (the blue
dots in the figure). The edges
of the dual lattice are dashed
to distinguish them from the
ones of the original square
lattice which are drawn solid

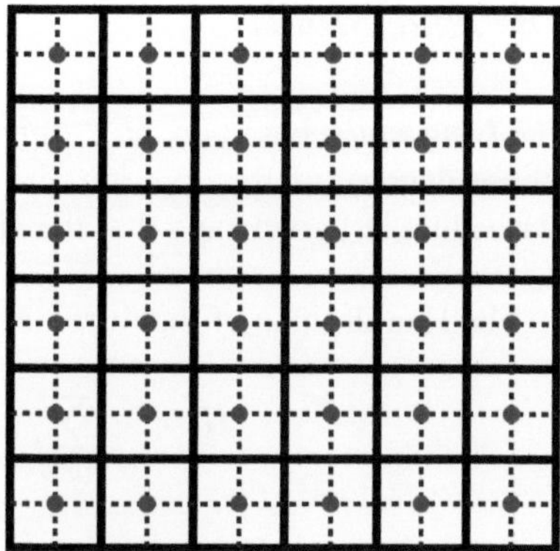

Fig. 4.5 The dual of the
2-dimensional hexagonal
lattice is the triangular lattice.
The centers of the original
plaquettes are the red dots,
and the edges of the dual
lattices are dashed

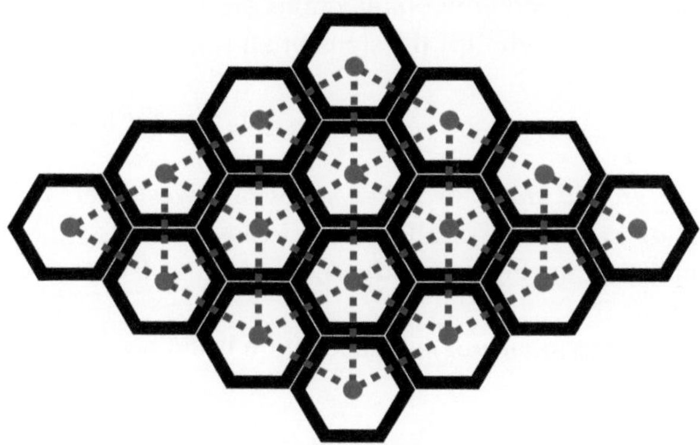

Dual Lattice If Λ is a periodic cubic lattice, its *dual lattice* Λ^* is the d-dimensional
lattice with $\Lambda_k^* = \Lambda_{d-k}$. Its sites are the central points of the fundamental cells i.e.

$$\boldsymbol{n}^* = \boldsymbol{n} + \frac{1}{2} \sum_{i=1}^{d} \boldsymbol{e}_i \tag{4.29}$$

and its edges are the segments connecting nearby dual sites: the edge between $\boldsymbol{n}^*$
and $\boldsymbol{n}^* + \boldsymbol{e}_i$ in Λ^* is dual (up to orientation) to a $(d-1)$-plaquette at $\boldsymbol{n} + \boldsymbol{e}_i$ in the
original lattice extended in the complementary directions $1, \ldots, \hat{i}, \ldots d$ (i.e. in the
directions normal to $\boldsymbol{e}_i$). See Fig. 4.4 for the $d = 2$ case. Clearly $(\Lambda^*)^* = \Lambda$. The
definition of dual lattice extends *mutatis mutandis* to lattices of other shapes: see
e.g. Fig. 4.5 for the dual of the 2-dimensional hexagonal lattice.

Remark 4.2 Thinking of the elements of Λ_k as singular k-chains, the map $\Lambda \to$
Λ^* is Poincaré duality. This observation is crucial for lattice gauge theories
(Sect. 4.8).

The Ising Model

The *Ising model* based on the d-dimensional lattice Λ is the discrete classical system whose degrees of freedom are one spin[11] σ_a at each site $a \in \Lambda$ which takes the two values ± 1: one says that the spin at a is *up* (resp. *down*) iff $\sigma_a = +1$ (resp. $\sigma_a = -1$).

The Hamiltonian of the ferromagnetic Ising model is

$$H = -J \sum_{\langle a,b \rangle} \sigma_a \, \sigma_b \equiv -J \sum_{l \in \Lambda_1} \sigma_{h(l)} \, \sigma_{t(l)}, \tag{4.30}$$

with a *positive* coupling constant $J > 0$. We stress that the sum is only over pairs of nearest neighbors: the interactions are *short-ranged*. The Hamiltonian H has a $\mathbb{Z}_2$ *symmetry* which flips the signs of all spins

$$\sigma_a \to -\sigma_a \qquad \text{for all } a \in \Lambda. \tag{4.31}$$

The *ground states* of H ($\equiv$ states of lowest energy) have all spins aligned: either all $\sigma_a = +1$ (up) or all $\sigma_a = -1$ (down). Thus we have *two* distinct ground states related by the $\mathbb{Z}_2$ symmetry (4.31). At $T = 0$ the system is in one of the two ground states.

The (spontaneous) magnetization is defined as

$$M = \lim_{\Lambda \uparrow \mathbb{Z}^d} \frac{1}{|\Lambda|} \sum_{a \in \Lambda} \langle \sigma_a \rangle_\Lambda \tag{4.32}$$

where $\langle \cdots \rangle_\Lambda$ is the expectation value with respect to a thermal probability distribution defined in the finite lattice Λ, and the limit is the thermodynamic limit to be defined more precisely momentarily. M is $\mathbb{Z}_2$ *odd*. A phase (state) is called *ordered* (resp. *disordered*) if $M \neq 0$ (resp. $M = 0$). The rationale of this terminology is clear: in a disordered state the spins point up and down in a "random" fashion, whereas in an ordered state the majority of the spins are aligned in an "orderly fashion" in a definite direction. The magnetization M measures "how much" the state is ordered, and is called the (magnetic) *order parameter. Order parameters* are the main tool to study ordered phases of a statistical system. The magnetization M is our first example of order parameter, a central and general notion in modern Statistical Mechanics to be discussed at length in this book.

---

[11] Properly speaking the spins take the half-integral values $\pm\frac{1}{2}$; however it is standard practice to change their normalization to ± 1.

As $T \to \infty$, i.e. $\beta \to 0$, the canonical distribution in a finite box Λ

$$\frac{1}{Z} \sum_{\{\sigma_m = \pm 1\}} \exp\left[\beta J \sum_{\langle n,n'\rangle} \sigma_n \, \sigma_{n'} \right] \tag{4.33}$$

goes into the product of independent probability distributions for the spin at each site, so there is no correlation between the several spins which then point in independent random directions; the system is *fully disordered* and therefore

$$M\big|_{T\to\infty} = 0. \tag{4.34}$$

Since at $T = 0$ we have $M = \pm 1$, while at $T = \infty$ the magnetization vanishes, $M = 0$, heuristically we expect that at some critical temperature T_c there is a phase transition between an ordered phase at low temperature $T < T_c$ and a disordered one at high temperature $T > T_c$. This picture will be made precise below. Note that the model depends only on the combination βJ, so we may set $J = 1$ with no loss.

We may consider a more general model depending on an extra parameter by switching on an external magnetic field B. This adds to the Hamiltonian an interaction linear in the spins

$$H = -J \sum_{\langle a,b\rangle} \sigma_a \, \sigma_b - B \sum_a \sigma_a. \tag{4.35}$$

For $B \neq 0$ there is a single ground state where all spins point up for $B > 0$ or down for $B < 0$. Then, heuristically, we do not expect any phase transition when $B \neq 0$; this fact will be proven below (the Yang-Lee theorem).

Remark 4.3 When $J < 0$ the Ising model is *anti-ferromagnetic*. In this situation energy considerations favor *misalignment* of spins: the energy is minimal when nearby spins point in *opposite* directions. Thus the anti-ferromagnetic model resists against magnetization in a weak external magnetic field. For a cubic lattice, we pass from a ferromagnetic model to an anti-ferromagnetic one by the spin redefinition

$$\sigma_n \rightsquigarrow (-1)^{|n|} \sigma_n, \qquad \text{where } |n| = n_1 + n_2 + \cdots + n_d, \tag{4.36}$$

hence we have still two ground states one with $\sigma_n = (-1)^{|n|}$ and one with $\sigma_n = -(-1)^{|n|}$. However in the anti-ferromagnetic case the choice of the shape of the lattice has dramatic effects. For instance, consider a $d = 2$ triangular lattice; let σ_1, σ_2 and σ_3 be the spins at the vertices of a fundamental triangle in the lattice (a *plaquette*). If σ_1 points in the opposite direction of σ_2, and σ_2 in the opposite direction of σ_3, the two nearest neighbor spins σ_1 and σ_3 will be aligned. One says that this plaquette (and the full system) is *frustrated* since the system's "desire" to misalign all its neighboring spin pairs is thwarted by the lattice geometry. Frustrated systems have a complicate geography of ground states. We shall say some more words about them in Chap. 7.

Boundary Conditions. Pure and Mixed Phases
To fully define the system in the finite lattice *box* $\Lambda \subset \mathbb{Z}^d$ of size L we need to specify the *boundary conditions* on the walls $\partial \Lambda$ of our box

$$\partial \Lambda \equiv \left\{ \boldsymbol{n} \equiv (n_1, \ldots, n_d) \in \Lambda \subset \mathbb{Z}^d : n_i \in \{1, L\} \text{ for some } i \right\}. \tag{4.37}$$

For computational purposes the most convenient condition is the *periodic* one (the one we used for the free systems in Chap. 3). However, to understand the physics of the ordered/disordered phases and to prove theorems, other boundary conditions are more natural and useful.

We define two boundary conditions, denoted respectively as $+$ and $-$, by fixing all spins on the boundary $\partial \Lambda$ of the box to be, respectively, *up* or *down*. Clearly the two boundary conditions $\pm$ break *explicitly* the $\mathbb{Z}_2$ symmetry by favoring spins up or, respectively, down. In particular, at $T = 0$ the $\pm$ boundary conditions select one out of the two ground states with magnetization

$$M_+\big|_{T=0} = +1, \quad \text{resp.} \quad M_-\big|_{T=0} = -1. \tag{4.38}$$

A correlation function in the finite box Λ with $\pm$ boundary condition has the form

$$\langle \sigma_{a_1} \cdots \sigma_{a_k} \rangle_\pm = \frac{1}{Z_\pm} \sum_{\{\sigma_n = \pm 1\}} \sigma_{a_1} \cdots \sigma_{a_k} \exp\left[\beta J \sum_{<b,c>} \sigma_b \sigma_c \right]\Bigg|_{\sigma_{m \in \partial \Lambda} = \pm 1} \tag{4.39}$$

where the subscript $\pm$ labels the above two special boundary conditions in the finite box. The (canonical) $\pm$ partition functions $Z_\pm$ are

$$Z_\pm = \sum_{\{\sigma_n = \pm 1\}} \exp\left[\beta J \sum_{<a,b>} \sigma_a \sigma_b \right]\Bigg|_{\sigma_{m \in \partial \Lambda} = \pm 1}, \tag{4.40}$$

The $\mathbb{Z}_2$ symmetry (4.31) interchanges the two boundary conditions $\pm$, so

$$Z_+ = Z_- = Z, \tag{4.41}$$

while

$$\langle \sigma_{a_1} \cdots \sigma_{a_k} \rangle_- = (-1)^k \langle \sigma_{a_1} \cdots \sigma_{a_k} \rangle_+. \tag{4.42}$$

We may think of two different possibilities:

TL1 the thermodynamic limit exists *and is unique*, that is, independent of the details of how we take the limit, in particular equal for all choices of boundary conditions. In this case the two probability distributions $\langle\cdots\rangle_+$ and $\langle\cdots\rangle_-$ agree in the limit, and then

$$\langle\sigma_a\rangle_+ = \langle\sigma_a\rangle_- = -\langle\sigma_a\rangle_+ = 0, \quad \text{for all } a \in \mathbb{Z}^d, \tag{4.43}$$

and the phase is *disordered.* This is the case at $T = \infty$. We note that uniqueness also entails *translational invariance* in the thermodynamic limit:

$$\langle\sigma_{a_1+b}\,\sigma_{a_2+b}\cdots\sigma_{a_s+b}\rangle = \langle\sigma_{a_1}\sigma_{a_2}\cdots\sigma_{a_s}\rangle \quad \text{for all } b \in \mathbb{Z}^d. \tag{4.44}$$

TL2 conversely: when the phase is *ordered* the thermodynamic limit *cannot be unique*. We have two *distinct* phases with probability distributions $\langle\cdots\rangle_+$ and $\langle\cdots\rangle_-$ where most of the spins point *up* or, respectively, *down:*

$$\langle\sigma_a\rangle_+ \equiv M_+ > 0, \qquad \langle\sigma_a\rangle_- \equiv M_- = -M_+ < 0, \tag{4.45}$$

It is intuitive that the distributions $\langle\cdots\rangle_\pm$ describe *pure* phases: this is obviously true at $T = 0$. In addition to the pure phases we have the *mixed* ones (just as in the ice-water mixture) whose expectation values have the form

$$\langle\cdots\rangle_\alpha = \alpha\,\langle\cdots\rangle_+ + (1-\alpha)\,\langle\cdots\rangle_-, \qquad 0 \le \alpha \le 1, \tag{4.46}$$

(cf. Eq. (4.8)). The mixed phase probability distributions are *convex combinations* of the pure phase ones. The pure phase distributions (hence their convex combinations) are translational invariant in the sense (4.44).

Cluster Property
To vindicate our intuition that the states $\langle\cdots\rangle_\pm$ are pure, we need a math definition of *pure phase*. There are several characterization of the *pure phases* (*states*). A convenient one is given by the *cluster property*:

Fact 4.2 (Cluster Property) *When $\langle\cdots\rangle$ is (the expectation value of) a* pure *phase the* cluster property *holds: for all operator insertion* $\sigma_{a_1}\cdots\sigma_{a_s}\,\sigma_{a_{s+1}+b}\cdots\sigma_{a_k+b}$

$$\lim_{|b|\to\infty} \langle\sigma_{a_1}\cdots\sigma_{a_s}\,\sigma_{a_{s+1}+b}\cdots\sigma_{a_k+b}\rangle = \langle\sigma_{a_1}\cdots\sigma_{a_s}\rangle\langle\sigma_{a_{s+1}}\cdots\sigma_{a_k}\rangle. \tag{4.47}$$

Conversely, if the phase is not pure the cluster property fails for some operator insertion $\sigma_{a_1}\cdots\sigma_{a_s}\,\sigma_{a_{s+1}+b}\cdots\sigma_{a_k+b}.$

Let us check that the cluster property fails in a mixed phase. Consider, for instance, the Ising model below the Curie temperature (where by definition

$\langle\cdots\rangle_+ \neq \langle\cdots\rangle_-$) in the $\mathbb{Z}_2$-symmetric mixed state

$$\langle\cdots\rangle = \frac{1}{2}\Big(\langle\cdots\rangle_+ + \langle\cdots\rangle_-\Big). \tag{4.48}$$

We have

$$\langle\sigma_a\sigma_b\rangle - \langle\sigma_a\rangle\langle\sigma_b\rangle = \langle\sigma_a\sigma_b\rangle_+. \tag{4.49}$$

As $|a - b| \to \infty$ the RHS goes to $\langle\sigma_a\rangle_+^2 = M_+^2 > 0$. We conclude that the cluster property does **not** hold in this mixed phase.

The cluster property yields a procedure to decompose a mixed phase into a convex combination of pure states. E.g. for the Ising model below the Curie temperature we reconstruct the pure phase distributions $\langle\cdots\rangle_\pm$ out of the $\mathbb{Z}_2$-symmetric mixed one $\langle\cdots\rangle$ by the cluster limit formula

$$\langle\sigma_{a_1}\cdots\sigma_{a_s}\rangle_+ = (-1)^s\,\langle\sigma_{a_1}\cdots\sigma_{a_s}\rangle_- =$$
$$= \Big(\lim_{|b|\to\infty}\langle\sigma_{a_1}\cdots\sigma_{a_s}\,\sigma_{a_1+b}\cdots\sigma_{a_s+b}\rangle\Big)^{1/2} \geq 0. \tag{4.50}$$

Existence of the Thermodynamic Limit

To make the previous scenario "rigorous", the first step is to establish that the *thermodynamic limit exists* for the Ising model, while the thermodynamic potentials have the appropriate "convexity" properties. All these facts can be established as rigorous math theorems, see Chaps. 3 and 5 of [2] for a list of precise math statements in different set-ups. We limit ourselves to a simple—but fully rigorous—argument.

By Eq. (4.50) it suffices to show the existence of the limit for boundary conditions leading to the $\langle\cdots\rangle$ symmetric state, i.e. for boundary conditions which preserve the $\mathbb{Z}_2$-symmetry. The simplest such boundary condition is the *free* condition, that is, in Eq. (4.39) we do not impose any condition on the spins in the boundary of the finite box but we sum over their two values ±1 just as for the spins in the bulk of Λ.

For the next argument it is convenient to consider a Hamiltonian more general than Eq. (4.30):

$$H = -\sum_{l\in\Lambda_1} J_l\,\sigma_{t(l)}\,\sigma_{h(l)} - \sum_{a\in\Lambda_0} B_a\,\sigma_a \tag{4.51}$$

where now the coupling constants J_l (resp. magnetic fields B_a) depend on the edge $l \in \Lambda_1$ (resp. site $a \in \Lambda_0$). We say that the generalized model (4.51) is *ferromagnetic* iff $J_l \geq 0$ for all links l and $B_a \geq 0$ for all sites a. It is convenient to think of the free boundary conditions on the lattice box $\Lambda \subset \mathbb{Z}^d$ as a generalized

Ising model (4.51) on the infinite lattice $\mathbb{Z}^d$ where we set to zero the couplings of all links which do not belong to Λ:

$$l \notin \Lambda_1 \quad \Rightarrow \quad J_l = 0. \tag{4.52}$$

Clearly the correlation functions of operators $\sigma_{a_1} \cdots \sigma_{a_s}$ with $a_i \in \Lambda$ are not affected by this trivial extension of the lattice from Λ to $\mathbb{Z}^d$. The canonical distribution for Λ finite is a *finite* polynomial in the spin variables $\{\sigma_a\}_{a \in \Lambda}$ of the form

$$
\begin{aligned}
\frac{e^{-\beta H}}{Z} &= \frac{1}{Z} \prod_{l \in \Lambda_1} \Big(\cosh(\beta J_l) + \sinh(\beta J_l)\, \sigma_{t(l)} \sigma_{h(l)} \Big) \times \\
&\qquad \times \prod_{a \in \Lambda_0} \Big(\cosh(\beta B_a) + \sinh(\beta B_a)\, \sigma_a \Big) = \\
&= \sum_{\ell \geq 0} \sum_{(a_1, \ldots, a_\ell) \in \Lambda_0^\ell} C_{a_1, \ldots, a_\ell}\, \sigma_{a_1} \sigma_{a_2} \cdots \sigma_{a_\ell},
\end{aligned}
\tag{4.53}
$$

where the coefficients $C_{a_1, \ldots, a_\ell}$ are *non-negative* when the model is ferromagnetic.

Ferromagnetic Inequalities We state a crucial property of the classical ferromagnetic systems called the *Griffiths inequalities*. Griffiths inequalities hold for a much more general class of ferromagnetic systems: for general statements see e.g. [2].

Theorem 4.3 (Griffiths Inequalities) *For the system* (4.51) *in a **finite** box* $\Lambda \subset \mathbb{Z}^d$ *with **free** boundary conditions, and **ferromagnetic** couplings* $J_l \geq 0$ *and* $B_a \geq 0$, *we have for arbitrary spin insertions:*

GI1   *first Griffiths inequality*

$$\langle \sigma_{a_1} \cdots \sigma_{a_s} \rangle \geq 0 \tag{4.54}$$

GI2   *second Griffiths inequality*

$$\langle \sigma_{a_1} \cdots \sigma_{a_r}\, \sigma_{b_1} \cdots \sigma_{b_s} \rangle - \langle \sigma_{a_1} \cdots \sigma_{a_r} \rangle \langle \sigma_{b_1} \cdots \sigma_{b_s} \rangle \geq 0. \tag{4.55}$$

Proof

GI1   By Eq. (4.53) the correlations functions

$$\langle \sigma_{a_1} \cdots \sigma_{a_s} \rangle \equiv \frac{1}{Z} \sum_{\{\sigma = \pm 1\}} e^{-\beta H} \sigma_{a_1} \cdots \sigma_{a_s} \tag{4.56}$$

are finite sums of terms of the form

$$
C_{b_1,\ldots,b_\ell} \sum_{\{\sigma=\pm 1\}} \sigma_{b_1} \cdots \sigma_{b_\ell}\, \sigma_{a_1} \cdots \sigma_{a_s}, \tag{4.57}
$$

with non-negative coefficients $C_{b_1,\ldots,b_\ell}$. The sum over the values ± 1 of the spins gives 1 iff in the list of sites $\{b_1, \ldots, b_\ell, a_1, \ldots, a_s\}$ each site appears with an even multiplicity, and zero otherwise. Thus (4.56) is non-negative.

GI2   We consider two identical copies of our generalized ferromagnetic system with spins $\sigma_a = \pm 1$ and $\tau_a = \pm 1$ and Hamiltonian

$$
\begin{aligned}
H(\{\sigma_a\}) &+ H(\{\tau_a\}) = \\[2mm]
&= -\sum_l J_l \left(\sigma_{h(l)}\, \sigma_{t(l)} + \tau_{h(l)}\, \tau_{t(l)} \right) - \sum_a B_a \left(\sigma_a + \tau_a \right) = \\[2mm]
&= -\frac{1}{2} \sum_l J_l \left(\xi_{h(l)}\, \xi_{t(l)} + \eta_{h(l)}\, \eta_{t(l)} \right) - \sum_a B_a\, \xi_a
\end{aligned} \tag{4.58}
$$

where

$$
\xi_a = \sigma_a + \tau_a, \qquad \eta_a = \sigma_a - \tau_a. \tag{4.59}
$$

ξ_a and η_a are random variables which take the values $0, \pm 2$; the two values ± 2 have equal probability. ξ_a and η_a are *not* independent: they are related by the identity (not summed over a!)

$$
\xi_a\, \eta_a = \left(\sigma_a + \tau_a \right)\left(\sigma_a - \tau_a \right) = \sigma_a^2 - \tau_a^2 \equiv 0. \tag{4.60}
$$

In this double system we consider the expectation

$$
\left\langle \xi_{a_1} \cdots \xi_{a_k} \eta_{b_1} \cdots \eta_{b_j} \right\rangle, \tag{4.61}
$$

it is a finite sum of terms of the form

$$
C_{\{a\}\{b\}\{c\}\{d\}} \sum_{\substack{\{\eta_n = \pm 2\},\\ \{\xi_n = \pm 2\}}}^{*} \xi_{a_1} \cdots \xi_{a_k}\, \eta_{b_1} \cdots \eta_{b_j}\, \xi_{c_1} \cdots \xi_{c_r}\, \eta_{d_1} \cdots \eta_{d_s}; \tag{4.62}
$$

where the symbol $\sum^{*}$ means that we sum only over configurations satisfying the constraint (4.60) and we omitted the terms with $\eta_n = 0$ and $\xi_n = 0$ since they vanish. The coefficient $C_{\{a\}\{b\}\{c\}\{d\}}$ is positive. By of Eq. (4.60) this term vanishes unless

$$
\{a_1, \ldots a_k, c_1, \ldots, c_r,\} \cap \{b_1, \ldots, b_j, d_1, \ldots, d_t\} = \varnothing. \tag{4.63}
$$

In order for the sum (4.62) to be non zero we need, in addition, that in each of the two disjoint lists (4.63) each site appears with an even multiplicity; when this is the case the sum (4.62) is

$$2^{(k+r+s+t)} \, C_{\{a\}\{b\}\{c\}\{d\}} > 0.$$
(4.64)

Thus the correlation (4.61) is the sum of non-negative quantities, hence non-negative:

$$\langle \xi_{a_1} \cdots \xi_{a_k} \eta_{b_1} \cdots \eta_{b_j} \rangle \geq 0 \quad \text{for all } \; a_1, \ldots, a_k, b_1, \cdots, b_j.$$
(4.65)

Now

$$\sigma_{a_1} \cdots \sigma_{a_j} - \tau_{a_1} \cdots \tau_{a_j} \equiv$$

$$\equiv \frac{1}{2^j} \left[(\xi_{a_1} + \eta_{a_1}) \cdots (\xi_{a_j} + \eta_{a_1}) - (\xi_{a_1} - \eta_{a_1}) \cdots (\xi_{a_j} - \eta_{a_j}) \right] =$$

$$= \frac{1}{2^j} \sum_{k=0}^{[(j-1)/2]} \sum_{\pi \in \mathfrak{S}_j} C^j_{k,\pi} \, \eta_{a_{\pi(1)}} \cdots \eta_{a_{\pi(2k+1)}} \xi_{a_{\pi(2k+2)}} \cdots \xi_{a_{\pi(j)}}$$
(4.66)

where the coefficients $C^j_{k,\pi}$ are *non-negative*. Now

$$\langle \sigma_{a_1} \cdots \sigma_{a_k} \sigma_{b_1} \cdots \sigma_{b_j} \rangle - \langle \sigma_{a_1} \cdots \sigma_{a_k} \rangle \langle \sigma_{b_1} \cdots \sigma_{b_j} \rangle =$$

$$= \langle \sigma_{a_1} \cdots \sigma_{a_k} \left(\sigma_{b_1} \cdots \sigma_{b_j} - \tau_{b_1} \cdots \tau_{b_j} \right) \rangle =$$
(4.67)

$$= \frac{1}{2^k} \sum_{k,\pi} C^j_{k,\pi} \left\langle \left(\xi_{a_1} + \eta_{a_1} \right) \cdots \left(\xi_{a_k} + \eta_{a_k} \right) \eta_{b_{\pi(1)}} \cdots \xi_{b_{\pi(j)}} \right\rangle \geq 0.$$

$$\square$$

Corollary 4.1 *Under the conditions of the **Theorem** the correlation functions*

$$\langle \sigma_{a_1} \cdots \sigma_{a_k} \rangle$$
(4.68)

increase when we increase any one coupling J_l or magnetic field B_a.

Remark 4.4 Before going to the math proof, we clarify the physical meaning of this **Corollary**. A positive coupling $J_l > 0$ favors the two nearby spins $\sigma_{h(l)}, \sigma_{t(l)}$ to be aligned, while the condition $B_a \geq 0$ favors the up alignment of σ_a. Making the couplings larger only forces the spins to be more aligned, and making the magnetic field larger makes the positive direction even more favorite. Therefore the (already positive) correlations of the spins may only become even more positive.

Proof One has

$$\frac{1}{\beta}\frac{\partial}{\partial J_l}\langle\sigma_{a_1}\cdots\sigma_{a_k}\rangle =$$

$$= \langle\sigma_{a_1}\cdots\sigma_{a_k}\sigma_{h(l)}\sigma_{t(l)}\rangle - \langle\sigma_{a_1}\cdots\sigma_{a_k}\rangle\langle\sigma_{h(l)}\sigma_{t(l)}\rangle \geq 0 \tag{4.69}$$

and

$$\frac{1}{\beta}\frac{\partial}{\partial B_b}\langle\sigma_{a_1}\cdots\sigma_{a_k}\rangle = \langle\sigma_{a_1}\cdots\sigma_{a_k}\sigma_b\rangle - \langle\sigma_{a_1}\cdots\sigma_{a_k}\rangle\langle\sigma_b\rangle \geq 0. \tag{4.70}$$

$\square$

Existence Now we are ready to prove the existence of the thermodynamic limit with free boundary conditions. What we have to show is that for all $k \in \mathbb{N}$ and all k-tuples $\{a_1, \ldots, a_k\} \in (\mathbb{Z}^d)^k$ the thermodynamic limit of the correlation $\langle\sigma_{a_1}\cdots\sigma_{a_k}\rangle$ exists. We consider an increasing sequence of finite boxes in $\mathbb{Z}^d$

$$\Lambda_{(1)} \subset \Lambda_{(2)} \subset \cdots \Lambda_{(n)} \subset \cdots \tag{4.71}$$

such that $\Lambda_{(n)} \uparrow \mathbb{Z}^d$ as $n \to \infty$ while $\Lambda_{(1)}$ is big enough to contain the insertion sites $\{a_1, \ldots, a_k\}$. Then we compute the correlation function $\langle\sigma_{a_1}\cdots\sigma_{a_k}\rangle_{(n)}$ for the generalized Ising model with Hamiltonian (4.51) defined in the n-th box $\Lambda_{(n)}$ with free boundary conditions.

Theorem 4.4

(1) For all operators insertions $\sigma_{a_1}\cdots\sigma_{a_k}$ the limit

$$\lim_{n\to\infty}\langle\sigma_{a_1}\cdots\sigma_{a_k}\rangle_{(n)} \equiv \langle\sigma_{a_1}\cdots\sigma_{a_k}\rangle \tag{4.72}$$

exists *and is translational invariant*

$$\langle\sigma_{a_1}\cdots\sigma_{a_k}\rangle = \langle\sigma_{a_1+b}\cdots\sigma_{a_k+b}\rangle, \quad \text{for all } b \in \mathbb{Z}^d. \tag{4.73}$$

(2) The limit

$$-\mathcal{F} = \lim_{\Lambda\to\mathbb{Z}^d}\frac{1}{|\Lambda|}\log Z_\Lambda, \tag{4.74}$$

where

$$Z_\Lambda = \sum_{\substack{a\in\Lambda:\\ \sigma_a=\pm 1}} e^{-\beta H(\{\sigma\})}, \tag{4.75}$$

exists *and is a **convex** function of* β

$$\frac{\partial^2}{\partial \beta^2}(-\mathcal{F}) \geq 0. \tag{4.76}$$

Proof

(1) The ferromagnetic Ising model in the n-th box Λ_n, with free boundary condition, is identified with the model (4.51) in the full infinite lattice $\mathbb{Z}^d$ where

$$J_l = \begin{cases} J > 0 & \text{if } l \in (\Lambda_n)_1 \\ 0 & \text{if } l \notin (\Lambda_n)_1. \end{cases} \tag{4.77}$$

We obtain the model defined in the next larger box $\Lambda_{(n+1)} \supset \Lambda_{(n)}$ by increasing the couplings J_l on the links $l \in \Lambda_{(n+1)} \setminus \Lambda_{(n)}$ from zero to $J > 0$. By Corollary 4.1 the correlation cannot decrease

$$\langle \sigma_{a_1} \cdots \sigma_{a_k} \rangle_{(n+1)} \geq \langle \sigma_{a_1} \cdots \sigma_{a_k} \rangle_{(n)}. \tag{4.78}$$

On the other hand

$$\langle \sigma_{a_1} \cdots \sigma_{a_k} \rangle_{(n)} \leq 1 \tag{4.79}$$

for all n. Since the sequence $\langle \sigma_{a_1} \cdots \sigma_{a_k} \rangle_{(n)}$ is monotonic non-decreasing and bounded above, it is convergent. Replacing the boxes $\Lambda_{(n)}$ with their translations by b we conclude that the correlation functions are translational invariant.

(2) Defining the canonical partition function Z_Λ as in Eq. (4.75)

$$\lim_{\Lambda \uparrow \infty} \frac{\partial}{\partial \beta} \frac{1}{|\Lambda|} \log Z_\Lambda = J \sum_{i=1}^{d} \langle \sigma_0 \sigma_{e_i} \rangle + B \langle \sigma_0 \rangle, \tag{4.80}$$

while at $\beta = 0$ we have $\log Z_\Lambda = |\Lambda| \log 2$ i.e. $-\mathcal{F}(\beta = 0) = \log 2$. Then

$$-\mathcal{F}(\beta) = \int_0^\beta \left[J d \langle \sigma_0 \sigma_{e_1} \rangle + B \langle \sigma_0 \rangle \right] d\beta' + \log 2 \tag{4.81}$$

and

$$\frac{\partial^2}{\partial \beta^2}(-\mathcal{F}) = J^2 d \sum_i \sum_a \left(\langle \sigma_0 \sigma_{e_1} \sigma_a \sigma_{a+e_i} \rangle - \langle \sigma_0 \sigma_{e_1} \rangle \langle \sigma_a \sigma_{a+e_i} \rangle \right) +$$

$$+ BJ \sum_i \sum_a \left(\langle \sigma_0 \sigma_a \sigma_{a+e_i} \rangle - \langle \sigma_0 \rangle \langle \sigma_a \sigma_{a+e_i} \rangle \right) + \qquad (4.82)$$

$$+ B^2 \sum_a \left(\langle \sigma_0 \sigma_a \rangle - \langle \sigma_0 \rangle \langle \sigma_a \rangle \right) \geq 0.$$

where we used Griffith's inequality **G2**.

$$\square$$

Remark 4.5 From the proof it is clear that the limit exists for *all shapes* of the finite boxes $\Lambda_{(n)}$ as long as $\Lambda_{(n)} \subset \Lambda_{(n+1)}$ and $\bigcup_n \Lambda_{(n)} = \mathbb{Z}^d$. In particular the thermodynamic limit is independent of the way we send the volume to infinity. (But it depends on the boundary condition on the spins).

No Spontaneous Magnetization at High Temperature
At infinite temperature, $\beta = 0$, the Ising model is totally disordered, and the spontaneous magnetization vanishes. One can show the much stronger result that there exists a finite temperature $\beta_0 > 0$, such that for all $\beta < \beta_0$ (i.e. $T > T_0$) there is no phase transition.

Fact 4.5 *Consider the ferromagnetic Ising model in presence of a magnetic field $B \in \mathbb{R}$ ($B = 0$ is not excluded!). There is a temperature T_0 such that for $T > T_0$ the free energy is an analytic function of $B \in \mathbb{R}$.*

In particular: for $T > T_0$ and $B = 0$ there is no spontaneous magnetization. We omit the proof; see §.5.2 of [2] and references therein. The idea of the proof is to check that the high temperature expansion in powers of β of the partition function has a finite radius of convergence around $\beta = 0$. The reader is invited to work out the details by herself.

4.3   The Yang-Lee Theorem

We apply the Yang-Lee theory to the d-dimensional Ising model in a magnetic field. The same argument covers many other systems, both continuous and discrete (cf. [2]). We start with a **Lemma** about the zeros of certain polynomials.

Lemma 4.1 *Let Δ_n be the set $\{1, 2, \ldots, n\}$, and let $A_{ij} = A_{ji}$ be a symmetric array of real numbers, where $i, j \in \Delta_n$ and $-1 \leq A_{ij} \leq 1$ for all i, j. The* *polynomial*

$$P(z) \stackrel{\text{def}}{=} \sum_{S \subset \Delta_n} z^{|S|} \prod_{i \in S} \prod_{j \in \Delta_n \setminus S} A_{ij}, \tag{4.83}$$

where the sum is over all subsets $S \subset \Delta_n$ (including $\varnothing \subset \Delta_n$ and Δ_n), is a real monic polynomial of degree n with the symmetry $P(z) = z^n P(1/z)$. All the zeros of $P(z)$ lay on the unit circle $|z| = 1$.

The proof is by induction on the degree n. We omit the details which can be found in §.5.1 of [2]. The reader may check the statement for polynomials of small degree.

We apply Lemma 4.1 to prove that the Ising model has no phase transition when $B \neq 0$. Physically this is intuitive: when $B \neq 0$ there is a unique ground state with all spins pointing up (or down) and the thermodynamical potentials should be smooth. At $B = 0$ we may or may not have a phase transition, depending of whether the model has spontaneous magnetization for the given value of βJ or not. For $T > T_0$ we know that there is no transition at $B = 0$ (cf. Fact 4.5), but we have still to understand what happens at low positive temperatures. When spontaneous magnetization is present, $\langle \sigma_a \rangle_+ > 0$ at $B = 0$ and the thermodynamical potential $\mathcal{F}(B)$ is non-analytic at $B = 0$; indeed

$$0 \neq 2\langle \sigma_a \rangle_+ = \langle \sigma_a \rangle_+ - \langle \sigma_a \rangle_- = \lim_{B \to 0^+} \langle \sigma_a \rangle - \lim_{B \to 0^-} \langle \sigma_a \rangle =$$

$$= -\frac{1}{\beta} \left(\frac{\partial \mathcal{F}}{\partial B} \bigg|_{B \to 0^+} - \frac{\partial \mathcal{F}}{\partial B} \bigg|_{B \to 0^-} \right), \tag{4.84}$$

so the *first* derivative of the thermodynamical potential is discontinuous in the magnetic field at $B = 0$. In other words: *in a ferromagnetic system (below the Curie temperature T_c) the phase transition is of first order in B*. Instead as a function of the temperature T along the line $B = 0$ we expect the phase transition at the Curie temperature T_c to be (at least) of *second order*, that is, we do not expect any latent heat to be involved in the order/disorder transition of the ferromagnetic systems.

We choose the additive constant in the Hamiltonian so that the energy of the ground states is zero:

$$H = -J \sum_{\langle a,b \rangle} (\sigma_a \sigma_b - 1) - B \sum_a (\sigma_a - 1), \tag{4.85}$$

where we assume $B \geq 0$. We write

$$z = \exp[-2\beta B]. \tag{4.86}$$

Working in a finite box $\Lambda \subset \mathbb{Z}^d$ with $|\Lambda|$ sites, we define

$$A_{a,b} = A_{b,a} = \begin{cases} \exp[-2\beta J] & \text{if } \langle a, b \rangle \\ 1 & \text{otherwise.} \end{cases} \tag{4.87}$$

A configuration of the system in the box Λ is specified by the subset $S \subset \Lambda$ of sites where the spins point *down;* in the complementary set $\Lambda \setminus S$ the spins point *up.* The canonical partition function (with the energy shifted as in (4.85)) in the finite box Λ is given by

$$Z_\Lambda(z) = \sum_{S \subset \Lambda} z^{|S|} \prod_{a \in S} \prod_{b \in \Lambda \setminus S} A_{a,b} \tag{4.88}$$

which is a monic polynomial in z of degree $|\Lambda|$. The polynomial $Z_\Lambda(z)$ has precisely the form (4.83) with $0 < A_{a,b} \leq 1$ for all a, b, so that for all boxes Λ, however big, the complex zeros of $Z_\Lambda(z)$ lay on the unit circle $|z| = 1$. By the Yang-Lee theory (Theorem 4.1), a phase transition may happen only at points on the positive real axis in the z-plane which are accumulation points of zeros of the partition functions $Z_\Lambda(z)$ as $\Lambda \uparrow \mathbb{Z}^d$. Since the zeros are on the unit circle, the only possible real positive accumulation point is $z = 1$, that is, $B = 0$.

4.4   The Transfer Matrix

The *transfer matrix* is a nice formalism for (classical) discrete system with finite range interactions. Informally, we may think of it as the Statistical Mechanical counterpart of the Hamiltonian formalism in Classical and Quantum Mechanics. For definiteness we introduce it in the context of the Ising model, the generalization to other discrete systems being obvious. Some reader may find the following discussion a bit "abstract": she is advised to jump to the next section where we present the explicit construction for the one-dimensional case; she may study that illuminating example carefully, and then return to the general story discussed in this section.

Slicing the Lattice Λ

The idea of the transfer matrix formalism is to build the lattice Λ by adding "one slice after the other". For simplicity we focus on the *cubic lattice* in d dimensions. Similar considerations apply to other regular lattices. Λ is the lattice with sites

$$\Lambda_0 = \{(k_1, \ldots, k_{d-1}, y) \in \mathbb{Z}^d : 1 \leq y \leq L, \ 1 \leq k_i \leq \tilde{L}\}, \tag{4.89}$$

together with the usual nearest neighbor edges of the cubic lattice (cf. Sect. 4.2). Now we subdivide the lattice Λ into a set of $(d-1)$-dimensional *slices,* each slice

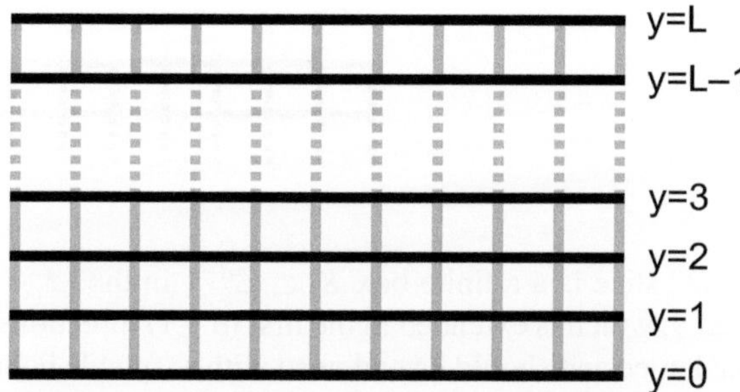

Fig. 4.6 Decomposing a two-dimensional $(L+1) \times \tilde{L}$ square lattice into $L+1$ "horizontal" slices K_y. The slice K_y at the fixed "vertical" position y consists of the sites $(k, y) \in \mathbb{R}^2$ $(k = 1, \ldots, \tilde{L})$ together with the "horizontal" edges connecting them (black edges in the figure). Two successive slices K_y and K_{y+1} are connected by the "vertical" edges (gray edges in the figure)

being a lattice in its own right. There are many distinct ways of doing this. For instance we may subdivide Λ into "horizontal" constant-y slices[12] K_y whose set of sites $(K_y)_0$ is

$$(K_y)_0 \equiv \left\{ (k_1, \ldots, k_{d-1}, y) \in \mathbb{Z}^d : 1 \le k_i \le \tilde{L}, \ i \neq d \right\} \quad y = 1, \ldots, L,$$

$$(4.90)$$

and whose edges are the "horizontal" edges of Λ connecting sites in the slice

$$(K_y)_1 \equiv \left\{ l \in \Lambda_1 : h(l) \in (K_y)_0, \ t(l) \in (K_y)_0 \right\}. \tag{4.91}$$

We see the original d-dimensional lattice Λ as the union of L $(d-1)$-dimensional lattice slices K_y $(y = 1, \ldots, L)$ with sites

$$\Lambda_0 = (K_1)_0 \cup (K_2)_0 \cup \cdots \cup (K_L)_0. \tag{4.92}$$

and edges

$$\Lambda_1 = \overbrace{(K_1)_1 \cup (K_2)_1 \cup \cdots \cup (K_L)_1}^{\text{"horizontal" edges}} \bigcup \overbrace{\left\{ \langle k, k + e_d \rangle : k, k + e_d \in \Lambda_0 \right\}}^{\text{"vertical" edges}}$$

$$(4.93)$$

See Fig. 4.6 for the two-dimensional case.

---

[12] Other choices of slice are allowed and often convenient. For instance in the two-dimensional square lattice we can consider the diagonal slices $n_1 + n_2 = $ const. which has some advantages pointed out by Baxter [6]; in the hexagonal lattice the most convenient choice of slices is the transverse brick-wall one [7].

Fig. 4.7 Adding an extra
layer of (gray) nodes at the
"time" $y + 1$ to the
2-dimensional cubic lattice of
size $L \times y$

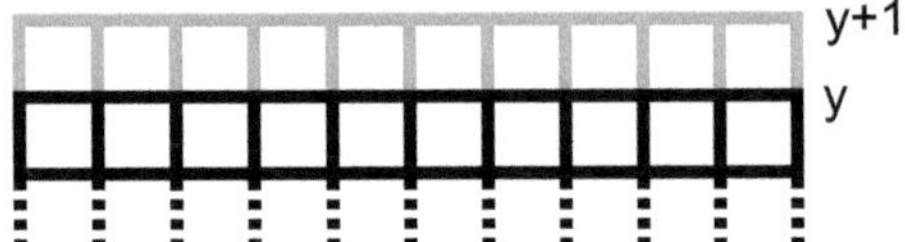

Each "horizontal" slice is a a finite box $K \subset \mathbb{Z}^{d-1}$ in the $(d-1)$-dimensional hyperplane of fixed y, which is extended in the first $(d-1)$ directions. K is a $(d-1)$-dimensional cubic lattice and should be endowed with a suitable boundary condition at its $(d-2)$-dimensional boundary ∂K (such as: free, periodic, $+$ or $-$, etc.).

We adopt the following **notation**: if A and B are any two lattices (graphs) with site and edge sets A_0, A_1 and, respectively, B_0, B_1, their *square product $A \boxtimes B$ is the lattice* with $(A \boxtimes B)_0 = A_0 \times B_0$ and $(A \boxtimes B)_1 = (A_1 \times B_0) \cup (A_0 \times B_1)$ or, in the language of chains, $\mathbb{Z}(A \boxtimes B)_k = \sum_{\ell=0}^{k} \mathbb{Z}A_\ell \otimes \mathbb{Z}B_{k-\ell}$. We denote by I_L the linear lattice over the sites $\{1, 2, \ldots, L\}$ with nearest neighbor links between sites i and $i + 1$. Then $\Lambda = K \boxtimes I_L$. By analogy with Euclidean QFT we shall (a bit abusively) refer to the first $(d-1)$ directions as (discretized) "space" and to the last one, y, as (discretized) imaginary[13] "time". We write $\sigma_i(y)$ for the spin variable at "time" y in the "spatial" position $i \in K$. In the Ising model each $\sigma_i(y)$ takes the two values ± 1.

Recursive Construction of the Canonical Ensemble
We think of the d-dimensional lattice Λ as being constructed by adding one $(d-1)$-dimensional slice K_y at each "time" $y \in \{1, \ldots, L\}$ (cf. Fig. 4.7). More precisely, our strategy is to construct the (non-normalized) canonical distributions $\exp[-\beta H_L]$ on the lattice $K \boxtimes I_L$ *recursively in the integer L*. To get the lattice $K \boxtimes I_{L+1}$ from $K \boxtimes I_L$ we have two add a new slice K_{L+1} and also the "vertical" edges connecting each node of the new slice at $y = L + 1$ to the corresponding node of the $y = L$ slice. Since the Hamiltonian is a sum of contributions from the edges, the new Hamiltonian H_{L+1} will differ from the old one H_L by two contributions: the one coming from the "horizontal" edges in the new slice K_{L+1}

$$-J \sum_{l \in (K_{L+1})_1} \sigma_{h(l)}(L+1)\, \sigma_{t(l)}(L+1) \tag{4.94}$$

and the one from the added "vertical" edges

$$-J \sum_{i \in (K_L)_0} \sigma_i(L)\, \sigma_i(L+1). \tag{4.95}$$

---

[13] That is, time Wick-rotated to Euclidean space-time signature.

Then one has the recursion relation

$$\exp[-\beta H_{L+1}] = S\,R\,\exp[-\beta H_L] \tag{4.96}$$

where

$$R = \exp\left[\beta J \sum_{i \in (K_L)_0} \sigma_i(L+1)\,\sigma_i(L)\right] \tag{4.97}$$

$$S = \exp\left[\beta J \sum_{l \in (K_{L+1})_1} \sigma_{h(l)}(L+1)\,\sigma_{t(l)}(L+1)\right]. \tag{4.98}$$

To solve the recursion relation (4.96) in an efficient way, we introduce a vector space V with an orthonormal basis indexed by the spin configurations on a lattice slice K. For a classical lattice model whose degree of freedom at each site $a \in \Lambda$ takes N values the Hilbert space V has dimension

$$\dim V = N^{|K_0|}. \tag{4.99}$$

E.g. for the Ising model a basis of V consists of $2^{|K_0|}$ elements: the orthonormal basis of V is chosen in the form

$$|\{s_{\boldsymbol{k}}\}\rangle \qquad \text{for } \boldsymbol{k} = (k_1, \ldots, k_{d-1}) \in K \subset \mathbb{Z}^{d-1}, \tag{4.100}$$

where each $s_{\boldsymbol{k}}$ takes the two values ± 1. The (fixed "time" y) spin operators σ_j are *diagonal* in this basis

$$\sigma_j |\{s_{\boldsymbol{k}}\}\rangle = s_j\,|\{s_{\boldsymbol{k}}\}\rangle, \qquad \forall\, j \in K, \tag{4.101}$$

and the elements of the basis are labeled by the eigenvalues $\{s_{\boldsymbol{k}}\}_{\boldsymbol{k} \in K}$ of the $\sigma_{\boldsymbol{k}}$'s. The expressions S and R in Eqs. (4.97) and (4.98) can be seen as linear operators (matrices) acting on V. S is diagonal in the spin-configuration basis (4.100)

$$S|\{s_{\boldsymbol{k}}^{(L+1)}\}\rangle = \exp\left[\beta J \sum_{l \in K_1} s_{h(l)}^{(L+1)}\,s_{t(l)}^{(L+1)}\right]|\{s_{\boldsymbol{k}}^{(L+1)}\}\rangle, \tag{4.102}$$

while

$$R|\{s_{\boldsymbol{k}}^{(L)}\}\rangle = \sum_{\{s_{\boldsymbol{k}}^{(L+1)}\}} \exp\left[\beta J \sum_{i \in K_0} s_i^{(L+1)}\,s_i^{(L)}\right]|\{s_{\boldsymbol{k}}^{(L+1)}\}\rangle, \tag{4.103}$$

where $\{s_k^{(y)}\}$ are the eigenvalues of the spin operators $\sigma_k(y)$ at "time" y (i.e. inserted in the y-th slice). The solution to the recursion relation (4.96) is then[14]

$$\exp[-\beta H_{L+1}] = \prod_{y=1}^{L} \langle\{s_k^{(y+1)}\}|\, SR\,|\{s_k^{(y)}\}\rangle =$$

$$= \langle\{s_k^{(L+1)}\}|\, SR\,|\{s_k^{(L)}\}\rangle\, \langle\{s_k^{(L)}\}|\, SR\,|\{s_k^{(L-1)}\}\rangle \times \cdots \tag{4.104}$$

Now suppose we want to compute the partition function $Z(\{s_k^{(1)}\}, \{s_k^{(L+1)}\})$ subjected to the boundary condition that the spin configurations at initial and final "times", $\{s_k^{(1)}\}$ and $\{s_k^{(L+1)}\}$, are fixed.[15] We have to sum over all possible configurations $\{s_k^{(y)}\}$ at all intermediate "times" $2 \le y \le L$. From Eq. (4.104) we see that these sums reproduce the definition of the product for matrices; hence

$$Z(\{s_k^{(1)}\}, \{s_k^{(L+1)}\}) = \langle\{s_k^{(L+1)}\}|\, (SR)^L\, S\,|\{s_k^{(1)}\}\rangle), \tag{4.105}$$

where $(SR)^L$ is the L-th power of the matrix SR. From our recursive construction of the lattice Λ by adding one slice after the other, we get that the linear operator $SR: V \to V$ has the effect of adding a lattice slice to the canonical ensemble. Said differently: the classical d-dimensional lattice system defined on Λ may be seen as the "evolution" in the "time" y of a **quantum** *lattice system in one less dimension* defined in the "spatial" lattice K with Hilbert space V and evolution operator SR.

The operator SR is not Hermitian. However it is conjugate to the Hermitian operator

$$T = S^{1/2} R\, S^{1/2}, \tag{4.106}$$

so that it becomes Hermitian (in facts real and symmetric) by a *diagonal* redefinition of the basis $|\{s_k\}\rangle \to S^{-1/2}|\{s_k\}\rangle$. In the new basis the spin operators σ_k remain diagonal.

Definition 4.1 The *transfer matrix* is the linear Hermitian operator $T: V \to V$ which "adds a time-slice to the lattice" or, said differently, which evolve the system from Euclidean "time" y to Euclidean "time" $y + 1$. The ℓ-th power of T then describes the Euclidean "time evolution" from y to $y + \ell$.

---

[14] To avoid cluttering, we omit the irrelevant factor $\langle\{s_k^{(0)}\}|\, R\,|\{s_k^{(0)}\}\rangle$ which may be absorbed in the definition of the boundary conditions.

[15] And a suitable boundary condition at the "spatial" boundary $(\partial K) \boxtimes I_L$.

From the above discussion we get that the transfer matrix of the d-dimensional Ising model is the linear operator with matrix elements

$$\langle \{s'_k\} | T | \{s_k\} \rangle =$$

$$= \exp\left[\tfrac{1}{2}\beta J \sum_{<i,j>} s'_i s'_j\right] \exp\left[\beta J \sum_h s'_h s_h\right] \exp\left[\tfrac{1}{2}\beta J \sum_{<l,m>} s_l s_m\right] \tag{4.107}$$

where we think of s_k as the eigenvalues of the spin operators $\sigma_k(y)$ and s'_k as the eigenvalues of $\sigma_k(y+1)$ in the next "time" slice. The spin configurations s'_k, s_k are required to obey the chosen boundary conditions at the "spatial" boundary ∂K. For simplicity we take periodic boundary conditions in the $d-1$ "spatial" directions, so that T is invariant by lattice translations in the 'spatial" directions.

Suppose we put the periodic boundary condition also in the "time" direction i.e. we impose the constraint

$$\sigma_k(y) = \sigma_k(y+L) \quad \text{that is,} \quad s_k^{(L+1)} \equiv s_k^{(1)} \quad \text{for all } k \in K. \tag{4.108}$$

The partition function with the "time periodic" boundary condition (4.108) is simply

$$Z_{\text{per}}(L) = \text{Tr}\left[T^L S\right] \equiv \sum_{\{s_k=\pm 1\}} \langle \{s_k^{(L+1)}\} | S^{1/2} T^L S^{1/2} | \{s_k^{(1)}\} \rangle \Big|_{s_k^{(L+1)} \equiv s_k^{(1)} \equiv s_k}$$

$$\tag{4.109}$$

Other kinds of boundary conditions at the initial and final "times", $y = 1$ and $y = L+1$, are described by boundary vectors $S^{-1/2}|B_{\text{in}}\rangle$, $S^{-1/2}|B_{\text{out}}\rangle \in V$

$$Z_{B_{\text{out}}, B_{\text{in}}}(L) = \langle B_{\text{out}} | T^L | B_{\text{in}} \rangle. \tag{4.110}$$

For instance the $\pm$ boundary condition is given by

$$|B_{\text{in}}\rangle = |B_{\text{out}}\rangle = |B, \pm\rangle \equiv |\{\pm, \pm, \cdots, \pm\}\rangle. \tag{4.111}$$

More generally, the boundary conditions are given by a density matrix

$$\eta = \sum_i |B_{\text{in}}, i\rangle\langle i, B_{\text{out}}| \in V \otimes V^\vee \tag{4.112}$$

through the formula

$$Z_\eta(L) = \text{Tr}[\eta\, T^L]. \tag{4.113}$$

The thermodynamic limit requires to send K to infinity, so that T becomes an operator in an infinite-dimensional Hilbert space, while also sending $L \to \infty$.

Let A be a finite subset of the "spatial" box K. We write $O_A(y)$ for the product of spin variables

$$O_A(y) \stackrel{\text{def}}{=} \prod_{k \in A} \sigma_k(y) \tag{4.114}$$

all inserted at the same "time" y, and O_A for the corresponding linear operator $\prod_{k \in A} \sigma_k$ acting on the vector space V. Then the correlation functions in the d-dimensional box $K \boxtimes I_L$, with periodic boundary conditions in the "time" direction, are (for $y_1 < y_2 < \cdots < y_s < L$)

$$\langle O_{A_s}(y_s) O_{A_{s-1}}(y_{s-1}) \cdots O_{A_1}(y_1) \rangle_{\text{per},\, L} =$$
$$= \frac{\text{Tr}[T^{(L-y_s)} O_{A_s} T^{(y_s-y_{s-1})} O_{A_{s-1}} \cdots T^{(y_2-y_1)} O_{A_1} T^{y_1} S]}{Z_{\text{per}}(L)}. \tag{4.115}$$

The way we defined it in Eq. (4.107), T is a real symmetric matrix (hence Hermitian), so its eigenvalues t_a are real and the eigenstates $|\psi_a\rangle$ are orthogonal

$$T|\psi_a\rangle = t_a|\psi_a\rangle, \qquad \langle\psi_a|\psi_b\rangle = \delta_{ab}. \tag{4.116}$$

T is a non-negative operator, i.e. $t_a \geq 0$: indeed $T = L^\dagger L$ where

$$\langle\{\tilde{s}_k\}| L |\{s_k\}\rangle =$$
$$= \prod_j \left[\sqrt{\cosh(\beta J)/2} + \sqrt{\sinh(\beta J)/2}\, \tilde{s}_j s_j\right] \exp\left[\tfrac{1}{2}\beta J \sum_{<l,l'>} s_l s_{l'}\right]. \tag{4.117}$$

T may be diagonalized in the form

$$T = \sum_a |\psi_a\rangle t_a \langle\psi_a|, \tag{4.118}$$

so the partition function with periodic boundary condition in the "time" direction is

$$Z_L = \text{Tr}[T^L S] = \sum_a t_a^L \langle\psi_a|S|\psi_a\rangle. \tag{4.119}$$

Let $t_{\max}$ be the largest eigenvalue of T, m its multiplicity and $|\phi_\alpha\rangle$ ($\alpha = 1, \ldots, m$) the corresponding eigenvectors. Then for large L

$$\text{Tr}[T^L S] = t_{\max}^L \left(\sum_{\alpha=1}^m \langle\phi_\alpha|S|\phi_\alpha\rangle + \text{exponentially small}\right) \quad L \gg 1, \tag{4.120}$$

and the thermodynamic limit is simply

$$\lim_{L\to\infty} \frac{1}{L} \log Z_L = \log t_{\max}.$$

(4.121)

Since we have proven that the thermodynamic limit exists, we know that

$$t_{\max} \approx e^{-|K_0|\mathcal{F}} \quad \text{for large } |K_0|.$$

(4.122)

Similarly, with other kinds of boundary conditions,

$$Z_{B_{\text{out}}, B_{\text{in}}}(L) = \langle B_{\text{out}}|T^L|B_{\text{in}}\rangle =$$

$$= t_{\max}^L \sum_{\alpha=1}^{m} \langle B_{\text{out}}|\phi_\alpha\rangle\langle\phi_\alpha|B_{\text{in}}\rangle + \text{subleading as } L \gg 1.$$

(4.123)

For generic boundary conditions $|B_{\text{in}}\rangle$, $|B_{\text{out}}\rangle$ the coefficient of $t_{\max}^L$ is non zero and we get the same thermodynamical potential

$$\frac{1}{L} \log Z_{B_{\text{in}}, B_{\text{out}}}(L) \approx \log t_{\max} \approx -|K_0|\mathcal{F}$$

(4.124)

as with periodic conditions. This condition is satisfied by all physically reasonable boundary conditions. What we mean by "reasonable"? Consider the thermodynamic limit of the thermal expectation value

$$\langle O_A(y)\rangle_{\text{ter.lim.}} = \lim_{L\to\infty} \frac{\langle B_{\text{out}}|\, T^{L/2-y}\, O_A\, T^{L/2+y}\,|B_{\text{in}}\rangle}{\langle B_{\text{out}}|\, T^L\,|B_{\text{in}}\rangle}.$$

(4.125)

If the genericity condition holds, the limit exists for all operators $O_A(y)$ and the expectation value is independent of y (invariant by translation in "time"). When the boundary states $|B_{\text{in}}\rangle$, $|B_{\text{out}}\rangle$ are "reasonable" the correlation is invariant also for translations in "space". This is automatic when there is a unique maximal eigenvector of T: since T is invariant by space translations, the unique maximal eigenvector should also be invariant. Below we shall prove "spatial" translational invariance in general.

Consider first the case where there is a *unique* eigenvector $|\phi_0\rangle$ of T with the largest eigenvector $t_{\max}$; we have

$$(t_{\max}^{-1} T)^L = |\phi_0\rangle\langle\phi_0| + \sum_{t_a < t_{\max}} |\psi_a\rangle(t_a/t_{\max})^L\langle\psi_a| \xrightarrow{L\to\infty} |\phi_0\rangle\langle\phi_0|$$

(4.126)

where the corrections for finite L are exponentially small as $L \gg 1$ when there is a gap in the spectrum between $t_{\max}$ and the other eigenvalues, otherwise they may go to zero just polynomially. Now, for all generic boundary states and $L \ggg 1$,

$$
\begin{aligned}
\langle O_A(y) \rangle &\equiv \left. \frac{\langle B_{\text{out}} | T^{L/2-y} O_A T^{L/2+y} | B_{\text{in}} \rangle}{\langle B_{\text{out}} | T^L | B_{\text{in}} \rangle} \right|_{L \ggg 1} \approx \\
&\approx \frac{\langle B_{\text{out}} | \phi_0 \rangle \langle \phi_0 | O_A | \phi_0 \rangle \langle \phi_0 | B_{\text{in}} \rangle}{\langle B_{\text{out}} | \phi_0 \rangle \langle \phi_0 | B_{\text{in}} \rangle} = \langle \phi_0 | O_A | \phi_0 \rangle,
\end{aligned}
\tag{4.127}
$$

while

$$
\begin{aligned}
&\langle O_A(y+x) O_B(y) \rangle - \langle O_A(y) \rangle \langle O_B(0) \rangle \approx \\
&\approx \langle \phi_0 | O_A (t_{\max}^{-1} T)^x O_B | \phi_0 \rangle - \langle \phi_0 | O_A | \phi_0 \rangle \langle \phi_0 | O_B | \phi_0 \rangle.
\end{aligned}
\tag{4.128}
$$

When $x \ggg 1$ we can again use Eq. (4.126) and replace $(t_{\max}^{-1} T)^x$ with $|\phi_0\rangle \langle \phi_0|$ up to small corrections. This shows that the cluster property holds; moreover the cluster limit is approached exponentially when there is a gap in the spectrum of T, that is, if in the sum in Eq. (4.126) $t_a \leq t_{\max} - \epsilon$ with $\epsilon > 0$. Comparing with the discussion of the cluster property in Sect. 4.2 we conclude:

Fact 4.6 *When the maximal eigenvalue of the transfer matrix T has multiplicity* 1 *the classical lattice system has a unique phase (which is then automatically pure).*

We stress that in the present context "uniqueness" is a subtle notion: all eigenvalues t of T with the property

$$
\lim_{|K_0| \to \infty} \frac{1}{|K_0|} \log t = -\mathcal{F}
\tag{4.129}
$$

become "maximal" in the thermodynamic limit. We are only interested in "maximal" eigenvalues whose eigenstates are invariant by translation in space.

The previous arguments show that in presence of multiple "maximal" eigenvalues (with translational invariant eigenvectors $|\phi_a\rangle$) we have several distinct phases σ

$$
\langle O_A \rangle_\sigma = \text{Tr}(\sigma O_A)
$$

$$
\text{where} \quad \sigma = \sum_{a=1}^{m} \sigma_a |\phi_a\rangle \langle \phi_a|, \quad \sigma_a \geq 0, \quad \sum_{a=1}^{m} \sigma_a = 1,
\tag{4.130}
$$

which are convex combination of pure ones $|\phi_a\rangle \langle \phi_a|$.

No Phase Transition in a Finite Box K We are ready to prove

Theorem 4.7

*(1) For all classical lattice systems with finite range interactions, and all **non-zero** temperatures, $\beta < \infty$, the transfer matrix T in a **finite** "spatial" box K has a*

unique maximal eigenvalue, *that is, in a finite "spatial" volume* $|K| < \infty$ *there cannot be any phase transition, except possibly at exactly zero temperature.*

(2) At exactly $T = 0$ *we can have different phases associated to different eigenvectors* $|\phi_a\rangle$ $(a = 1, \ldots, m)$ *of* $T(\beta = \infty)$ *all associated to the same maximal eigenvalue* t_{max}. *For (say) periodic boundary conditions in the "space" directions, all such eigenvalues* $|\phi_a\rangle$ *are invariant by translation in the "space directions".*

Remark 4.6 The vectors $|\phi_a\rangle$ in item **(2)** correspond to the ground states of the system. For instance in the Ising model (with periodic boundary conditions) we have two of them: all spins up and all spins down. This variational characterization explains physically why they are translation invariant (for the math argument, see below). The proof of the **Theorem** will provide more details on the $\beta = \infty$ physics.

The **Theorem** is a consequence of the Perron-Frobenius theorem [8]:

Theorem 4.8 (Perron-Frobenius)

(1) Let $A = (A_{ij})$ *be an* $n \times n$ *matrix whose entries are* ***positive*** $A_{ij} > 0$ *for* ***all*** $i, j = 1, \ldots, n$. *Then* A *has a positive eigenvalue* $\lambda_{\max}$ *such that all other eigenvalues* λ *satisfy*

$$|\lambda| < \lambda_{\max}. \tag{4.131}$$

The eigenvalue $\lambda_{\max}$ *is* ***simple***, *and its eigenvector* $(v_1, \ldots, v_n)^t$ *has strictly positive entries* $v_i > 0$ *for* $i = 1, \ldots, n$. *There is no other eigenvector with non-negative entries.*

(2) Suppose A *has* ***non-negative*** *entries* $A_{ij} \geq 0$ *and in addition is* ***irreducible*** *in the Frobenius sense, that is, there is no non-trivial linear subspace spanned by a proper subset of the standard basis vectors of* $\mathbb{R}^n$ *which is invariant under* A. *Then there is a single real positive eigenvalue* $\lambda_{\max}$, *whose eigenvector has strictly positive entries, such that all other eigenvalues satisfy* $|\lambda| \leq \lambda_{\max}$.

Proof of Theorem 4.7

(1) The entries of the (finite lattice) transfer matrix T_{ij} are of the form

$$T_{ij} = \exp[-\tfrac{1}{2}\beta E(i)] \, \exp[-\beta \, U(i, j)] \, \exp[-\tfrac{1}{2}\beta E(j)] \tag{4.132}$$

where $U(i, j)$ (resp. $E(i)$) is the interaction between a slice of spins in the i-th configuration and a nearby slice of spins in the j-th configuration (resp. the energy of a slice of spins in the i-th configuration). Hence for $\beta < \infty$ all entries of T are *strictly* positive, while at $\beta = \infty$ we have $T_{ij} \geq 0$. Moreover T is Hermitian and non-negative, hence all its eigenvalues are real non-negative. Then when $\beta < \infty$ there is a unique maximal eigenvalue $\lambda_{\max}$ whose eigenvector $|\phi_0\rangle$ has positive entries. Thus at $\beta < \infty$ we cannot have phase transitions in a finite "spatial" volume $|K| < \infty$.

(2) Set the temperature strictly to zero i.e. $\beta = \infty$; now $T_{ij} \geq 0$. If T is irreducible in the Frobenius sense, we get the same conclusion as in **(1)**. Let us assume T is not irreducible in Frobenius' sense. Since T is Hermitian, if it is reducible it is *totally* reducible, that is, by a mere permutation of the standard basis vectors we put T in a block-diagonal form

$$
T = \begin{pmatrix}
T_1 & 0 & \cdots & \cdots & 0 \\
0 & T_2 & \cdots & \cdots & 0 \\
0 & 0 & T_3 & \cdots & 0 \\
\vdots & \vdots & \vdots & \ddots & \vdots \\
0 & \cdots & \cdots & 0 & T_s
\end{pmatrix}
\tag{4.133}
$$

that is,

$$
V = \bigoplus_k V_k \quad \text{and} \quad T V_k \subseteq V_k, \quad T|_{V_k} = T_k,
\tag{4.134}
$$

where each block T_k is irreducible in the Frobenius sense and has non-negative entries. Hence we can apply the Perron-Frobenius theorem independently in each block. We conclude that each block T_k has a unique maximal eigenvalue $\lambda_{k,\max}$, and all other eigenvalues of T_k are strictly smaller (being real and non-negative). T is invariant under "space" translations (say for the periodic boundary conditions in the $(d-1)$ spatial directions). This implies that each T_k is separately invariant under "space" translations, and so it is the unique eigenvector $|\phi_k\rangle$ associated to each eigenvalue $\lambda_{k,\max}$. Moreover the entries of $|\phi_k\rangle$ are strictly positive in the V_k summand and zero in $V_{h \neq k}$. We have one phase at $\beta = 0$ per each block T_k such that $\lambda_{k,\max} = t_{\max}$.

$\square$

Remark 4.7 For classical lattice models the statement that there are no phase transitions at non-zero temperature in a finite box already follows from the Yang-Lee theory. For instance, for the one-dimensional Ising model (where we define the Hamiltonian so that the energy of the ground state is zero) the finite-volume partition function is a polynomial in $z \equiv \exp(-2\beta J)$ hence an entire function. It is easy to compute the complex zeros of the partition functions Z_N and check that they accumulate only at $z = 0$ as $N \to \infty$, see Problem 4.1. In the next section we recover this result from the exact expression of the free energy.

4.5   Ising Model in One Dimension

Consider a one-dimensional classical discrete system with finite-range interactions. The "space" is zero-dimensional, i.e. a point, and the transfer matrix T is a $N \times N$ matrix where N is the number of states of the discrete degree of freedom ("spin") on this point. E.g. in the $d = 1$ Ising model T is a 2×2 matrix. Since "space" is automatically finite, Theorem 4.7 says

Corollary 4.2 *In a one-dimensional classical discrete system with finite-range interactions there are no phases transitions at $\beta < \infty$; in particular the thermo-dynamical potential $\mathcal{F}$ is analytic in β for $0 < \beta < \infty$. There is a phase transition at exactly zero temperature iff the number of ground states of the system is > 1.*

Let us check this prediction for the one-dimensional Ising model which has two vacua: spin *up* and spin *down*. The transfer matrix is

$$T = \begin{pmatrix} e^{\beta J} & e^{-\beta J} \\ e^{-\beta J} & e^{\beta J} \end{pmatrix}, \qquad J > 0. \tag{4.135}$$

For $\beta < \infty$, the matrix $e^{-\beta J}T$ satisfies the condition of part **(1)** of the Perron-Frobenius theorem, while for $\beta = \infty$ it has the totally reducible Frobenius form (4.133) with two diagonal blocks corresponding to the two vacua *up* and *down*.

The two eigenvalues (resp. eigenvectors) of T are

$$\lambda_{\max} = 2 \cosh(\beta J), \qquad\qquad |\phi\rangle = \tfrac{1}{\sqrt{2}} \begin{pmatrix} 1 \\ 1 \end{pmatrix} \tag{4.136}$$

$$\lambda_{\min} = 2 \sinh(\beta J) \qquad\qquad |\psi\rangle = \tfrac{1}{\sqrt{2}} \begin{pmatrix} 1 \\ -1 \end{pmatrix} \tag{4.137}$$

and the free energy per site is

$$-\beta F = \log \cosh(\beta J) + \log 2 \tag{4.138}$$

which is analytic in $z \equiv \beta J$ along the real axis (precisely: for $|\mathrm{Im}\, z| < \pi/2$) as expected from the general theory. Correspondingly, the magnetization is zero

$$\langle \sigma \rangle = \lim_{L \to \infty} \frac{\langle \phi | \begin{pmatrix} 1 & 0 \\ 0 & -1 \end{pmatrix} T^L |\phi\rangle}{(2 \cosh(\beta J))^L} = 0. \tag{4.139}$$

We can switch on a non-zero magnetic field; the new transfer matrix is

$$T(B) = \begin{pmatrix} e^{\beta(J+B)} & e^{-\beta J} \\ e^{-\beta J} & e^{\beta(J-B)} \end{pmatrix} \tag{4.140}$$

whose largest eigenvalue is

$$\lambda_{\max}(B) = \exp(\beta J)\cosh(\beta B) + \sqrt{\exp(-2\beta J) + \exp(2\beta J)\sinh^2(\beta B)} \tag{4.141}$$

which is analytic in βJ and βB and positive for $\beta J, \beta B \in \mathbb{R}$ and *even* in B

$$\lambda_{\max}(B) = \lambda_{\max}(-B) \tag{4.142}$$

so that

$$\text{(spontaneous magnetization)} \equiv \frac{\partial \log \lambda_{\max}(B)}{\partial B}\bigg|_{B=0} = 0, \tag{4.143}$$

as expected on general grounds.

4.6 Classical Discrete Systems in $d \geq 2$

We consider a classical discrete system defined on a regular lattice Λ in dimension $d \geq 2$. On each site $k \in \Lambda_0$ there is one degree of freedom, denoted as σ_k, which takes a finite number m of values. We write s_k for the (eigen)value of σ_k, and loosely call both σ_k and s_k the "spin" at k.

When $d \geq 2$ the thermodynamic limit ($|K_0| \to \infty$ and $L \to \infty$) is highly non-trivial. In a finite box T has a unique maximal eigenvalue,[16] and the partition function Z is analytic in β. The second statement is elementary: Z is a sum of exponentials. In the limit $|K_0| \to \infty$ (say with periodic boundary conditions in "space") T becomes a non-negative Hermitian operator whose matrix elements in the spin-eigenstate basis, $\langle \{s'_k\} | T | \{s_k\} \rangle$, are *non-negative*. By part **(2)** of the Perron-Frobenius Theorem 4.8, T is totally reducible. Then the vector space V has a decomposition

$$V = \bigoplus_a V_a, \tag{4.144}$$

preserved by T, $T V_a \subset V_a$, such that $T|_{V_a}$ is irreducible in the Frobenius sense with respect to the spin-eigenstate basis. By definition V_a is spanned by a set of spin-eigenstate basis elements $|\{s_k\}\rangle$ of V. Since all spin operators σ_k are diagonal in this basis, cf. Eq. (4.101), we have

$$\sigma_j V_a \subset V_a \quad \text{for all } j \in V_a, \tag{4.145}$$

that is, *acting with **finitely many** spin operators on a subspace V_a we remain in the subspace V_a*, that is,

$$\langle \psi_a | O_A | \psi_b \rangle = 0 \quad \text{for } |\psi_a\rangle \in V_a, \ |\psi_b\rangle \in V_b, \text{ and } a \neq b. \tag{4.146}$$

In each V_a we have a **unique** eigenvector $|\phi_a\rangle$ of T associated to the maximal eigenvalue of $T|_{V_a}$, which is invariant under "spatial" translations and is an

---

[16] To make the thermodynamic limit smooth, we define T using a Hamiltonian H shifted so that the energy of the ground state is zero.

eigenvector of the "time" translation T. Each V_a whose maximal eigenvalue is equal to $t_{\max}$ in the sense (4.129) defines a *pure* phase with probability distribution

$$\langle O_A \rangle_a \equiv \lim_{L \to \infty} \frac{\langle \phi_a | T^L O_A | \phi_a \rangle}{\langle \phi_a | T^L | \phi_a \rangle} = \langle \phi_a | O_A | \phi_a \rangle. \tag{4.147}$$

$|\phi_a\rangle$ is a *pure state* since the cluster property holds for all finite products of spins O_A, O_B

$$\lim_{|A-B| \to \infty} \Big(\langle O_A O_B \rangle_a - \langle O_A \rangle_a \langle O_B \rangle_a \Big) \to 0 \tag{4.148}$$

in view of the fact that

$$\langle \phi_b | T^L O_A | \phi_a \rangle = 0 \quad \text{for } a \neq b \quad \text{and any } L, \tag{4.149}$$

so that we can apply the argument around Eq. (4.128) to each Frobenius block separately. We conclude:

Fact 4.9 *The pure phases of a classical discrete system correspond to the subspaces of V which are invariant by translation, preserved by the algebra of finite operators O_A ($\equiv$ polynomials in the "spins" σ_k), and contain a unique eigenvector of the "renormalized" transfer matrix $T/t_{\max}$ with eigenvalue 1 (whose entries are automatically non-negative)* [9].

The Physical Picture Let us explain what is going on in a more intuitive fashion. Consider, say, the Ising model in a $d \geq 2$ cubic lattice. As long as we are in finite volume, T has a strictly positive matrix element

$$\langle \{s_k'\} | T | \{s_k\} \rangle > 0 \tag{4.150}$$

between *any two* spin configurations $\{s_k\}$ and $\{s_k'\}$. As the volume of the "spatial" box K goes to infinity, some matrix elements may go to zero. For instance: let T_0 be the Ising transfer matrix defined with the *natural* Hamiltonian shifted so that the ground state has energy zero. Its matrix element between the configurations all spins up and all spins down is

$$\langle \{s_k' = -1\} | T_0 | \{s_k = +1\} \rangle = \exp\big(-2\beta J |K_0| \big). \tag{4.151}$$

For any finite "spatial" box $|K_0| < \infty$ this matrix element is strictly positive and hence there is a unique Perron-Frobenius eigenstate $|\phi_0\rangle$. At strictly the thermodynamic limit $|K_0| = \infty$ this matrix element vanishes, and the possibility of multiple phases (states) opens up.

In more physical terms: minus the logarithm of the matrix element (4.151) is the *energy barrier* between the two spin configurations; more precisely it is the energy of a *domain wall* separating the two configurations. As long as the barrier is finite, we can pass through it by the thermal analogue of the "tunneling effect". Since the energy of the domain wall is proportional to its volume $|K_0|$, the height of the barrier goes to $+\infty$ in the infinite volume limit, and the "tunneling" between the two configurations stops.

In a general model when two "fixed-time" configurations $|\Sigma_1\rangle$ and $|\Sigma_2\rangle$ are separated by an infinite barrier (in the thermodynamic limit) i.e.

$$\langle \Sigma_1 | T_0 | \Sigma_2 \rangle = 0, \tag{4.152}$$

they lead to distinct equilibrium states *unless* there is an alternate path connecting the two configurations which passes through *finitely-many* intermediate configurations separated by *finite* barriers, that is, unless there is a $k \in \mathbb{N}$ such that

$$\langle \Sigma_1 | T_0^k | \Sigma_2 \rangle > 0 \quad \text{in the thermodynamic limit} \tag{4.153}$$

i.e. unless the two configurations $|\Sigma_1\rangle$, $|\Sigma_2\rangle$ belong to the same Frobenius irreducible component.

The analysis of the possible phases using energy considerations for domain walls which separate spin configurations is made systematic by the Peierls argument.

4.7   The Peierls Argument

We have seen that in $d = 1$ the Ising model has no phase transition and no spontaneous magnetization (no ordered phase) for $\beta < \infty$. We also known that for all dimension d there exist a temperature T_d such that for $T > T_d$ the Ising model is disordered. We remain with the question: *What happens at low temperatures when* $d \geq 2$?

Theorem 4.10 (Peierls) *In the $d \geq 2$ dimensional Ising model there exists a $\beta_0 < \infty$ such that for all $\beta > \beta_0$ we have spontaneous magnetization i.e.*

$$\langle \sigma_n \rangle_+ > 0, \tag{4.154}$$

where $\langle \cdots \rangle_+$ is the thermal probability distribution defined by taking the thermodynamical limit with the $+$ boundary condition.[17]

Thus the Ising system models a physical ferromagnet in all dimensions $d \geq 2$.

While the proof holds in all $d \geq 2$, for clarity we shall mainly work in the $d = 2$ case where we can draw nice pictures. For simplicity we assume the lattice to be cubic, leaving the generalization to other shapes to the reader.

---

[17] We proved that the limit exists.

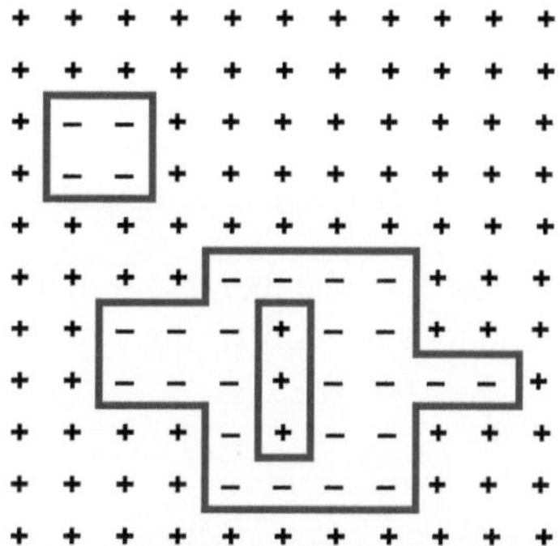

Fig. 4.8 A typical configuration of the 2-dimensional Ising model at low temperature with $+$ boundary condition. The blue lines are the *phase boundaries* i.e. the union of all links in the dual lattice which cross links of the original lattice connecting sites with spins $+1$ and -1

We have seen in Sect. 4.3 that a Ising configuration is defined by the region $S \subset \Lambda$ of sites with spins pointing down. We consider the *phase boundary* $\partial S \subset \Lambda^*$ defined as follows: $\partial S \equiv \gamma$ is the union of all $(d-1)$-plaquettes of Λ^* which are dual (i.e. cross in the middle) the edges of Λ which connect nearest neighbor sites $\langle i, j \rangle$ with $\sigma_i \sigma_j = -1$. As its name suggests, the phase boundary separates regions in Λ where the spins point up from regions where the spins point down: see Fig. 4.8 for an example in the $d = 2$ case.

Defining the Hamiltonian so that a ground state has energy zero, the energy of a configuration is

$$H(\partial S) = 2J|\partial S| \tag{4.155}$$

where $|\partial S|$ is the volume of the phase boundary ∂S given by the number of dual $(d-1)$-plaquettes it contains. Clearly we can write the partition function (with the $+$ boundary condition) as a sum over all possible phase boundaries ∂S.

We consider the configurations such that the phase boundary ∂S contains the closed component γ. The probability of such a configuration is

$$\mathsf{Pr}(\gamma) = \frac{\sum_{\partial S \supset \gamma} e^{-\beta H(\partial S)}}{\sum_{\partial S} e^{-\beta H(\partial S)}} \tag{4.156}$$

where ∂S runs over all phase boundaries or in other words over all spin configurations with $+$'s at infinity.

Lemma 4.2 $\mathsf{Pr}(\gamma) \leq e^{-2\beta J|\gamma|}$.

Proof Let $\gamma \subset \partial S$; we write $\partial S^{\sharp}$ for the configuration obtained by flipping the spins inside γ. This eliminates the component γ from ∂S

$$\partial S^{\sharp} = \partial S \setminus \gamma. \tag{4.157}$$

From (4.155)

$$\frac{e^{-\beta H(\partial S)}}{e^{-\beta H(\partial S^{\sharp})}} = e^{-2\beta J|\gamma|}. \tag{4.158}$$

For fixed γ the correspondence $\partial S \leftrightarrow \partial S^{\sharp}$ is a 1-to-1 map from the phase boundaries ∂S containing γ to the phase boundaries $\partial S'$ such that

$$\partial S' \cap \gamma = \varnothing. \tag{4.159}$$

We have the inequalities

$$\Pr(\gamma) = \frac{\sum_{\partial S \supset \gamma} e^{-\beta H(\partial S)}}{\sum_{\partial S'} e^{-\beta H(\partial S')}} \leq \frac{\sum_{\partial S \supset \gamma} e^{-\beta H(\partial S)}}{\sum_{\partial S' \cap \gamma = \varnothing} e^{-\beta H(\partial S')}} =$$

$$= \frac{\sum_{\partial S \supset \gamma} e^{-\beta H(\partial S)}}{\sum_{\partial S^{\sharp}} e^{-\beta H(\partial S^{\sharp})}} = e^{-2\beta J|\gamma|}. \tag{4.160}$$

$\square$

Before taking the thermodynamic limit we may identify the system with the $+$ boundary condition as the Ising model in the full $\mathbb{Z}^d$ lattice where all spins σ_j with j not contained in the finite box Λ are frozen in the $+1$ value. We write $\langle \cdots \rangle_{+,\Lambda}$ for the expectation value in this set-up.

Proposition 4.1 *For β sufficiently large (low temperature) and $d \geq 2$*

$$0 \leq 1 - \langle \sigma_i \rangle_{+,\Lambda} \leq e^{-\beta J}, \tag{4.161}$$

that is, in the $+$ phase we have a spontaneous magnetization M_+ with

$$1 - e^{-\beta J} \leq M_+ \leq 1. \tag{4.162}$$

Hence at low temperature we have an ordered *phase.*

Proof Only configurations with $\sigma_i = -1$ contribute to $\langle (1 - \sigma_i) \rangle_+$. In any such configuration with phase boundary ∂S there is a minimal component $\gamma(\partial S) \subset \partial S$ which incloses the site i. Then

$$1 - \langle \sigma_i \rangle_+ = 2\frac{\sum_{\gamma} \sum_{\partial S:\gamma(\partial S)=\gamma} e^{-\beta H(\partial S)}}{\sum_{\partial S} e^{-\beta H(\partial S)}} \leq 2\frac{\sum_{\gamma} \sum_{\partial S:\supset\gamma} e^{-\beta H(\partial S)}}{\sum_{\partial S} e^{-\beta H(\partial S)}} =$$

$$= 2\sum_{\gamma} \Pr(\gamma) \leq 2\sum_{\gamma} e^{-2\beta J|\gamma|} \tag{4.163}$$

Let $N(|\gamma|)$ be the number of phase boundaries of area $|\gamma|$ which enclose the site i. Our task is to estimate $N(|\gamma|)$. We may think of constructing the phase boundary γ (say for the $d = 2$ lattice) by adding one edge at the time in the following way: we start from a point $*$ in the center of a plaquette, and add a dual edge starting at $*$ and going in one of the 4 possible directions; at the second endpoint of this edge we add another dual edge in one of the possible 3 directions, ending in the center of a third plaquette, where we also have 3 possible choices for the next edge of γ, and so on for $|\gamma|$ steps. The number of closed contours of length $|\gamma|$ passing through $*$ is then bounded above by $ab^{|\gamma|}$ for some constants a, b depending on the lattice (in d dimensions we can choose $b = 2d - 1$). Since the site i is enclosed by γ, the point $*$ cannot be more than $|\gamma|/2$ lattice lengths away from i, so $*$ is contained in a cube of side $|\gamma|$ in d dimensions, and we can choose $*$ in at most $|\gamma|^d$ ways. In conclusion

$$N(|\gamma|) \leq a|\gamma|^d \, b^{|\gamma|} \tag{4.164}$$

for some constants a, b which depend on the shape of the lattice. Then

$$1 - \langle \sigma_i \rangle_+ \leq 2a \sum_{r=2d}^{\infty} r^d c^r e^{-2\beta J r} \leq e^{-\beta J} \tag{4.165}$$

for β sufficiently large. $\qquad\qquad\qquad\qquad\qquad\qquad\qquad\qquad\qquad\qquad\qquad\qquad\quad\square$

Note that the critical value β_c is at most equal to the largest value of β for which the sum in (4.165) converges, that is, we have the (rough) bound

$$\beta_c \, J \leq \frac{1}{2} \log c. \tag{4.166}$$

In $d = 2$, using the natural value $c = 3$, this gives $\beta_c \, J \lesssim 0.549306$ while the exact value of the critical temperature is $\beta_c J \approx 0.440684$.

Remark 4.8 The argument does not work in $d = 1$ where the phase boundary γ consists of points, with $|\gamma|$ the number of points. $N(|\gamma|)$ is unbounded for $|\gamma| = 2$ since the two points can be anywhere (one on the left and one on the right of i).

4.8 $d = 2$ Ising Model: Order-Disorder Duality

The $d = 2$ Ising model with no magnetic field $B = 0$ is exactly solvable, that is, one can get an exact expression for its free energy in the thermodynamic limit in terms of quadratures. There are many different techniques which allow us to do that, see [5] for a review of several of them. To keep our treatment elementary, in this book we analyze the deep physical properties of the $d = 2$ Ising model without reference to its exact solution (which, of course, confirms the results of our *a priori* analysis).

To make the story more transparent, and simplify the computations, it is helpful to reformulate the Ising model in a deeper geometric language.

The Language of Gauge Theories on Lattices

Informally a lattice Λ is a discretization of its ambient manifold M (usually $\mathbb{R}^d$). The rough idea is that sending the length ϵ of the edges to zero we "recover" the continuum theory on the smooth manifold M. Dynamically this "continuum limit" of the lattice model is quite subtle [10], but for the moment we are interested only in the geometric structures on Λ not on physical processes.

Suppose the ambient manifold M is the base of a bundle $\mathcal{V} \to M$ with a compact structure group G and fibers $\mathcal{V}_m \simeq V$ over the points $m \in M$. For instance, $\mathcal{V}$ may be a principal G-bundle or a vector bundle whose fibers carry a finite-dimensional unitary G-representation V. Let A be a G-connection on $\mathcal{V} \to M$.

We see $\Lambda \subset M$ as a lattice "discretization" of M. By restricting $\mathcal{V}$ to the discrete subset Λ we get a fiber $\mathcal{V}_a \simeq V$ (with G-action) over each site $a \in \Lambda \subset M$ of our lattice. Consider two sites $t(l)$, $h(l)$ connected by the link $l \subset M$. We have a map

$$U_l : \mathcal{V}_{t(l)} \to \mathcal{V}_{h(l)} \quad \phi_{t(h)} \mapsto \exp\left(\int_l A\right)\phi_{t(h)} \in \mathcal{V}_{h(l)}, \quad U_l \in G, \qquad (4.167)$$

given by the *parallel transport* ($\equiv$ holonomy element) along the edge l with the G-connection A. If l is the oriented link at a in the i-th direction e_i, we informally think of U_l as

$$1 + \epsilon A_i(a) + O(\epsilon^2). \qquad (4.168)$$

Parallel transport along the inverse link is given by U_l^{-1}. Now consider the parallel transport along the oriented boundary of a 2-plaquette in the i, j-directions[18]

$$\Phi_{i,j}(a) \equiv U(a + e_j)_{e_j}^{-1} U(a + e_i + e_j)_{e_i}^{-1} U(a + e_i)_{e_j} U(a)_{e_i}, \qquad (4.169)$$

see Fig. 4.9. Informally we may think of $\Phi_{i,j}(a)$ as $1 + \epsilon^2 F(a)_{ij} + O(\epsilon^3)$ where the 2-form F is the *curvature* of the connection A on M.

These considerations lead us to define the *lattice gauge theories* as follows [10]. Let G be a *compact group* which may be a finite group or a Lie group of positive dimension. Over each site $a \in \Lambda$ we put a copy of the G-space V. Depending on the gauge system, V may be a vector space carrying a finite-dimensional unitary representation of the group G or a copy of the group itself. Even if V is not a vector space, we assume the existence of a G-invariant map $V \times V \to \mathbb{C}$ that we write (abusively) as $\phi^* \psi$ for $\phi, \psi \in V$. To each oriented link l we associate the element

---

[18] In Eq. (4.169) we write $U(t(l))_{h(l)-t(l)}$ for $U_l, l \in \Lambda_1$.

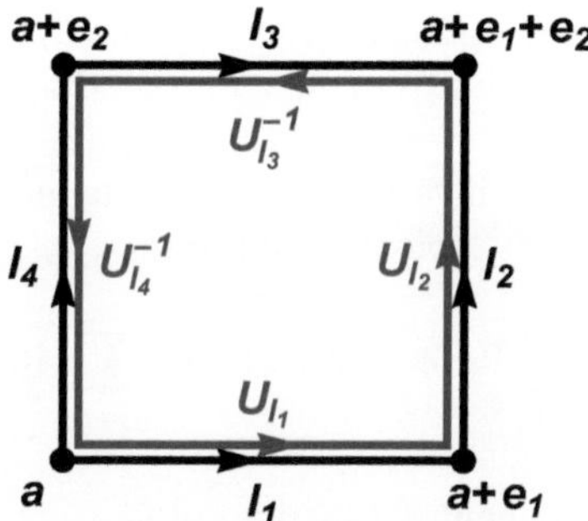

Fig. 4.9   The 2-plaquette at site $\boldsymbol{a}$ in the directions 1,2. The nodes of the plaquette are drawn purple, and the *positively oriented* links l_1,l_2,l_3,l_4 on the boundary of the plaquette in black. The holonomy element $\Phi(\boldsymbol{a})_{1,2}$ at site $\boldsymbol{a}$ in the 1, 2 directions is the holonomy along the closed oriented loop drawn in blue: $\Phi(\boldsymbol{a})_{1,2} = U_{l_1} U_{l_2} U_{l_3}^{-1} U_{l_4}^{-1}$

$U_l \in G$ which gives the parallel transport from the tail site $t(l)$ of l to its head site $h(l)$. As discussed before, U_l may be seen as the discretized version of the gauge field. The parallel transport (holonomy) along the (closed) oriented boundary of a 2-plaquette $\square$

$$\Phi_{\square} = \overrightarrow{\prod_{l \in \partial \square}} U_l \tag{4.170}$$

is then the discrete version of the curvature (field-strength) of the connection as we see in Fig. 4.9. On each lattice site $\boldsymbol{a} \in \Lambda_0$ we have a degree of freedom $\phi_a \in V$ which we call "spin". To specify a configuration of the gauge model we have to give the values of both the spin degrees of freedom $\{\phi_a\}_{a \in \Lambda_0}$ and the gauge ones $\{U_l\}_{l \in \Lambda_1}$. The gauge transformation $\{g_a \in G\}_{a \in \Lambda_0}$ acts on the degrees of freedom $\{\phi_a, U_l\}$ as

$$\phi_a \rightarrow g_a \, \phi_a, \quad U_l \rightarrow g_{h(l)} \, U_l \, g_{t(l)}^{-1}. \tag{4.171}$$

Under a gauge transformation the group element $\Phi_{\square}(\boldsymbol{a})$, defined as the parallel transport from the site $\boldsymbol{a}$ to $\boldsymbol{a}$ along the boundary of a plaquette $\square$ with $\boldsymbol{a}$ its first vertex, transforms as

$$\Phi_{\square}(\boldsymbol{a}) \rightarrow g_a \, \Phi_{\square}(\boldsymbol{a}) \, g_a^{-1}. \tag{4.172}$$

A gauge configuration $\{U_l\}$ is *pure gauge* if we can put it in the form $U_l = 1$ for all $l \in \Lambda_1$ by a gauge transformation (4.171). In this case

$$U_l = g_{h(l)}^{-1} \, g_{t(l)} \tag{4.173}$$

for some g_a's in G. Standard arguments show[19] that $\{U_l\}$ *is pure gauge iff* $\Phi_\square = 1$ *for all plaquettes* $\square$. This follows from the corresponding well-known statement for G-connections on manifolds.

A (isotropic, translational invariant, "local") Hamiltonian which is invariant under the gauge symmetry has typically the form

$$H(\phi_a, U_l) = -J \sum_{l \in \Lambda_1} \mathrm{Re}(\phi_{t(l)}^* U_l \, \phi_{h(l)}) + \sum_{\square \in \Lambda_2} \mathrm{Re} \, \chi(\Phi_\square) \tag{4.174}$$

where $\chi(\cdot)$ is a *character* ($\equiv$ central function) of G.

The Ising Hamiltonian (4.30) has the form (4.174) with

$$G = \mathbb{Z}_2, \quad V = \mathbb{Z}_2, \quad \phi_a \equiv \sigma_a, \quad \phi_a^* \phi_b = \sigma_a \sigma_b \in \mathbb{R}, \tag{4.175}$$

and

$$U_l = 1 \quad \text{for all } l \in \Lambda_1. \tag{4.176}$$

In other words, in the Ising model the gauge connection U_l is a *background*[20] which is *pure gauge* ($\equiv$ trivial). However we are free to write this background in *any gauge we like* without changing the Thermodynamics, that is, we can assign to each link l an element $U_l \in \mathbb{Z}_2$ in *any way* we please subjected only to the condition that the plaquette element $\Phi_\square$ in Eq. (4.170) is 1 for all $\square \in \Lambda_2$. Under this condition on the $\{U_l\}$, the system is still our original Ising model since we can return to the original Hamiltonian with $U_l \equiv 1$ by a gauge transformation.

Electromagnetic and Abelian Dualities

There are many advantages in the gauge point of view, which allows us to get a geometric understanding of many properties of the Ising model and other lattice systems. First of all we have the gauge freedom in the definition of the couplings

$$H = -J \sum_{l \in \Lambda_1} U_l \, \sigma_{t(l)} \, \sigma_{h(l)}, \quad \{U_l\} \text{ any pure gauge background}, \tag{4.177}$$

which can be used to define additional interesting observables (see below). What is more, we can infer properties of the Ising model from know properties of the Abelian gauge theories, such as electromagnetism. The physical electromagnetism in 4-dimensional space-time has a duality which interchanges electric and magnetic

---

[19] The statement applies to lattices which discretize manifold with trivial fundamental group.

[20] By a *background* we mean a configuration which is kept *fixed* in the ensemble, that is, we do not sum over it when computing the partition function.

fields, or in the form notation[21]

$$F \leftrightarrow *F, \tag{4.178}$$

where $F = dA$ is the field-strength 2-form and $*$ is the Hodge dual which maps k-forms into $(d-k)$-forms in d-dimensions (so 2-forms into 2-forms in 4 dimensions).

Electromagnetic duality is a special instance of a more general class of *Abelian dualities* [11]. In d-dimensions[22] a k-form Abelian gauge field $A^{(k)}$ has the gauge symmetry[23]

$$A^{(k)} \rightsquigarrow A^{(k)} + d\xi^{(k-1)}, \tag{4.179}$$

where $\xi^{(k-1)}$ is an arbitrary $(k-1)$-form parameter. The local gauge-invariant quantity is the $(k+1)$-form field-strength

$$F^{(k+1)} = dA^{(k)}, \tag{4.180}$$

and the "free" action is

$$S = \frac{1}{2} \int F^{(k+1)} \wedge *F^{(k+1)}. \tag{4.181}$$

The field-strength $F^{(k+1)}$ satisfies the Bianchi identities and equations of motion

$$dF^{(k+1)} = 0, \qquad d*F^{(k+1)} = 0. \tag{4.182}$$

The Abelian duality interchanges the roles of these two equations, and describes the *same physical system* in terms of *different degrees of freedom,* namely in terms of a *magnetic*[24] *dual* $(d-k-2)$ gauge potential $B^{(d-k-2)}$ with dual field-strength

$$G^{(d-k-1)} = dB^{(d-k-2)} \equiv *F^{(k+1)}, \tag{4.183}$$

---

[21] As written the formula applies to Euclidean electromagnetism (i.e. electromagnetism in a spacetime whose metric is positive-definite) not to the physical one in Minkowski space.

[22] We take spacetime to be Euclidean (that is, endowed with a positive-definite metric) both for simplicity and because this is the relevant situation for the applications to Statistical Mechanics which is tantamount to Euclidean QFT [10].

[23] For a zero-form gauge field $A^{(0)}$ the gauge symmetry is replaced by a Peccei-Quinn symmetry $A^{(0)} \rightsquigarrow A^{(0)} + \text{const.}$ which leaves invariant the "field strength" $F^{(1)} = dA^{(0)}$ [11].

[24] The original field $A^{(k)}$ is called the *electric potential.* Which one of the two dual potentials $A^{(k)}$ and $B^{(d-k-2)}$ is the electric (resp. magnetic) one is a matter of convention.

so that Eqs. (4.182) take the equivalent dual form

$$\mathrm{d} * G^{(d-k-1)} = \mathrm{d}G^{(d-k-1)} = 0. \tag{4.184}$$

One passes from one formulation to the dual one by a functional Fourier transform (see Sect. 5.5) which for free systems reduces to a Legendre transform of the action. We sketch the argument: write[25]

$$S = \int \left(\frac{1}{2} F^{(k+1)} \wedge * F^{(k+1)} + \mathrm{i}\, B^{(d-k-2)} \wedge \mathrm{d}F^{(k+1)} \right) \tag{4.185}$$

where $F^{(k+1)}$ is a *unrestricted* $(k+1)$-form and the $(d-k-2)$-form $B^{(d-k-2)}$ is a Lagrange multiplier enforcing the Bianchi identity $\mathrm{d}F^{(k+1)} = 0$ whose (local) solution is $F^{(k+1)} = \mathrm{d}A^{(k)}$ for an "electric" potential k-form $A^{(k)}$, showing that the action (4.185) *on-shell* is equivalent to the original action (4.181). Alternatively we can integrate by parts the second term in (4.185), and consider the equations of motion for the unrestricted field $F^{(k+1)}$

$$0 = \frac{\delta S}{\delta F^{(k+1)}} = * F^{(k+1)} + \mathrm{i}(-1)^{kd}\mathrm{d}B^{(d-k-2)}. \tag{4.186}$$

Plugging this equation back into (4.185), we get

$$S = \frac{1}{2} \int \mathrm{d}B^{(d-k-2)} \wedge * \mathrm{d}B^{(d-k-2)}, \tag{4.187}$$

showing that the "electric" model (4.181) is physically equivalent to the dual "magnetic" theory (4.187) of a (free) $(d-k-2)$-form gauge field $B^{(d-k-2)}$. A more detailed analysis, which applies also to interacting systems (and takes into due account Dirac quantization of charge) is described in Chap. 1 of [11]. There one shows that the dualities form a much richer web of equivalences.

From the above discussion it is clear that the Abelian dualities are especially deep and powerful when the dimension d is even, $d = 2n$, and the gauge-fields are $(n-1)$-forms $A^{(n-1)}$. In this case the dual gauge fields are also $(n-1)$-forms $B^{(n-1)}$ and the duality relates two descriptions with the "same physics". For $n = 2$ this gives back the usual electromagnetic duality in 4 dimensions. For $n = 1$ we get a duality between a two-dimensional system of zero-form fields $A^{(0)}$, invariant under the Peccei-Quinn symmetry $A^{(0)} \rightsquigarrow A^{(0)} + \text{const.}$, and a dual system of the same kind. This two-dimensional duality is usually called *T-duality* [12] (see also [13]).

---

[25] The imaginary unit i in front of the second term is due to the fact that we are working in a spacetime of Euclidean ($\equiv$ positive) signature.

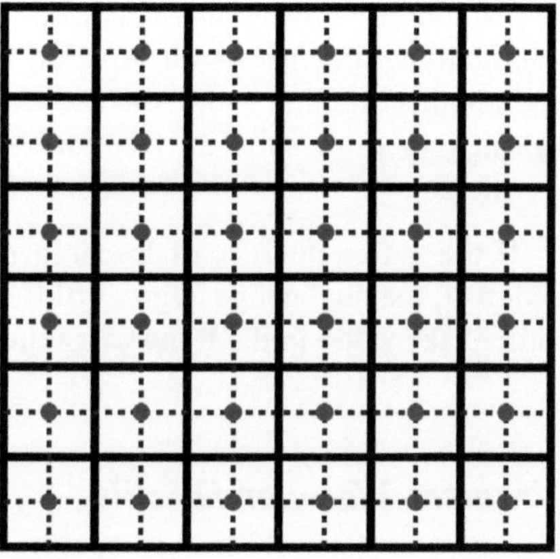

Fig. 4.10 Duality in $d = 2$ square lattice. Generalized electromagnetic duality replaces the original lattice (drawn in black) with its dual (drawn in blue)

Abelian Dualities on a Lattice

The generalized electromagnetic dualities of continuum Abelian gauge theories extend in a straightforward way to Abelian gauge theories defined on a lattice. Since the k-forms are dual to k-chains, and k-chains on a lattice are well-defined as linear combinations of oriented k-plaquettes, in view of Eq. (4.178) the generalized electromagnetic duality in a d-dimensional Abelian lattice theory *should interchange k-plaquettes with dual $(d - k)$-plaquettes*. We already know such a lattice duality operation: replacing the lattice Λ with its dual Λ^* has the effect

$$\mathbb{Z}\Lambda_k \leftrightarrow \mathbb{Z}\Lambda^*_{d-k} \tag{4.188}$$

cf. Sect. 4.2 and Fig. 4.10.

Fact 4.11 *For an Abelian lattice gauge theory, the two electromagnetic dual descriptions live in dual lattices.*

The lattice 0-form (Abelian) degrees of freedom (called *spins*) live on the sites $a \in \Lambda_0$; the 1-form degrees of freedom U_l live on the links and are called *gauge fields*. In general the (Abelian) k-form degrees of freedom live on k-plaquettes and their dual degrees of freedom live on $(d - k - 2)$-plaquettes of the *dual lattice*. In particular:

(a) In two dimensions an Abelian spin system is dual to another Abelian spin system;
(b) in two dimensions an (Abelian) gauge system is dual to a topological system with no local degrees of freedom (since $\Lambda^*_{-1} = \varnothing$);
(c) in three dimension an Abelian spin system is dual to a gauge theory and viceversa;
(d) in three dimensions a 2-form Abelian gauge theory is dual to a topological theory;

(e) in four dimension an Abelian spin system is dual to a 2-form gauge theory;
(f) in four dimensions an Abelian gauge theory is electromagnetically dual to another Abelian gauge theory;
(g) in four dimension an Abelian 3-form gauge theory is dual to a topological theory.

If the gauge degrees of freedom in the k-plaquettes take value in the Abelian group G, the dual gauge degrees of freedom in the $(d-k-2)$-plaquettes of the dual lattice take value in the Pontryagin dual group $\mathsf{Hom}(G, U(1))$.

Kramers-Wannier Duality

We apply the machinery of generalized electromagnetic duality to the $d = 2$ Ising model. As we have seen in Eq. (4.175) the Ising model is a spin (0-form) gauge theory with gauge group $\mathbb{Z}_2$. This is an Abelian group, and we have a duality with another spin theory; $\mathbb{Z}_2$ is its own Pontryagin dual, so the dual model is the Ising model again, i.e. *the Ising model is self-dual*. This self-duality of the Ising model is known as the *Kramers-Wannier duality* after the authors who first described it (using quite different methods). The duality holds for any two-dimensional lattice. We discuss the case of the square lattice, leaving the hexagonal/triangular lattices to the diligent reader (the lazy reader may have a look to [5]).

The Ising Model as a Gauge System   Let Λ be any graph not necessarily regular. The Ising Hamiltonian on Λ is as

$$H = -J \sum_{l \in \Lambda_1} \sigma_{h(l)} \sigma_{t(l)}. \tag{4.189}$$

We rewrite the action as

$$H = -J \sum_{l \in \Lambda_1} U_l, \qquad U_l \equiv \sigma_{h(l)}\, \sigma_{t(l)} \tag{4.190}$$

and we replace the original degrees of freedom $\{\sigma_a\}$ living on the sites with the new gauge degrees of freedom $\{U_l\}$ living on the links l. However the $\{U_l\}$ are not unconstrained variables. Indeed from (4.190) we see that $\{U_l\}$ is a pure gauge configuration. Hence for all closed path γ in Λ we have

$$\overrightarrow{\prod_{l \in \gamma}} U_l = 1. \tag{4.191}$$

If this condition is satisfied for all closed paths γ in Λ, the gauge model in the left part of Eq. (4.190) is equivalent to the original Ising model (with nearest neighbor interactions) defined on the graph Λ. Indeed: if the U_l's satisfy the conditions (4.191), we can reconstruct the original spin configuration σ_a out of the U_l up to simultaneous multiplication of all spins by ± 1 (which is a symmetry). This is achieved as follows: fix a reference site $*$ and call σ_* the spin there. The spin at any other site $a \in \Lambda$ is then

$$\sigma_a = \sigma_* \prod_{l \in \gamma(a)} U_l \tag{4.192}$$

where $\gamma(a)$ is a *path* ($\equiv$ an oriented concatenation of links) which connects the site a to the reference one $*$. The conditions (4.191) just say that the definition (4.192) is well-posed, i.e. independent of choices. When working with the conceptually more natural boundary conditions, the $\pm$ ones, the overall sign ambiguity gets totally fixed by choosing as reference point $*$ a site whose spin is frozen to ± 1 by the boundary condition. In the language of lattice gauge theory the $\{U_l\}$ form a $\mathbb{Z}_2$ gauge connection (on the lattice), and the conditions (4.195) say that $\{U_l\}$ is pure gauge i.e. each U_l is the product of two $\mathbb{Z}_2$-valued degrees of freedom $\sigma_{h(l)}$, $\sigma_{t(l)}$ living on the endpoints (sites) of the edge l:

$$U_l = \sigma_{h(l)} \, \sigma_{t(l)}. \tag{4.193}$$

Replacing this expression in the Hamiltonian (4.190) we get back the original Ising Hamiltonian with, say, $\sigma_* = +1$.

Recall that a graph is called a *tree* if it does not contain any non-trivial closed path. If Λ is a tree, the constraint (4.191) is empty. Hence the Ising model on any tree is equivalent to the trivial gauge model with Hamiltonian $H = -J \sum_l U_l$. Working with free boundary conditions, we conclude

Fact 4.12 *The Ising partition function on a tree Λ is*

$$Z = \sum_{\{U_l = \pm 1\}} \prod_{l \in \Lambda_1} e^{\beta J U_l} = (2 \cosh(\beta J))^{|\Lambda_1|} \tag{4.194}$$

This formula holds for the one-dimensional regular lattice $\mathbb{Z}$ which is a tree, cf. Sect. 4.5.

Remark 4.9 In the above discussion we have not payed attention to the boundary conditions since they are inessential for the partition function in the thermodynamic limit. The reader is invited to repeat the argument more carefully for the usual boundary conditions (periodic, $\pm$, free, etc.).

The Ising Model in a Square Two-Dimensional Lattice It is immediate to see that the constraint (4.191) holds for all closed paths in the $d = 2$ square lattice if and only if it holds for all elementary plaquettes. Then there is one constraint per

2-plaquette

$$\overrightarrow{\prod_{l\in\partial\square}} U_l = 1 \quad \text{for all } \square \in \Lambda_2. \tag{4.195}$$

In conclusion we can rewrite the 2-dimensional Ising partition function in the form

$$Z = \sum_{\{U_l=\pm 1\}} \exp\left[\beta J \sum_{l\in\lambda_1} U_l\right] \prod_{\square\in\Lambda_2} \delta_{\mathbb{Z}_2}\left(\prod_{l\in\partial\square} U_l\right) \tag{4.196}$$

where $\delta_{\mathbb{Z}_2}(a)$ is the δ-function on the $\mathbb{Z}_2$-group (written multiplicatively) i.e.

$$\delta_{\mathbb{Z}_2}(1) = 1, \qquad \delta_{\mathbb{Z}_2}(-1) = 0, \tag{4.197}$$

which enforces the *pure gauge condition* (4.195) on the gauge configuration $\{U_l\}$.

Replacing the multiplicative notation with the additive one, i.e. writing $U_l = e^{i\pi u_l}$ where $u_l \in \mathbb{Z}/2\mathbb{Z}$, and using the delta-function on $\mathbb{Z}/2\mathbb{Z}$ (now in additive notation)

$$\delta_{\mathbb{Z}_2}(x) \equiv \frac{1}{2} \sum_{m=0}^{1} e^{i\pi m x} = \begin{cases} 1 & x = 0 \bmod 2 \\ 0 & x = 1 \bmod 2, \end{cases} \tag{4.198}$$

we get

$$Z = \frac{1}{2^{|\Lambda_2|}} \sum_{\{u_l\in\mathbb{Z}/2\mathbb{Z}\}} \exp\left[\beta J \sum_{l\in\Lambda_1} \exp(i\pi u_l)\right] \times$$
$$\times \sum_{\{m_\square=0,1\}} \exp\left[i\pi \sum_{\square\in\Lambda_2} \left(m_\square \sum_{l\in\partial\square} u_l\right)\right]. \tag{4.199}$$

The $m_\square$'s may be seen as degrees of freedom, valued in $\mathbb{Z}/2\mathbb{Z}$, which live on the sites a^* of the dual lattice Λ^* (identified with the 2-plaquettes $\square$ of the original lattice Λ): we can rearrange the sum in the exponential in the form

$$\sum_{\square\in\Lambda_2} m_\square \sum_{l\in\partial\square} u_l = \sum_{l\in\Lambda_1} u_l\left(m_{h(l^*)} + m_{t(l^*)}\right) \equiv \sum_{l^*\in\Lambda_0^*} u_{l^*} \sum_{a^*\in\partial l^*} m_{a^*} \bmod 2, \tag{4.200}$$

where l^* is the (unique) link of Λ^* which intersects the link l in Λ and we set $u_{l^*} \equiv u_l$. We write

$$\mu_{a^*} = \exp(i\pi m_{a^*}) \in \{\pm 1\}. \tag{4.201}$$

Now the $\{\mu_{a^*}\}$ are spin variables living on the sites of the dual lattice Λ^* which take the two values ± 1 (with equal probability). The partition function Z becomes

$$Z = \frac{1}{2^{|\Lambda_2|}} \sum_{\{\mu_{a^*}=\pm 1\}} \sum_{\{u_{l^*}=1,0\}} e^{\beta J \sum_{l^*} \exp(i\pi u_{l^*})} \prod_{l^* \in \Lambda_1^*} \left(\mu_{h(l^*)}\mu_{t(l^*)}\right)^{u_{l^*}}. \qquad (4.202)$$

The sum over the u_{l^*} is elementary; we get

$$Z = \frac{1}{2^{|\Lambda_2|}} \sum_{\{\mu_{a^*}=\pm 1\}} \prod_{l^* \in \Lambda_1^*} \left(e^{\beta J} + e^{-\beta J}\, \mu_{h(l^*)}\mu_{t(l^*)}\right). \qquad (4.203)$$

Setting

$$e^{-2\beta J} \equiv \tanh(\beta^* J), \qquad (4.204)$$

and using

$$\cosh(\beta^* J) + \sinh(\beta^* J)\, \mu_{h(l^*)}\mu_{t(l^*)} \equiv \exp\left[\beta^* J \mu_{h(l^*)}\mu_{t(l^*)}\right], \qquad (4.205)$$

we can rewrite (4.203) as

$$Z = \frac{1}{2^{|\Lambda_2|}} \left(\frac{e^{\beta J}}{\cosh(\beta^* J)}\right)^{|\Lambda_1|} \sum_{\{\mu_{a^*}=\pm 1\}} \exp\left[\beta^* J \sum_{l^* \in \Lambda_1^*} \mu_{h(l^*)}\mu_{t(l^*)}\right]. \qquad (4.206)$$

We have not payed attention to boundary conditions, so this formula is correct up to boundary corrections. It becomes *exact* asymptotically as $\Lambda \uparrow \mathbb{Z}^2$ (thermodynamic limit). In this limit $|\Lambda_2| = |\Lambda_0|$ and $|\Lambda_1| = 2|\Lambda_0|$.

We conclude that, up to a trivial factor, we get back the partition function of the Ising model but at a *different temperature* $\beta \rightsquigarrow \beta^*$. The two equations (4.204), (4.206) can be written more symmetrically as

$$\sinh(2\beta^* J)\sinh(2\beta J) = 1 \qquad (4.207)$$

$$Z(\beta) = \sinh(2\beta^* J)^{-|\Lambda_0|}\, Z(\beta^*), \qquad (4.208)$$

or, in terms of the thermodynamic potential $\mathcal{F} \equiv -\log Z$,

$$\mathcal{F}(\beta) = \mathcal{F}(\beta^*) + \log\sinh(2\beta^* J). \qquad (4.209)$$

Equations (4.207) and (4.208) show that repeating the procedure twice we get back the original model at the original temperature, i.e. $(\beta^*)^* = \beta$.

The dual Ising model is defined on the dual lattice Λ^* and its degrees of freedom are the *new* spin variables μ_{a^*} replacing the original ones σ_a. We say that β^* is the dual (inverse) temperature of β. Note that as $\beta \to 0$, we have $\beta^* \to \infty$ and viceversa:

Corollary 4.3 *The duality* $\beta \leftrightarrow \beta^*$ *maps high temperature (the disorder phase) into low temperature (the ordered phase).*

Remark 4.10 Had we kept into due account the boundary conditions, we would have found that the fixed boundary conditions $\pm$ is mapped into the free one,[26] and that the partition function with periodic boundary conditions on both "space" and "time"

$$Z_{pp} \equiv \mathrm{Tr}[T_p^L], \qquad T_p : \left[\begin{array}{l}\text{transfer matrix with periodic}\\ \text{boundary conditions}\\ \text{in the "space" direction}\end{array}\right. \tag{4.210}$$

is equal to

$$(\text{prefactor})\left(\mathrm{Tr}[T_p^{*L}] + \mathrm{Tr}[T_a^{*L}] + \mathrm{Tr}[I\,T_p^{*L}] + \mathrm{Tr}[I\,T_a^{*L}]\right), \tag{4.211}$$

where T_p^* (resp. T_a^*) is the dual transfer matrix with periodic (resp. antiperiodic) conditions and I is the operator that makes $\mu_{a^*} \to -\mu_{a^*}\ \forall a^*$ in the "time" slice.

Discussion The duality means that the physics at the two temperatures β and β^* is the *same:* indeed, for all insertions of spin variables,[27]

$$\left\langle \sigma_{a_1}\,\sigma_{a_2}\cdots\sigma_{a_s}\right\rangle_\beta = \left\langle \mu_{a_1^*}\,\mu_{a_2^*}\cdots\mu_{a_s^*}\right\rangle_{\beta^*} \tag{4.212}$$

where $\langle\cdots\rangle_\beta$ is the probability distribution for the Ising model at the inverse temperature β in the thermodynamic limit.

We arrive to the astonishing conclusion that order and disorder are inherently equivalent. We shall discuss why this is not a contradiction momentarily. This duality is called the *order-disorder duality* or the *Kramers-Wannier* duality after its discovers [14, 15]. We stress that it is just a first example of a large class of "electro-magnetic dualities" which, abstractly speaking, are just functional Fourier transforms in configuration space.

Let us extract some physical consequences of the duality: we know that at high temperature the model is in the *disordered phase* (zero magnetization) while in the low temperature regime it is in an *ordered* phase. Therefore there must be at least one temperature T_c at which we have a phase transition. A priori one may wonder that there may be a chain of $n > 1$ critical temperatures where we have phase

---

[26] This is the discrete version of the fact that T-duality interchanges Dirichlet and Newmann boundary conditions.

[27] Here we use $a^* = a + \frac{1}{2}\sum_i e_i$, see Eq. (4.29).

transitions between $n + 1$ distinct phases: but this scenario is very unlikely. There is no evidence of a third intermediate phase between the high and low temperature ones, and there is no plausible physics such an extra phase may have. On the other hand, if $\beta_c > 0$ is a critical temperature (a point where $\mathcal{F}(\beta)$ is not analytic), so is β_c^*. Hence, if there is only one phase transition, it must happen at the fixed point of the map $\beta \to \beta^*$, i.e. at the *self-dual temperature.* From Eq. (4.207) we get

$$\beta_c J = \frac{1}{2} \operatorname{arcsinh}(1) \approx 0.440687. \tag{4.213}$$

The exact solution of the two-dimensional Ising model confirms that there is a unique transition at this precise temperature [5].

Remark 4.11 One can consider a slightly more general Ising model in which the couplings along the horizontal and vertical links are different. The dual is another Ising model with the same property. See e.g. [5].

Order-Disorder Operators

The Kramers-Wannier duality interchanges the ordered and the disordered phases, and also the original spin operators $\{\sigma_a\}$ with the dual spin operators $\{\mu_{a^*}\}$ living in the dual lattice. We stress that each set of variables yields a *complete* description of the Ising model on its own. At low temperature the phase is *ordered,* and in a pure state the order parameter $\langle \sigma_a \rangle = \pm M(\beta) \neq 0$, while at hight temperature the phase is *disordered* and $\langle \sigma_a \rangle = 0$. Applying the duality we get (in a pure state)

$$\langle \mu_{a^*} \rangle = \begin{cases} \pm M(\beta^*) \neq 0 & \beta < \beta_c \\ 0 & \beta > \beta_c. \end{cases} \tag{4.214}$$

The operator μ_{a^*}, with the property that its expectation is non-zero in the disordered phase, is called the *disorder operator* of the Ising model. The distinction between order σ_a and disorder μ_{a^*} operators is a matter of convention: the physics is identical in opposite regimes of temperature when expressed in dual variables. Disorder is just order described in terms of the dual algebra of operators.

We wish to construct the disorder operators μ_{a^*} in terms of the *original degrees of freedom* of the Ising model $\{\sigma_a\}$. To this end, let us see what is the effect of inserting the operator μ_{a^*} in the a^*-th site of the dual lattice (i.e. in the center of the plaquette at a in the original lattice Λ). We focus on the neighborhood of this plaquette: the following is a purely local analysis. Equation (4.199) gets replaced by

$$Z \langle \mu_{a^*} \rangle = \frac{1}{2^{|\Lambda_2|}} \sum_{\{u_l \in \mathbb{Z}/2\mathbb{Z}\}} \exp\left[\beta J \sum_{l \in \Lambda_1} \exp(i\pi u_l) \right] \times$$

$$\times \sum_{\{m_{b^*} = 0, 1\}} e^{i\pi m_{a^*}} \exp\left[i\pi \sum_{l^* \in \Lambda_1^*} u_l \left(m_{h(l^*)} + m_{t(l^*)} \right) \right] \tag{4.215}$$

Fig. 4.11 The definition of
the disorder operator μ using
Dirac strings. The site in the
dual lattice where μ is
inserted is the pink dot. The
dashed paths in the dual
lattice denoted η_1, η_2 and η_3,
drawn respectively in blue,
purple and red, are three
different but gauge equivalent
choices of the Dirac string

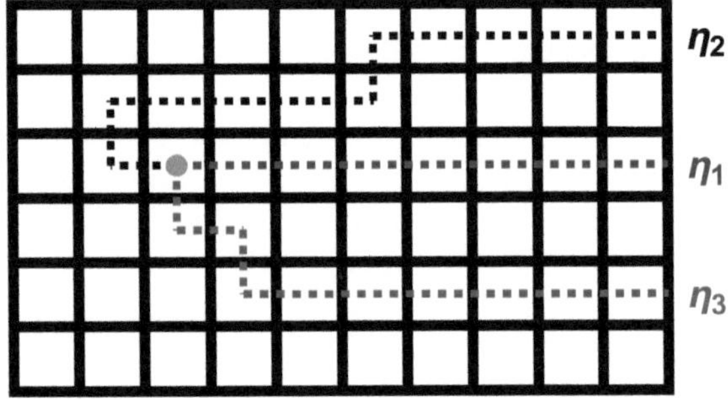

Summing over the m_{b^*}'s with the help of Eq. (4.198), we get the modified constraint

$$\prod_{l \in \partial b^*} U_l \equiv \exp\left[\sum_{l \in \partial b^*} i\pi u_l \right] = \begin{cases} -1 & b^* = a^* \\ 1 & \text{otherwise,} \end{cases} \tag{4.216}$$

that is, inserting μ_{a^*} at the center of the plaquette a^* we create a non-trivial flux
of the $\mathbb{Z}_2$ magnetic field trough the plaquette. Thus μ_{a^*} is the 2d version of the 3d
operator which creates a magnetic monopole (here for the gauge group $\mathbb{Z}_2$; the Dirac
monopole has instead $G = U(1)$). This is natural in view of our description of the
Kramers-Wannier duality as a discrete analog of electric-magnetic duality.

Next we construct the gauge configurations $\{U_l\}$ with the property (4.216) in the
infinite lattice $\mathbb{Z}^2$. They are all gauge equivalent, so it suffices to consider the very
simplest ones. The analogy with the Dirac monopole suggests that we can construct
examples of these configurations using a *Dirac string*. A Dirac string η is a *path*
(oriented concatenation of edges) in the dual lattice Λ^* which starts at the dual site
a^* and goes to infinity, see Fig. 4.11. A Dirac string is a magnetic flux tube which
"absorbs" the flux created by the μ_{a^*} insertion: by flux conservation it can end only
at (dual) sites where a disorder operator is inserted. Having chosen a path η starting
at a^*, a gauge configuration $\{U_l\}$ which satisfies (4.216) is

$$U_l = \begin{cases} -1 & \text{if the link } l \in \Lambda_1 \text{ is crossed by } \eta \\ +1 & \text{otherwise} \end{cases} \tag{4.217}$$

Comparing with the general Hamiltonian (4.177) we realize that inserting the
operator μ_{a^*} is equivalent to computing the canonical partition function with the
modified Hamiltonian obtained by flipping the sign $J \to -J$ of the coupling
between nearest neighbor spins which are connected by a link crossed by η, i.e. the
couplings of spins on the two sides of η are now *anti*-ferromagnetic. The expectation
of the operator μ_{a^*} is then

$$\langle \mu_{a^*} \rangle = \frac{\tilde{Z}(\eta)}{Z}, \tag{4.218}$$

where Z is the canonical partition function for the original Hamiltonian, and $\tilde{Z}(\eta)$ the one for the Hamiltonian modified by a flip of the couplings along the path η starting at $\boldsymbol{a}^*$. The RHS is well-defined, that is, is independent of the chosen η. This is the Kadanoff-Ceva construction of the disorder operators $\mu_{\boldsymbol{a}^*}$ [16].

Naively it may seem that $\mu_{\boldsymbol{a}^*}$ corresponds to a physical object extended along the path η. However it is not so: the local physics in a region R along the path η is not modified by the flip in sign of the couplings. The physics of the system with the given choice of η is gauge equivalent to the one with a different choice of η (as long the two have the same starting point $\boldsymbol{a}^*$), and we may always choose the support of η to be far away from R. Hence the Dirac string has no local effects on the physics: it is "invisible" ($\equiv$ no local observable can detect its presence). Only the local physics in the vicinity of the starting point of η—i.e. at the site $\boldsymbol{a}^*$ where we insert the disorder operator $\mu_{\boldsymbol{a}^*}$—is affected by the modification of the couplings. Indeed now the four spins at the vertices of the plaquette $\boldsymbol{a}^*$ are *frustrated*: if we minimize the energy contribution of three sides of the plaquette, the contribution from the forth one will be maximal.

We conclude that the $\mu_{\boldsymbol{a}^*}$'s are *bona fide* local operators which affect the physics only locally at their insertion points $\boldsymbol{a}^* \in \Lambda^*$. In the 2d Ising model we have two sets of operators $\{\sigma_{\boldsymbol{a}}\}$ and $\{\mu_{\boldsymbol{a}^*}\}$, and each set suffices by itself to describe all the degrees of freedom of the system. The operators in each set are local with respect to each other, *but the $\mu_{\boldsymbol{a}^*}$'s are **not local** with respect to the $\sigma_{\boldsymbol{a}}$'s*. This is already manifest from the Kadanoff-Ceva construction: to make $\{\mu_{\boldsymbol{a}^*}\}$ we have to flip the coupling of *infinitely-many* $\sigma_{\boldsymbol{a}}$. A more intrinsic viewpoint is to study how the order-disorder correlation function

$$\langle \sigma_{\boldsymbol{a}}\, \mu_{\boldsymbol{b}^*} \rangle \tag{4.219}$$

varies when we carry the operator $\sigma_{\boldsymbol{a}}$ along a closed path γ in Λ which encircles the dual site $\boldsymbol{b}^*$: the parallel transport of the order operator along the path γ is given by

$$\prod_{l \in \gamma} U_l = -1, \tag{4.220}$$

since γ contains the $\mu_{\boldsymbol{b}^*}$ insertion point $\boldsymbol{b}^*$. Thus, when the order operator gets back to its original site $\boldsymbol{a}$, it has picked up an extra minus sign

$$\langle \sigma_{\boldsymbol{a}}\, \mu_{\boldsymbol{b}^*} \rangle \rightsquigarrow -\langle \sigma_{\boldsymbol{a}}\, \mu_{\boldsymbol{b}^*} \rangle, \tag{4.221}$$

that is, *a thermal correlation function containing both order and disorder operators is **not univalued** as a function of their insertion points $\boldsymbol{a}, \boldsymbol{b}^*$*. This non-univaluedness of the correlation functions is the definition of *relative non-locality* between the σ's and the μ's: notice that the order operator may be far away from the disorder one, and the path γ may stay quite apart from $\boldsymbol{b}^*$ and yet the sign of $\langle \sigma_{\boldsymbol{a}}\, \mu_{\boldsymbol{b}^*} \rangle$ flips. In other words: the order operator feels the presence of the disorder one even remaining at an arbitrarily large distance from it, and this is precisely what we mean by *mutual non-locality* of two operators.

The fact that μ_{a*}'s are mutually local *between themselves* is a consequence of duality since the σ_a's are by definition mutually local. The fact that the two sets are not mutually local with respect to each other is needed for consistency: *any operator local with respect to the σ_a's will detect the same notion of order*. Since the μ_{a*}'s see a different notion of order, they should be non-local with respect to the σ_a's.

The prediction from the duality is (4.214). Let us check that this is correct using Eq. (4.218). In the ordered phase most spins σ_a point up. Most links with anti-ferromagnetic coupling along η will then contribute a factor $\mathrm{e}^{-2\beta J}$ to $\tilde{Z}/Z$. More precisely: in the thermodynamic limit there are infinitely many *frustrated links* in any configuration and hence

$$\langle \mu_{a*} \rangle = \frac{\tilde{Z}(\eta)}{Z} \approx \mathrm{e}^{-2\beta J |\eta|} \to 0 \quad \text{in the thermodynamic limit.} \tag{4.222}$$

Conversely in the disordered phase the spins are randomly distributed so that $\tilde{Z}/Z$ is non-zero. At strictly $\beta = 0$ (infinite temperature) the ratio is just 1.

More interesting is the correlations of two μ's. Now a Dirac string η which starts from the first μ may terminate in the second one (which absorbs the $\mathbb{Z}_2$ magnetic flux carried by the Dirac string), so we need to flip the coupling of only finitely many links see Fig. 4.12.

Deep in the ordered phase (i.e. away from the critical temperature) each link crossed by η will contribute a factor $\approx \mathrm{e}^{-2\beta J}$ so that the correlation $\langle \mu_{a*} \mu_{b*} \rangle$ will go exponentially to zero as the distance $|a^* - b^*|$ between the two insertion sites goes to infinity. In particular

$$\lim_{|a^*-b^*| \to \infty} \langle \mu_{a*} \mu_{b*} \rangle = \langle \mu_{a*} \rangle \langle \mu_{b*} \rangle \equiv 0, \tag{4.223}$$

showing that the cluster property is strongly satisfied.

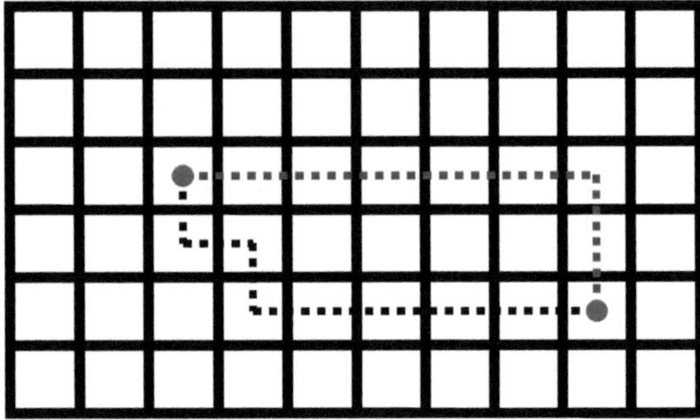

Fig. 4.12 The correlation $\langle \mu_{\tilde{a}} \mu_{\tilde{b}} \rangle$ The disorder operators are inserted in the sites in the dual lattice denoted by the red dots. We can choose the Dirac string to have the two red dots as endpoints. The two Dirac string drawn dashed (blue and purple) are gauge equivalent

Ising Fermions

We define two new operators to be inserted at the site $\boldsymbol{a} \in \Lambda$

$$\psi_{\boldsymbol{a}}^{\pm} = \sigma_{\boldsymbol{a}}\,\mu_{\boldsymbol{a}\pm}, \quad \text{where} \quad \boldsymbol{a}^{\pm} \equiv \boldsymbol{a} \pm \tfrac{1}{2}\sum_i \boldsymbol{e}_i \in \Lambda^* \tag{4.224}$$

We consider a thermal correlation where we insert two such operators. If we carry the first operator around the second one, the correlation function will return to its original value since we pick up a minus sign when carrying the operator $\sigma_{\boldsymbol{a}}$ around $\sigma_{\boldsymbol{b}}\,\mu_{\boldsymbol{b}\pm}$ and another minus sign when carrying $\mu_{\boldsymbol{a}\pm}$ around it. Thus the operators (4.224) are now *mutually local*.

However consider now carrying the first operator along half a circle around the second one. This has the effect of permuting the positions of the two operators. We may do the job in such a way that the order operators cross just one frustrated link, that is, we pick up just *one* minus sign. This means that the ψ's, seen as operators, *anticommute*. Comparing with the discussion of second quantization in Chap. 3, we see that they are *fermions*. Indeed one can show that the $d = 2$ Ising model is a (free) theory of fermions (for more details see e.g. [5] and references therein).

4.9 Continuous Symmetries in $d = 2$: Mermin-Wagner Theorem

The Peierls argument shows that in the $d \geq 2$ Ising model we have a non-zero spontaneous magnetization at low non-zero temperature. In absence of magnetic field the Ising model has a $\mathbb{Z}_2$ symmetry up $\leftrightarrow$ down, and this symmetry is *spontaneously broken* at low temperature, meaning that the correlation functions in a pure state are not invariant under the symmetry of the theory. For instance, the magnetization $\langle \sigma_{\boldsymbol{a}} \rangle$ is $\mathbb{Z}_2$ odd and it is sent to minus itself by the symmetry, so that any non-zero value of the magnetization implies a spontaneous breaking of $\mathbb{Z}_2$.

Spontaneous breaking of a symmetry is a central notion in physics. The Perron-Frobenius argument shows that classical one-dimensional systems (discrete or continuous) do not break any symmetry (at non-zero temperature). In $d \geq 2$ we can construct models which break spontaneously any (finite) discrete symmetry at low but non-zero temperature, the Ising model being a primary example of this story. *What about **continuous** symmetries* ($\equiv$ the symmetry group G is a non-trivial connected Lie group)?

Spontaneous breaking of a continuous symmetry is possible but only in dimensions $d \geq 3$. This is the statement of the Mermin-Wagner theorem equivalent to the Coleman theorem (which is stated in the language of QFT).

Theorem 4.13 (Mermin-Wagner [17], Coleman [18]) *A continuous symmetry cannot spontaneously break in $d = 2$, that is, when $d = 2$ the correlation functions in **any** phase are invariant under all continuous symmetry of the Hamiltonian.*

The **Theorem** refers to an arbitrary (connected) Lie symmetry G. By restricting to a one-parameter subgroup of G we can assume that the symmetry group is $U(1)$ with no loss. By a reparametrization of the degrees of freedom we can always reduce to variables of the form $\{\theta_a, \chi_a^\alpha\}_{a \in \Lambda_0}$ where θ_a is an angle on which $U(1)$ acts by translations

$$\theta_a \rightsquigarrow \theta_a + \alpha \mod 2\pi \quad \forall a \in \Lambda_0, \tag{4.225}$$

and the χ_a^α's are invariant under $U(1)$. The χ_a^α's play no role in the argument, and we shall disregard them. We remain with a system (called the *XY model* or the *rotor model*) where the degrees of freedom are one element $e^{i\theta_a}$ of $U(1)$ at each site $a \in \Lambda_0$ of our 2-dimensional regular lattice Λ. The Hamiltonian of the "ferromagnetic" XY-model is

$$H = -\sum_{l \in \Lambda_1} F(\theta_{h(l)} - \theta_{t(l)}) \tag{4.226}$$

where $F(x)$ is a periodic function of period 2π with $F(-x) = F(x)$, whose maximum is at $x = 0 \mod 2\pi$. The most common choice is $F(x) = J \cos(x)$ with $J > 0$. The XY model differs from the Ising model only by the replacement of the finite group $\mathbb{Z}_2$ with the continuous one $U(1)$. However this replacement has a dramatic physical consequence: now the magnetization vanishes for all $\beta < \infty$

$$\langle e^{i\theta_a} \rangle = 0 \tag{4.227}$$

in *all* phases (the XY model has distinct high and low temperature phases [19]).

Let us prove the above claim. By definition, if the $U(1)$ symmetry was spontaneously broken we could find an operator O such that its variation under the $U(1)$ symmetry, δO, has non-zero expectation value $\langle \delta O \rangle \neq 0$, that is, such that δO is an *order parameter* for the $U(1)$ symmetry. To simplify the notation we assume that O has support at a single site a, i.e. it is of the form $O \equiv f(\theta_a)$ for some periodic function $f(x)$. The extension of the argument to the general case is straightforward.

Ward Identities

In this paragraph Λ is a finite regular lattice in d-dimensions that we take cubic for definiteness. Integration by parts yields the identity

$$
0 = \oint \prod_c \frac{d\theta_c}{2\pi} \frac{\partial}{\partial \theta_b} \left(f(\theta_a) \prod_{l \in \Lambda_1} \exp\left[\beta F(\theta_{h(l)} - \theta_{t(l)})\right] \right) =
$$

$$
= \oint \frac{d\theta_c}{2\pi} \prod_{l \in \Lambda_1} \exp\left[\beta F(\theta_{h(l)} - \theta_{t(l)})\right] \times
$$

$$
\times \left(\delta_{a,b}\, f'(\theta_a) - \beta\, f(\theta_a) \sum_{i=1}^{d} \left[F'(\theta_{b+e_i} - \theta_b) - F'(\theta_b - \theta_{b-e_i}) \right] \right)
$$

$$
\tag{4.228}
$$

We introduce the notations[28]

$$
J(a)_i \equiv F'(\theta_a - \theta_{a-e_i}), \tag{4.229}
$$

$$
\nabla_j J(a)_i \equiv J(a + e_j)_i - J(a)_i. \tag{4.230}
$$

Then the identity (4.228) becomes[29]

$$
\delta_{a,b} \left\langle f'(\theta_a) \right\rangle = \beta\, \nabla_i \left\langle J(b)_i\, f(\theta_a) \right\rangle \quad \forall\, b \in \Lambda \tag{4.231}
$$

which is (a special instance of) the *Ward identity*. This identity is the discretized version of the conservation law of the Noether current [20] associated to the shift symmetry (4.225), and the Ward identities we are after are the lattice version of the usual Ward identities in continuous space field theory [20].

Consider now a smaller cubic box $\Lambda' \subset \Lambda$ centered at the insertion site a of size $2N$ (see Fig. 4.13 for the $d = 2$ case). We sum the identity (4.231) over all points $b \in \Lambda'$. On the RHS we get a pure boundary term

$$
\left\langle f'(\theta_a) \right\rangle = \beta \sum_{b \in \partial \Lambda'} \left\langle J(b)_n\, f(\theta_a) \right\rangle \tag{4.232}
$$

where $J(b)_n$ is the component normal to the boundary, that is, if $b_i = a_i \pm N$ we have $J(b)_n \equiv \pm J(b \pm e_i)_i$. In the limit $|\Lambda| \to \infty$ Eq. (4.232) becomes an identity between expectation values in the thermodynamic limit valid for all choices

---

[28] ∇_j is the forward partial derivative in the j-th direction on the lattice Λ while $J(a)_i$ is the i-th component of the lattice version of the conserved "Noether" current associated to the $U(1)$ symmetry ("electric current" in the standard terminology).

[29] Sum over the repeated index i implied.

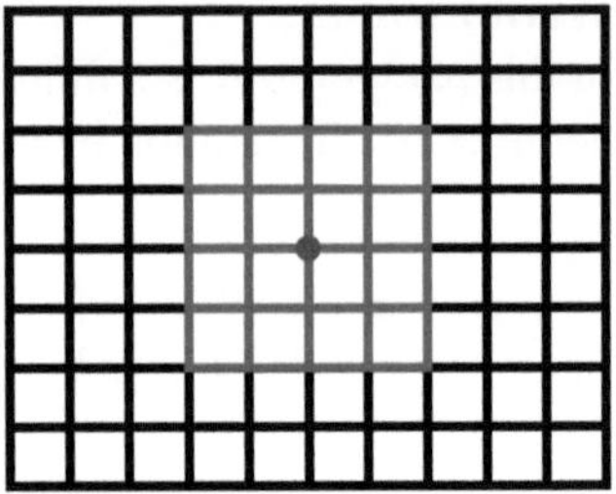

Fig. 4.13 The sublattice $\Lambda' \subset \Lambda$ is drawn in red and contains the insertion site a (blue dot) in its center (here $N = 2$)

of boundary conditions in the bigger box Λ, i.e. in all equilibrium states. The RHS is independent of the size of the box Λ': taking $\Lambda' \uparrow \mathbb{Z}^2$ sends $|b| \to \infty$. If the LHS of (4.232) is non-zero, i.e. if the $U(1)$ symmetry is broken in the particular phase, the correlation

$$\langle J(b)_n\, f(\theta_a)\rangle \tag{4.233}$$

can go to zero at most as $|b|^{-(d-1)}$ as $|b| \to \infty$. We consider the minimal Hamiltonian, i.e. $F'(x) = -\sin x$; the general case is reduced to this one by considering the Fourier series of $F'(x)$:

$$\langle J(b)_n\, f(\theta_a)\rangle = -\langle \sin(\theta_b - \theta_{b-e_n})\, f(\theta_a)\rangle \approx$$
$$\approx -\langle (\theta_b - \theta_{b-e_n})\, f(\theta_a)\rangle. \tag{4.234}$$

Thus $\langle J(b)_n\, f(\theta_a)\rangle$ is the discretized derivative of $-\langle \theta_b\, f(\theta_a)\rangle$ with respect to the position of the first insertion. Since the LHS goes to zero not more rapidly that the distance $|a - b|$ to the power $-(d - 1)$ we must have

$$\langle \theta_b\, f(\theta_a)\rangle = \begin{cases} O(|a - b|^{-(d-2)}) & \text{for } d \geq 3 \\ O(\log |a - b|) & \text{for } d = 2, \end{cases} \tag{4.235}$$

that is, $\langle \theta_b\, f(\theta_a)\rangle$ goes to zero at infinity as the Coulomb potential in d dimensions. But in $d = 2$ the Coulomb potential *grows* logarithmically with distance, and this behavior is not consistent with the cluster property. In $d = 2$ the cluster property then implies that the RHS of (4.232) is zero, and so must be the LHS. Thus no there is no spontaneous breaking of continuous symmetries in $d = 2$. For $d \geq 3$ there is no constraint to symmetry breaking, except that the correlation (4.235) cannot decay with the distance r more rapidly than the power $r^{-(d-2)}$ in presence of breaking.

Corollary 4.4 (Goldstone Theorem) *In arbitrary dimension $d \geq 3$, suppose that a continuous symmetry is broken with order parameter $\langle \delta O \rangle \neq 0$. Then the two-*

point correlation $\langle O(x)\,O(0)\rangle$ goes to zero only as a power $|x|^{-(d-2)}$ of the distance (i.e. the correlation length[30] ξ is infinite).

4.10 Critical Exponents and Universality Classes

We close this chapter by returning to the issue of the possible kinds of phase transitions that we hinted at in Sect. 4.1. The modern theory classifies the critical phenomena in *universality classes*.

Close to a critical point the order parameters and the response functions (such as heat capacity, compressibility, magnetic susceptibility, etc.)[31] have special behaviors which are typically singular. Based on both theoretical and experimental motivations, these behaviors are expressed in the form of *power laws* whose exponents are called *critical exponents*. These exponents characterize the *universality classes* of statistical systems. All these assertions can be fully justified using the Renormalization Group approach to critical phenomena [5, 21] a topic which is beyond this textbook.

Many of the "abstract" statements of this section will become concrete in Chap. 5 where we compute several of the relevant quantities using various approximation schemes and/or phenomenological descriptions.

Critical Exponents

We review the critical exponents more frequently used. The symbols α, β, γ, δ, ν, and η are nowadays standard in the literature.

We start from the connected two point function of the order parameter $\sigma(r)$

$$\langle \sigma(r)\,\sigma(0)\rangle_{\text{conn.}} \overset{\text{def}}{=} \langle \sigma(r)\sigma(0)\rangle - \langle \sigma(r)\rangle\langle \sigma(0)\rangle \tag{4.236}$$

in a pure state. We know that it goes to zero as $r \equiv |r| \to \infty$ by the cluster property of pure states. At a generic point in the space of control parameters, the pure state connected correlation functions go to zero *exponentially* up to a power, i.e. we have roughly a behavior of the form

$$\langle \sigma(r)\,\sigma(0)\rangle_{\text{conn}} \sim \text{const.}\, r^{-a} e^{-r/\xi} \qquad \text{for } r \gg \xi, \tag{4.237}$$

since the correlation functions are invariant under rotations. The length scale ξ (which is a function of the control parameters) is called the *correlation length*. The

---

[30] For the *correlation length* see Sects. 4.10 and 5.7.

[31] Linear response theory and the generalized susceptibilities ($\equiv$ response functions) will be treated in detail in Chap. 6.

exact expression of the connected 2-point function may be written in the form

$$\langle \sigma(r)\,\sigma(0)\rangle_{\mathrm{con}} = \frac{1}{r^{d-2-\eta}}\, f(r/\xi) \tag{4.238}$$

where $f(r/\xi)$ is some function which vanishes exponentially for large r and tends to a constant as $r \to 0$. In Eq. (4.238) d is the spatial dimension of the model, and η is its *critical exponent*. It is customary to write the exponent as $-(d-2-\eta)$ because in the Gaussian model (i.e. the free theory) the exponent is $-(d-2)$ and η is the so-called "anomalous" part of the exponent which yields its deviation from the "naive" free value $-(d-2)$ produced by non-trivial interactions (see Sect. 5.7 for more).

Equation (4.238) says that the typical length scale of the fluctuations of the order field $\sigma(r)$ is $\xi(T)$. The cluster property yields

$$\lim_{r\to\infty} \langle \sigma(r)\,\sigma(0)\rangle = \begin{cases} 0 & \text{for } T > T_c \\ \langle\sigma\rangle^2 \neq 0 & \text{for } T < T_c. \end{cases} \tag{4.239}$$

To set a long-range order in a disordered system, the correlations must have a long range. Therefore, when we approach the transition from above, $T \to T_c^+$, the correlation length must diverge $\xi \to \infty$. Dually, approaching the critical temperature from below, $T \to T_c^-$, the fluctuations in the order field $\sigma(r)$ should have long range to be able to destroy the long-range order. Therefore $\xi \to \infty$ when we approach T_c from both sides. To understand the nature of the phase transition of the particular system at hand, we study the leading behavior of ξ as we approach T_c from the left and from the right

$$\xi(T) = \begin{cases} \xi_+\, t^{-\nu}(1 + O(t)) & T > T_c \\ \xi_-\,(-t)^{-\nu}(1 + O(t)) & T < T_c, \end{cases} \tag{4.240}$$

where

$$t \equiv \frac{T - T_c}{T_c}, \tag{4.241}$$

and ξ_+, ξ_- are constants. ν is the *critical exponent* of the correlation length. It is a deep result of the Renormalization Group approach that the exponent ν is the same above and below the critical temperature while the overall constants $\xi_\pm$ are typically different on the two sides.

There are other important critical exponents. Consider a ferromagnetic system in the ordered phase where the spontaneous magnetization M is non-zero (at zero magnetic field $B = 0$). As we approach the critical temperature T_c from below the magnetization goes to zero. Again we are interested in understanding the behavior of M at criticality. This behavior defines a new critical exponent called β

$$M \simeq M_0(-t)^\beta, \quad \text{as } t \to 0^-. \tag{4.242}$$

We can also consider the dependence of M from the magnetic field B while keeping the temperature fixed at its critical value T_c; this defines the critical exponent δ

$$M(B, T_c) \simeq m_0 \, B^{1/\delta} \quad \text{as } B \to 0^+. \tag{4.243}$$

The *magnetic susceptibility* is the response function of the system to an external magnetic field B

$$\chi(T) = \left. \frac{\partial M(B, T)}{\partial B} \right|_{B=0} \tag{4.244}$$

(see Chap. 6 for the general theory of response functions). The magnetic susceptibility is singular at a magnetic phase transition

$$\chi(T) = \begin{cases} \chi_+ t^{-\gamma} & T > T_c \\ \chi_-(-t)^{-\gamma} & T < T_c, \end{cases} \tag{4.245}$$

where, again, the critical exponent γ is the same on the two sides. Another quantity which (in general) has a singular behavior at $T = T_c$ is the heat capacity $C_V(T)$

$$C_V(T) = \begin{cases} C_+ t^{-\alpha} & T > T_c \\ C_-(-t)^{-\alpha} & T < T_c, \end{cases} \tag{4.246}$$

which defines the critical exponent α. The definitions of the standard critical exponents are summarized in Table 4.1.

The exponents α, β, γ, δ, ν and η are not all independent. They satisfy a number of algebraic relations (whose coefficients may depend on the dimension d of the system). We quote them here without proof. Some of these relations will

Table 4.1   Summary of the most widely used critical exponents

Exponent	Definition	At
α	$C_V \propto \lvert T - T_c \rvert^{-\alpha}$	$B = 0$
β	$M \propto (T_c - T)^{-\beta}$	$B = 0, T < T_c$
γ	$\chi \propto \lvert T - T_c \rvert^{-\gamma}$	$B = 0$
δ	$\lvert M \rvert \sim \lvert B \rvert^{1/\delta}$	$T = T_c$
ν	$\xi \propto \lvert T - T_c \rvert^{-\nu}$	$B = 0$
η	$\langle \sigma(r)\,\sigma(0) \rangle \propto r^{-(d-2+\eta)}$	$T = T_c, B = 0$

be illustrated in Chap. 5 in the context of the Landau-Ginzburg theory and further analyzed in Chap. 6 with the help of linear response theory.

$$\alpha + 2\beta + \gamma = 2 \tag{4.247}$$

$$\alpha + \beta\delta + \delta = 2 \tag{4.248}$$

$$\nu(2 - \eta) = \gamma \tag{4.249}$$

$$\alpha + \nu d = 2. \tag{4.250}$$

To illustrate the meaning of these relations we sketch a proof of (4.249) (say for the Ising model):

$$\text{const. } t^{-\gamma} = \chi = \frac{\partial}{\partial B}\langle \sigma_0 \rangle\big|_{B=0} = \frac{1}{\beta}\sum_{i\in\Lambda_0}\langle \sigma_i\sigma_0\rangle \approx \frac{1}{\beta}\int d^d r\,\langle \sigma(r)\sigma(0)\rangle \approx$$

$$\approx \frac{\text{const.}}{\beta}\int \frac{d^d r}{r^{d-2-\eta}}\,e^{-r/\xi} = \text{const. }\xi^{2-\eta} = \text{const. }t^{-(2-\eta)\nu}. \tag{4.251}$$

Universality Classes

The critical behavior of a system near a transition point is independent of most details of the underlying microscopic model. Generically it depends only on the space dimensionality d of the system and its symmetry group G; by this we mean that two *generic* models with the same d and G will have the same critical behavior. Two models with the same critical behavior are said to belong to the *same universality class*. A proper theory of the universality classes requires the use of the Renormalization Group which is part of Quantum Field Theory or, in the present application to critical phenomena, of Statistical Field Theory [5, 10, 21], topics which are beyond this textbook.

The fact that critical phenomena organize in universality classes implies that we can limit ourselves to study the "minimal model" of each class, i.e. the simpler one. This is a great advantage.

From the previous discussion it is clear that the critical exponents α, β, γ, δ, ν and η are properties of the universality class, not of the particular microscopic model. The Renormalization Group analysis shows that there are other invariants which depend only on the universality class and not on the specific model. Simple examples of such universality invariants are the *invariant ratios*

$$\xi_+/\xi_-, \quad C_+/C_-, \quad \chi_+/\chi_- \tag{4.252}$$

For more general universality invariants, see e.g. [5].

Critical Dimensions

Each class of statistical models has two critical dimensions d_1 and d_2. d_1 is defined by the condition that in dimension $d \le d_1$ the system has no phase transition at positive temperature. For instance: for the Ising model $d_1 = 1$; the same holds for other discrete systems with finite symmetry groups. For systems with continuous symmetries we have $d_1 \geq 2$ by the Mermin-Wagner theorem. The upper critical dimension d_2 is the dimension such that for $d > d_2$ the system behaves at criticality as a free system (more precisely: as expected from mean field theory, see Chap. 5). For the Ising model the upper critical dimension is 4 while for the tri-critical Ising model $d_2 = 6$: both statements will be demonstrated and clarified in Chap. 5.

Appendix: Proofs in Yang-Lee Theory

We start from a **Lemma** in complex analysis [1]:

Lemma 4.3 *Consider the series*

$$\sum_{l=0}^{\infty} b_l(V)z^l = S(V, z) \tag{4.253}$$

where $|b_l(V)| \leq A\sigma^{-l}$ for some positive constants A, σ. For all real z between $-\sigma$ and σ assume $\lim_{V\to\infty} S(V, z)$ to exist. Then

(a) $\lim_{V\to\infty} b_l(V) \equiv b_l(\infty)$ exists;
(b) $\lim_{V\to\infty} S(V, z) = \sum_{l=0}^{\infty} b_l(\infty)z^l$ for all $|z| < \sigma$ (where the series converges).

Proof $b_0(\infty) = \lim_{V\to\infty} S(V, 0)$ exists. Let us prove that $\lim_{V\to\infty} b_1(V)$ exists. For $0 < \epsilon < \sigma/2$, there exists a volume V_0 such that for all volumes $V, W > V_0$

$$|S(V, \epsilon) - S(W, \epsilon)| < \epsilon^2, \qquad |b_0(V) - b_0(w)| < \epsilon^2. \tag{4.254}$$

Since

$$\left|\sum_{l=2}^{\infty} b_l(V)\epsilon^l\right| \leq \frac{A\epsilon^2}{1 - \epsilon\sigma^{-1}} \leq 2A\epsilon^2 \geq \left|\sum_{l=2}^{\infty} b_l(W)\epsilon^l\right|, \tag{4.255}$$

while

$$\epsilon \, b_1(V) = S(V, \epsilon) - b_0(v) - \sum_{l=2}^{\infty} b_l(V)\epsilon^l, \tag{4.256}$$

and one has

$$\epsilon|b_1(V) - b_1(w)| < (2 + 4A^2)\epsilon^2, \tag{4.257}$$

and $\lim_{V\to\infty} b_1(V)$ exists; the same argument shows that all $\lim_{V\to\infty} b_l(V)$ exist. Point (b) is then standard. $\qquad\square$

Proof of Theorem 4.1 We follow [1]. At finite volume we have

$$Q(V, \zeta) = \prod_{i=1}^{d} \left(1 - \frac{\zeta}{\zeta_i} \right) \tag{4.258}$$

where the $\zeta_i = \zeta_i(v)$ are the (complex) zeros of the polynomial. Suppose that there is a disk $D \subset R$ centered at the point $\zeta = \eta$ on the real axis, and set $z = \zeta - \eta$. Then

$$Q(V, z) = \prod_{i=1}^{d} \left(1 - \frac{z}{z_i} \right) \frac{z_i}{\zeta_i} \tag{4.259}$$

where $z_i \equiv \zeta_i - \eta$ are the roots of the translated polynomial. Since $\log Q(V, z)$ is analytic in D, we have

$$\frac{1}{V} \log Q(V, z) = \sum_{l=0}^{\infty} b_l(V) z^l \tag{4.260}$$

with

$$b_l(V) = -\frac{1}{V l} \sum_{j=1}^{d} \frac{1}{z_i^l} \quad \text{for } l \geq 1, \tag{4.261}$$

and

$$b_0(V) = \frac{1}{V} \sum_{i=1}^{d} \log \frac{z_i}{\zeta_i} \tag{4.262}$$

Let σ be the radius of D; since D does not contain roots, $|z_i| \geq \sigma$. Then by (4.261)

$$|b_l(z)| \leq \frac{d}{V} \frac{1}{l \sigma^l} \quad \text{for } l \geq 1. \tag{4.263}$$

d/V is bounded by the constant c. So we can apply Lemma 4.3 to D and prove Theorem 4.1 for $\zeta \in D$. By similar arguments we extend the theorem into a disk $D' \subset R$ whose center is in D. Repeating the process we cover the full region R. $\qquad\square$

Problems

4.1 Consider the one-dimensional Ising model with Hamiltonian normalized to that

$$H = -J \sum_{i=1}^{L} (\sigma_i \sigma_{i+1} - 1)$$

with periodic boundary condition $\sigma_{i+L} \equiv \sigma_i$. Let $Z_L(z)$ be the canonical partition function as a function of $z = \exp(-2\beta J)$. Find its $2[L/2]$ zeros in the complex z-plane. Show that $z = 0$ is the only *real* accumulation point of zeros of $Z_L(z)$ as $L \to \infty$ (thermodynamic limit).

4.2 One-Dimensional Potts Model

Consider a model defined on the one-dimensional lattice $\mathbb{Z}$ where on each site i we have a discrete degree of freedom ξ_i which takes the n values $\{1, 2, \ldots, n\}$. The Hamiltonian is

$$H = -J \sum_i \delta_{\xi_i, \xi_{i+1}} \qquad J > 0$$

where $\delta_{\xi_i, \xi_{i+1}}$ is Kronecker's delta equal 1 if $\xi_i = \xi_{i+1}$ and 0 otherwise. Compute the free energy in the thermodynamic limit and discuss the phases of the system.

4.3 One-Dimensional Gaussian Model

On each site of the periodic one-dimensional lattice of period N, $\Lambda \equiv \mathbb{Z}/N\mathbb{Z}$, there is a continuous degree of freedom x_i taking value in $\mathbb{R}$. The Hamiltonian is

$$H = \frac{1}{2} \sum_{i=1}^{N} (x_{i+1} - x_i)^2.$$

Compute the partition function at finite N. Then take the thermodynamic limit.

4.4 Ising Model on a Caley Tree

A Caley tree of valency z is an infinite graph without closed loops ($\equiv$ a tree graph) whose vertices have all valency z (i.e. z nearest neighbors). We the Ising model with Hamiltonian

$$H = -J \sum_{<i,j>} \sigma_i \sigma_j \qquad \sigma_i = \pm 1, \quad J > 0$$

(the sum is over pair of nearest neighbor sites). Compute the free energy using the free boundary condition. What changes with other boundary conditions?

4.5 Two-Dimensional Gaussian Model

Same Hamiltonian as in Problem 4.3 but with a two-dimensional square lattice (with nearest neighbor interactions along all links). Compute the partition function in a periodic lattice of sizes M and N. Then take the thermodynamic limit.

4.6 Work out the details of the duality between the Ising models defined on a two-dimensional triangular lattice and a hexagonal (regular) lattice.

4.7 Duality in Two-Dimensional Spin Models

Consider a periodic 2-dimensional square lattice Λ. On each site i of Λ we have a spin taking values in $\mathbb{Z}/m\mathbb{Z}$ (m an integer > 2) which we see as a discrete angle θ_i taking the values $2\pi k/m$ with k an integer defined mod m. The Hamiltonian is

$$H = -J \sum_{<i,j>} \cos(\theta_i - \theta_j)$$

(sum over nearest neighbors). Construct the dual model on the dual lattice Λ^*.

References

1. C.N. Yang, T.D. Lee, Statistical theory of equations of state and phase transitions. I. Theory of condensation. Phys. Rev. **87**, 404–409 (1952)
2. D. Ruelle, *Statistical Mechanics. Rigorous Results* (Benjamin, 1969)
3. J.L. Schiff, *Normal Families*. Universitext (Springer, 1993)
4. A.J. Berlinsky, A.B. Harris, *Statistical Mechanics. An Introductory Graduate Course* (Springer, 2019)
5. G. Mussardo, *Statistical Field Theory. An Introduction to Exactly Solved Models in Statistical Physics*, 2nd edn. Oxford Graduate Texts (Oxford University Press, 2020)
6. R.J. Baxter, *Exactly Solved Models in Statistical Mechanics* (Academic Press, 1982)
7. G.H. Wannier, Antiferromagnetism. The triangular Ising net. Phys. Rev. **79**, 357–364 (1950)
8. F. Gantmacher, *The Theory of Matrices*, vol. 2 (AMS, 2000)
9. J. Glimm, A. Jaffe, *Quantum Physics. A Functional Integral Point of View* (Springer, 1981)
10. C. Itzykson, J.-M. Drouffe, *Statistical Field Theory*, vols. 1 and 2. Cambridge Monographs in Mathematical Physics (Cambridge University Press, 1989)
11. S. Cecotti, *Supersymmetric Field Theories. Geometric Structures and Dualities* (Cambridge University Press, 2015)
12. S. Cecotti, *Introduction to String Theory*. Theoretical and Mathematical Physics (Springer, 2023)
13. S. Cecotti, S. Ferrara, L. Girardello, Hidden noncompact symmetries in string theory. Nucl. Phys. B **308**, 436 (1988)
14. H.A. Kramers, G.H. Wannier, Statistics of the two-dimensional ferromagnet. Part I. Phys. Rev. **60**, 252 (1941)
15. H.A. Kramers, G.H. Wannier, Statistics of the two-dimensional ferromagnet. Part II. Phys. Rev. **60**, 263 (1941)
16. L.P. Kadanoff, H. Ceva, Determination of an operator algebra for the two-dimensional Ising model. Phys. Rev. **B3**, 3918 (1971)
17. N. Mermin, H. Wagner, Absence of ferromagnetism or antiferromagnetism in one- or two-dimensional isotropic Heisenberg models. Phys. Rev. Lett. **17**, 1133–1136 (1966)

18. S. Coleman, There are no Goldstone bosons in two dimensions. Comm. Math. Phys. **31**, 259–264 (1973)
19. J.M. Kosterlitz, D.J.Thouless, Ordering, metastability and phase transitions in two-dimensional systems. J. Phys. **C6**, 1181–1203 (1972)
20. M.E. Peskin, D.V. Schroeder, *An Introduction to Quantum Field Theory* (Perseus Books, 1995)
21. D.J. Amit, *Field Theory, the Renormalization Group, and Critical Phenomena*. International Series in Pure and Applied Physics (McGraw-Hill, 1978)

Chapter 5
Approximate Methods and Landau Theory

The computation of the partition function and other relevant quantities for an interacting system with a huge number of degrees of freedom is typically very complicated. A few exactly solvable models exist (notably the 2d Ising model), but for most systems of physical interest, including the Ising models in $d \geq 3$, we have to revert to approximate techniques and effective methods. We are in particular interested in approximation schemes which allow us to study the physical behavior in the vicinity of a critical point, in order to get a more detailed (if approximate) understanding of these crucial and tricky phenomena. The simplest such scheme is the Landau theory of phase transitions and its refinements to be introduced later in this chapter.

5.1 Mean Field Theory

The simpler and more widely applicable approximation scheme is the so-called *mean field* (MF) approach which is ubiquitous in Theoretical Physics (under different names in the diverse subjects).

The physical idea is to replace the complicate effects on one degree of freedom due to its interactions with each one of the others with the "average" effect of all degrees of freedom. This process can be described (and somehow justified) in different physical languages. We describe its logic mathematically: one writes a (physically well motivated) *ansatz* for the answer which depends on a few undetermined parameters which measure the collective effect of the several degrees of freedom on one of them, and then optimize the *ansatz* by fixing these parameters by a *variational principle* i.e. by imposing that the trial thermodynamic potential is maximal/minimal according to the convexity properties of the potential at hand. In other words, the method combines the variational characterization of the exact thermodynamic potential as the function which extremizes the appropriate

229

S. Cecotti, *Statistical Mechanics*, UNITEXT for Physics,
https://doi.org/10.1007/978-3-031-67874-5_5

functional[1] inside its natural space of functions, and the physical insight which provides a "good" finite-dimensional sub-family of trial potentials where to look for the extremum.

To better illustrate the ideas beyond the method, we start by reconsidering the Ising model in this (approximate) approach.

The Ising Model Again

We consider the ferromagnetic Ising model on a lattice Λ of dimension d with Hamiltonian

$$H = -J \sum_{l \in \Lambda_1} \sigma_{h(l)} \sigma_{t(l)} - B \sum_{a \in \Lambda_0} \sigma_a. \tag{5.1}$$

We do not specify the shape of the lattice, but only its *coordination number z*, i.e. the number of nearest neighbors of each site. For instance: for the cubic lattice $z = 2d$, for the $d = 2$ triangle lattice $z = 6$, etc.

We know from Chap. 4 that in the thermodynamic limit (or in a finite box with periodic boundary condition) the equilibrium states (*pure* as well *mixed*) are translational invariant. In particular the one-spin expectation value

$$\langle \sigma_a \rangle = m, \tag{5.2}$$

is independent of the site $a \in \Lambda_0$ and equal to the *magnetization*

$$m = \text{magnetization} \overset{\text{def}}{=} \lim_{\Lambda \to \infty} \frac{1}{|\Lambda_0|} \sum_{a \in \Lambda_0} \langle \sigma_a \rangle. \tag{5.3}$$

When we consider the collective effect on one spin σ_a of all the others, it is natural to think that, on the average, it is given by the magnetic field produced by all the spins in the material, which is proportional to the magnetization m. In formulae: we start from the identity

$$\sigma_{h(l)} \sigma_{t(l)} = -m^2 + m(\sigma_{h(l)} + \sigma_{t(l)}) + (\sigma_{h(l)} - m)(\sigma_{t(l)} - m), \tag{5.4}$$

valid for all $m \in \mathbb{R}$, and rewrite the Hamiltonian in the form

$$H = J|\Lambda_1|m^2 - (zmJ + B) \sum_{a \in \Lambda_0} \sigma_a - J \sum_{l \in \Lambda_1} (\sigma_{h(l)} - m)(\sigma_{t(l)} - m). \tag{5.5}$$

The last term is quadratic in the fluctuations of the spin variables σ_a around their mean value m. If the fluctuations were small, this term would be negligible with respect to the others. The *mean field approximation* consists in neglecting

---

[1] For instance, it maximizes entropy at constant U or it minimizes U at constant S.

the terms of higher order in the fluctuations, so that each spin σ_a interacts only with the average magnetic field collectively produced by all spins. Clearly such an approximation is reliable only in physical regimes where the fluctuations are small enough. Therefore one has to evaluate *a posteriori* the size of these fluctuations in order to validate or reject the results of the mean field computation.

The mean field approximate Hamiltonian is then

$$H_{\mathrm{MF}} \equiv H_{\mathrm{MF}}(m) = J\big|\Lambda_1\big|m^2 - B_{\mathrm{eff}} \sum_{a\in\Lambda_0} \sigma_a \tag{5.6}$$

where the effective magnetic field

$$B_{\mathrm{eff}} \equiv z\,Jm + B \tag{5.7}$$

is the sum of the external field B plus the magnetic field $z\,Jm$ induced by the spins themselves. Now the computation of the canonical partition function is trivial

$$Z_{\mathrm{MF}}(m) = \mathrm{e}^{-\beta Jm^2|\Lambda_1|} \prod_{i\in|\Lambda_0|} \left(\sum_{\sigma_a=\pm 1} \mathrm{e}^{\beta B_{\mathrm{eff}}\sigma_a} \right) = \left(2\mathrm{e}^{-z\beta Jm^2/2}\cosh(\beta B_{\mathrm{eff}})\right)^{|\Lambda_0|} \tag{5.8}$$

where we used $2|\Lambda_1| = z|\Lambda_0|$. The mean field "free energy" per site is

$$F_{\mathrm{MF}}(m) = zJm^2/2 - kT \log\big(2\cosh(\beta B_{\mathrm{eff}})\big) \tag{5.9}$$

see the plots in Fig. 5.1 for $B = 0$. We wrote "free energy" in quotes because for now $F_{\mathrm{MF}}(m)$ is just a one-parameter family of trial free energies parametrized by m.

To fix the mean field Hamiltonian $H_{\mathrm{MF}}(m)$ it remains to determine the parameter m. From a physics viewpoint a necessary condition that m should satisfy is the so-called *self-consistency condition:* the magnetization computed using the

Fig. 5.1 The mean field free energy F_{MF} of the Ising model above and below the critical temperature, $T_c = zJ/k$ at $B = 0$ (see main text)

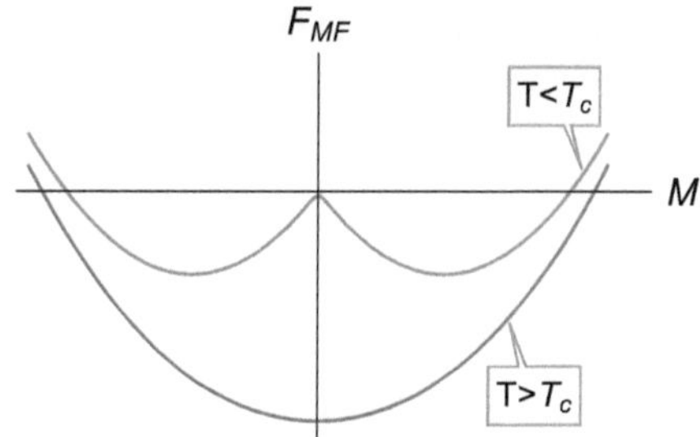

approximate ensemble distribution

$$\frac{e^{-\beta H_{\mathrm{MF}}(m)}}{Z(m)_{\mathrm{MF}}} \tag{5.10}$$

should give back the input magnetization m

$$m \equiv \langle \sigma_i \rangle = \frac{1}{Z_{\mathrm{MF}}} \sum_{\{\sigma_a = \pm 1\}} \sigma_i\, e^{-\beta H_{\mathrm{MF}}} = -\frac{\partial F_{\mathrm{MF}}}{\partial B} = \tanh(z\,\beta J m + \beta B) \tag{5.11}$$

an equation which determines m as a function of β, J, and B. The significance of this equation is to fix the parameter m in the effective Hamiltonian $H_{\mathrm{MF}}(m)$ so that (5.10) is the best approximation to the exact canonical ensemble distribution.

As already mentioned, there is a better interpretation of Eq. (5.11) which is variational in nature and follows directly from the fundamental principles. We know that the free energy is *minimized* by the *exact* canonical probability distribution; hence the optimal distribution in the approximate family

$$\left\{ \frac{e^{-\beta H_{\mathrm{MF}}(m)}}{Z(m)_{\mathrm{MF}}}, \quad m \in \mathbb{R} \right\} \tag{5.12}$$

is the one which yields the minimal free energy, that is, we need to impose

$$\frac{\partial F_{\mathrm{MF}}}{\partial m} = zJ\left(m - \tanh(\beta B_{\mathrm{eff}}) \right) = 0 \tag{5.13}$$

$$\frac{\partial^2 F_{\mathrm{MF}}}{\partial m^2} = zJ\left(1 - \frac{z\beta J}{\cosh^2(\beta B_{\mathrm{eff}})} \right) \geq 0. \tag{5.14}$$

Equation (5.13) gives back the "consistency" equation. However this is merely a necessary condition: we need also the convexity condition (5.14) to hold (otherwise the putative approximate distribution will violate the Griffiths inequalities and hence will be certainly wrong). If both equations hold, m is a *local* minimum of $F_{\mathrm{MF}}(m)$: we still have to check that it gives the *absolute* minimum. Local minima *do not* describe equilibrium states: they are merely (possibly long-lived) *meta-stable states* as we shall see later in this chapter. Note that when (5.13) holds, (5.14) reduces to

$$1 - z\beta J(1 - m^2) \geq 0. \tag{5.15}$$

For the rest of this section we set the external magnetic field to zero $B = 0$. We have

$$|m| = \tanh(\beta z J |m|) \leq \beta z J |m| \tag{5.16}$$

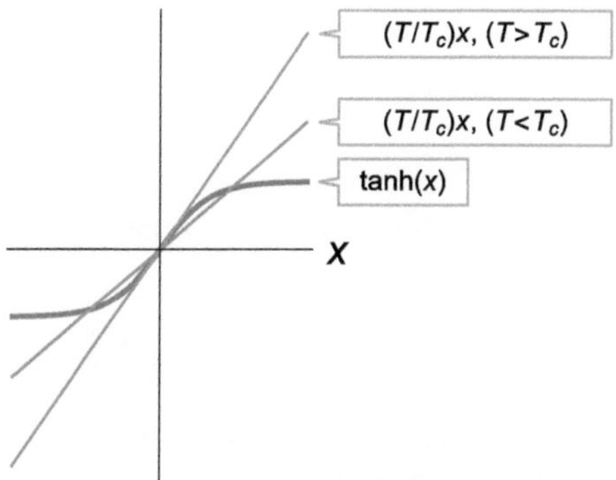

Fig. 5.2 The graphical solution of the consistency equation

so that when $\beta z J \le 1$, that is, $kT \ge kT_c \equiv z J$, the only solution to Eq. (5.13) is $m = 0$. This means that at temperatures higher than the *critical temperature*

$$T_c \equiv z J / k \tag{5.17}$$

the magnetization vanishes and the model is in a *disordered* phase. This is consistent with the general theorem that states that at sufficient high temperature classical discrete systems are disordered (cf. Sect. 4.2). The equation

$$\tanh x = (T/T_c)x \tag{5.18}$$

$(x \equiv \beta z J m)$ can be solved graphically by looking at the intersection points between the two curves $y = \tanh x$ and $y = (T/T_c)x$, cf. Fig. 5.2. We see that when $T < T_c$ the consistency condition has three solutions $m = 0$ and $m = \pm\mu \ne 0$. From (5.15) we learn that $m = 0$ is a local maximum (see also Fig. 5.1), and hence it is not a valid solution to the variational problem. We remain with the two solutions $m = \pm\mu$ which are physically equivalent since they are related by the $\mathbb{Z}_2$ symmetry of the Ising system. $m = \pm\mu$ are absolute minima of F_{MF} hence valid solutions.

   In conclusion: the mean field theory predicts a disordered phase at high temperature and an ordered one at low temperature for all values of $z \ge 1$. The mean field theory depends on the dimension d only on through the coordination number z: e.g. for a cubic lattice $z = 2d$. Thus mean field fails badly in $d = 1$ where no phase transition is present: this is due to the fact that in $d = 1$ the fluctuations are quite large and neglecting them is totally unjustified. For $d \ge 2$ the picture emerging from mean field theory is *qualitatively* correct but quantitatively poor. Let us compare the $d = 2$ critical temperatures (for the square lattice) obtained in different approaches:

$$\beta_c J \Big|_{\mathrm{MF}} = 0.25, \quad \beta_c J \Big|_{\mathrm{Peierls}} \lesssim 0.5493, \quad \beta_c J \Big|_{\mathrm{exact}} \approx 0.4406. \tag{5.19}$$

However one can show (see Sect. 5.7) that the mean field approach becomes exact in the limit $d \to \infty$ where the fluctuations are suppressed.

We introduce the variable

$$t = \frac{T - T_c}{T_c} \tag{5.20}$$

and rewrite the consistency equation in the form

$$m = -\frac{B}{kT_c} + (1 + t)\operatorname{arctanh} m \tag{5.21}$$

At $B = 0$ and close to the critical point $t \approx 0$, the spontaneous magnetization is small; expanding the previous equation

$$m = (1 + t)\operatorname{arctanh} m = (1 + t)\left[m + \frac{1}{3}m^3 + \frac{1}{5}m^5 + \cdots\right] \tag{5.22}$$

and solving recursively for m, we get

$$m = (-3t)^{1/2}\big(1 + O(t)\big). \tag{5.23}$$

Thus *the critical exponent of magnetization in mean field theory is $\beta = \frac{1}{2}$*, for all dimensions d, very different from the exact one which for $d = 2$ is $1/8$.

To compute the magnetic susceptibility we take the derivative of (5.21):

$$\chi \stackrel{\text{def}}{=} \frac{\partial m}{\partial B} = -\frac{1}{kT_c} + \frac{1 + t}{1 - m^2}\chi \tag{5.24}$$

For $B = 0$ and $T > T_c$, $m = 0$ hence

$$\chi = \frac{1}{kT_c t}, \quad T > T_c. \tag{5.25}$$

For $B = 0$ but $T < T_c$ we get

$$\chi = \frac{1}{2kT_c}(-t)^{-1}\big(1 + O(t)\big). \tag{5.26}$$

The critical exponent γ in the mean field approximation is 1, while in $d = 2$ the exact value is $7/4$. Finally, let us check that the *heat capacity* is discontinuous at the critical temperature. The (mean field) internal energy

$$U = \frac{\partial}{\partial \beta}(\beta F) = zJ\left(\frac{m^2}{2} - m\tanh(\beta zJm)\right) = -zJ\frac{m^2}{2} \tag{5.27}$$

$$\approx \begin{cases} 0 & t > 0 \\ \frac{3zJ}{2}t & t < 0 \end{cases} \quad \text{for } t \text{ small}$$

is *continuous at* $t = 0$, while its derivative (the *heat capacity*)

$$C_V \equiv \frac{\partial U}{\partial t} \approx \begin{cases} 0 & t > 0 \\ \frac{3zJ}{2T_c} \equiv \frac{3k}{2} & t < 0 \end{cases} \tag{5.28}$$

is *discontinuous* so that the phase transition is of *second order in the temperature*, as expected. We leave to the reader the computations of other interesting quantities.

Remark 5.1 For future comparison we emphasize again that in the mean-field approximation the internal energy U of the ferromagnetic Ising model is identically zero for $T > T_c$. The mean-field entropy (per site)

$$S_{\mathrm{MF}} = \beta \frac{\partial}{\partial \beta}(\beta F_{\mathrm{MF}}) - \beta F_{\mathrm{MF}} = -\beta z J m^2 + \log\left(2\cosh(z\beta J M)\right) =$$

$$\tag{5.29}$$

$$= \log 2 - z\beta J m^2 - \frac{1}{2}\log(1 - m^2)$$

is identically $\log 2$ for $T > T_c$, which is the maximal possible value since each site carries a degree of freedom with 2 states. Thus $S(U)_{\mathrm{MF}}$ is singular as a thermodynamic potential in the disorder phase and the mean-field microcanonical ensemble is a bit subtle.

Ising Model: The Bethe-Peierls Approximation
The mean field approximation is very crude and for small d is particularly unreliable. It predicts a phase transition in $d = 1$ which can be rigorously ruled out on general grounds as well as by explicit exact computations (see Chap. 4). The Bethe-Peierls approximation is a refinement of the mean field approach, that is, a better family of trial probability distributions to which we apply the variational strategy.

In the original mean field approach we focused on a single spin σ_a and considered the rest of the system as a "gray box" whose only effect is to produce an effective Hamiltonian for the only degree of freedom we are considering. As a result, we got a statistical distribution for the two values of the degree of freedom σ_a

$$\frac{e^{\beta B_{\mathrm{eff}}\sigma_a}}{Q} \tag{5.30}$$

where B_{eff} is the effective magnetic field produced at a by the rest of the system and $Q \equiv 2\cosh(\beta B_{\mathrm{eff}})$ is the normalization factor. In absence of external fields, $B \equiv 0$, B_{eff} is proportional to the magnetization m (cf. Eq. (5.7)) so we can use B_{eff} as our order parameter. This set-up corresponds to the left part of Fig. 5.3 (drawn in $d = 2$).

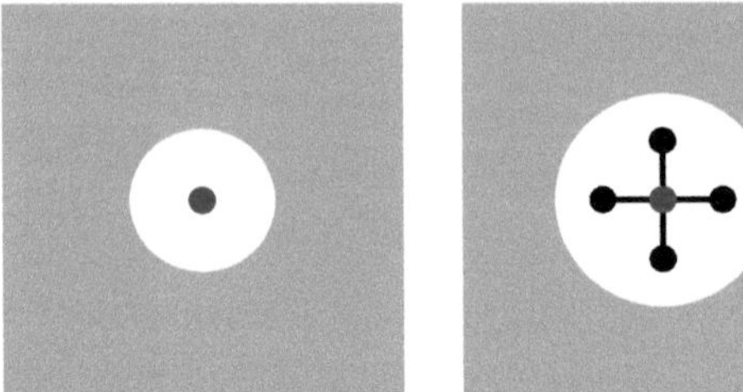

Fig. 5.3 Comparison between mean field and Bethe-Peierls approximations ($d = 2$ Ising). LEFT: in mean field we single out one spin (blue dot) and see the rest of the system as a gray box whose effect is to produce an effective Hamiltonian for the blue spin. RIGHT: Bethe-Peierls approximation. We focus on the central spin (blue dot) and its z nearest neighbors (purple dots). The rest of the lattice is a gray box which yields an effective Hamiltonian for purple spins

The Bethe-Peierls approach consists in focusing on $1 + z$ degrees of freedom, namely a central spin σ_0 and its z nearest neighbors:[2] σ_i ($i = 1, \ldots, z$) see the right side of Fig. 5.3. The improved ensemble distribution which replaces (5.30) is

$$\frac{1}{Q} \exp\left[\beta(J\sigma_{i_0} + B_{\text{eff}}) \sum_{i=1}^{z} \sigma_i \right]. \tag{5.31}$$

The consistency requirement is now

$$\langle \sigma_0 \rangle = \langle \sigma_i \rangle = \text{mean magnetization in the full lattice.} \tag{5.32}$$

We have

$$\frac{1}{2} Q \langle \sigma_0 + 1 \rangle = \left(e^{\beta(J+B_{\text{eff}})} + e^{-\beta(J+B_{\text{eff}})} \right)^z \tag{5.33}$$

$$\frac{1}{2} Q \langle \sigma_i + 1 \rangle = \left(e^{\beta(J+B_{\text{eff}})} + e^{-\beta(J+B_{\text{eff}})} \right)^{z-1} e^{\beta(J+B_{\text{eff}})} +$$
$$+ \left(e^{\beta(-J+B_{\text{eff}})} + e^{-\beta(-J+B_{\text{eff}})} \right)^{z-1} e^{\beta(-J+B_{\text{eff}})} \tag{5.34}$$

equating the last two expressions divided by $(e^{\beta(J+B_{\text{eff}})} + e^{-\beta(J+B_{\text{eff}})})^{z-1}$ we get

$$e^{\beta(J+B_{\text{eff}})} + e^{-\beta(J+B_{\text{eff}})} = e^{\beta(J+B_{\text{eff}})} + e^{\beta(B_{\text{eff}}-J)} \left(\frac{\cosh[\beta(J - B_{\text{eff}})]}{\cosh[\beta(J + B_{\text{eff}})]} \right)^{z-1}, \tag{5.35}$$

---

[2] We assume that the lattice is such that the nearest neighbors of σ_0 are not nearest neighbors between themselves, as it happens in the cubic lattice. This will not be true for a triangular lattice.

that is, the self-consistency condition is

$$\exp\left(\frac{2\,\beta B_{\mathrm{eff}}}{z-1}\right) = \frac{\cosh[\beta(J+B_{\mathrm{eff}})]}{\cosh[\beta(J-B_{\mathrm{eff}})]}. \tag{5.36}$$

The original mean field approximation wrongly predicted a phase transition in one-dimension when the coordination number $z = 2$. In this case Eq. (5.36) becomes

$$e^{2\beta B_{\mathrm{eff}}}\left(e^{\beta(J-B_{\mathrm{eff}})} + e^{-\beta(J-B_{\mathrm{eff}})}\right) = e^{\beta(J+B_{\mathrm{eff}})} + e^{-\beta(J+B_{\mathrm{eff}})} \;\Rightarrow\; e^{4\beta B_{\mathrm{eff}}} = 1, \tag{5.37}$$

so the order parameter B_{eff} is always zero for $d = 1$: this approximation does not predict (correctly) a phase transitions at positive temperature. Setting

$$y = \exp\left(\frac{2\,B_{\mathrm{eff}}}{z-1}\right), \tag{5.38}$$

the self-consistency condition becomes a polynomial equation in y of degree z

$$P(y) \equiv y^z - e^{2\beta J}\,y^{z-1} + e^{2\beta J}\,y - 1 = 0. \tag{5.39}$$

Note that the polynomial $P(y)$ has the symmetry property

$$P(y) = -y^z\,P(1/y) \tag{5.40}$$

so that, if y is a root, also $1/y$ is a root, that is, if B_{eff} is a solution so is $-B_{\mathrm{eff}}$. Of course this is just the $\mathbb{Z}_2$ symmetry of the Ising model at $B = 0$, and we may focus on solutions with $B_{\mathrm{eff}} \geq 0$ i.e. $y \geq 1$. Note that $y = 1$ is *always* a root of $P(y)$, and if the degree z is even (as it is the case for the cubic lattice in any dimension) also $y = -1$ is a root.

Theorem 5.1 (Descartes' Rule of Signs (1637) [1]) *The number of positive real zeros of a real polynomial $P(y)$ is equal to the number of changes of sign in front of each monomial (written in order of decreasing degrees and omitting terms with zero coefficient) or it is less than this number by an **even** integer.*

The polynomial (5.39) has three changes of sign, so it has either 3 or 1 positive roots. In the second case the positive root must be $y = 1$. In the first case they are

$$y = 1, \quad y = y_+ \geq 1, \quad \text{and} \quad y = y_- \equiv 1/y_+. \tag{5.41}$$

Hence there is at most one positive value for the self-consistent magnetization B_{eff}. We have to characterize the polynomials (5.39) with 3 positive roots, i.e. values of the coefficient $e^{2\beta J}$ for which a positive magnetization is possible. If a polynomial $P(y)$ has three positive roots $1/y_+, 1, y_+ > 1$, its first derivative $P'(y)$ must have two positive simple zeros: one in the range $1/y_+ < y < 1$ and one in the range $1 < y < y_+$; by Descartes' rule $P'(y)$ has either no positive zero or two positive zeros. If $P'(y)$ has two positive roots $P'(1) < 0$. On the other hand, if $P'(1) < 0$ the polynomial is negative for $y = 1 + \epsilon$, and since is positive for $y \gg 1$, it should have a root $y_+ > 1$, hence 3 positive roots by Descartes' rule. We conclude:

Lemma 5.1 *The self-consistency condition* (5.39) *has a non-trivial solution with* $y \equiv \exp(\beta B_{\mathit{eff}}) > 1$ *if and only if* $P'(1) < 0$. *Since*

$$P'(1) = z - (z-2)e^{2\beta J} < 0 \quad \Rightarrow \quad e^{2\beta J} > \frac{z}{z-2} \tag{5.42}$$

the non-trivial solution is present in the low temperature regime

$$kT < \frac{2J}{\log[z/(z-2)]}. \tag{5.43}$$

In particular the critical temperature is

$$\beta_c J = \frac{1}{2} \log\left(\frac{z}{z-2}\right). \tag{5.44}$$

For $d = 1$, i.e. $z = 2$, we recover $\beta_c = \infty$, i.e. $T_c = 0$, which is the correct result. For the $d = 2$ cubic lattice, i.e. $z = 4$, we get

$$\beta_c J \approx 0.3465 \tag{5.45}$$

which is a better approximation to the correct value than the result of the plain mean field theory, cf. (5.19).

5.2 The q-State Potts Model

Up to now we used the Ising model as our pet laboratory to understand the statistical mechanics of phase transitions. The Ising phase transition is of second order (in temperature) without latent heat. To illustrate the physics of phase transitions *with* latent heat in a simple situation, we now consider a generalization of the Ising model which presents a first order transition and the related phenomena (such as *supercooling* and *superheating*). The Potts model will give us as a further example of mean field approximation in preparation for (old-fashioned) Landau theory of critical phenomena which describes phase transitions of diverse orders.

The q-state Potts model is the generalization of the Ising model where at each site i of the d-dimensional lattice Λ there is a discrete degree of freedom ξ_i (call it "spin") which takes $q \in \mathbb{N}$ distinct values, say $\xi_i = 1, 2, \ldots, q$, with Hamiltonian

$$H = -J \sum_{\langle i,j \rangle} \delta_{\xi_i, \xi_j} \equiv -J \sum_{l \in \Lambda_1} \delta_{\xi_{h(l)}, \xi_{t(l)}} \equiv -J \sum_{l \in \Lambda_1} \left(\sum_{a=1}^{q} \delta_{\xi_{h(l)}, a} \, \delta_{\xi_{t(l)}, a} \right),$$

$$(5.46)$$

where $\delta_{i,j}$ is the Kronecker delta. In the "ferromagnetic" case $J > 0$, so that the interactions tend to align the "spins" to a common value $a_0 \in \{1, \ldots, q\}$. Therefore we have q ground states in which all spins have the same value a_0. The system has a symmetry $\mathfrak{S}_q$ given by permutations of the q possible spin values $\{1, 2, \ldots, q\}$. Note that this group is Abelian for $q \leq 2$ and non-Abelian for $q \geq 3$. For $q = 2$ the Potts model is equivalent to the Ising model up to a redefinition of the coupling constant J. The q ground states form an orbit of the symmetry group $\mathfrak{S}_q$. On a vacuum the symmetry group $\mathfrak{S}_q$ is broken down to the subgroup $\mathfrak{S}_{q-1}$ which fixes the elements $a_0 \in \{1, \ldots, q\}$ and permutes the other $(q - 1)$ elements of the set $\{1, \ldots, q\}$. Therefore we may expect a phase transition between a disordered phase at high temperature, where the symmetry group $\mathfrak{S}_q$ is unbroken, and an ordered phase at low temperature where the symmetry breaks down to a subgroup $\mathfrak{S}_{q-1}$.

We solve the Potts model in the mean field approximation. We know that this approximation becomes reliable when the dimension d of the lattice is large enough, so the new phenomena we shall discover *do take place* under the appropriate conditions.

As order parameters we introduce the q "magnetizations"

$$m_a = \frac{1}{|\Lambda_0|} \sum_{i \in \Lambda_0} \langle \delta_{\xi_i, a} \rangle, \qquad a = 1, 2 \ldots, q. \qquad (5.47)$$

Note that $m_a = N_a/|\Lambda_0|$ where N_a is the number of spins in state a. In particular $0 \leq m_a \leq 1$. The m_a's are not independent since there is a relation

$$m_1 + m_2 + \cdots + m_q = 1, \qquad (5.48)$$

and we have only $(q - 1)$ independent "magnetizations"

$$\Delta_a = m_a - m_{a+1}, \quad a = 1, 2, \ldots, q - 1. \qquad (5.49)$$

The actual order parameters are the Δ_a's not the m_a's. Indeed in the disordered phase the symmetry $\mathfrak{S}_q$ is unbroken, and all m_a's are equal (and non-zero) while $\Delta_a = 0$ for all a. In an ordered phase at least one Δ_a is not zero. In a ground state

$$m_a = \begin{cases} 1 & a = a_0 \\ 0 & \text{otherwise} \end{cases} \qquad \Delta_a = \begin{cases} 1 & a = a_0 \\ -1 & a = a_0 - 1 \\ 0 & \text{otherwise.} \end{cases} \qquad (5.50)$$

The m_a's transform in the defining (reducible) representation of $\mathfrak{S}_q$ and the Δ_a's in the irreducible $(q-1)$-dimensional representation.

Mean Field Approximate Solution

The Hamiltonian (5.46) can be identically rewritten as

$$H = J|\Lambda_1| \sum_{a=1}^{q} m_a^2 - zJ \sum_{i\in\Lambda_0} \sum_{a=1}^{q} m_a\, \delta_{\xi_i,a} - $$
$$- J \sum_{l\in\Lambda_1} (\delta_{\xi_{h(l)},a} - m_a)(\delta_{\xi_{t(l)},a} - m_a). \tag{5.51}$$

Again, the mean field approximation consists in neglecting the term quadratic in the fluctuations. Neglecting it, and summing over the ξ_i's, we get the partition function

$$Z_{\mathrm{MF}} = \exp\left[-\beta J|\Lambda_1| \sum_{a=1}^{q} m_a^2\right] \left(\sum_{a=1}^{q} \exp\left[z\beta J m_a\right]\right)^{|\Lambda_0|} \tag{5.52}$$

compare with Eq. (5.8) for the Ising model. Let

$$F_{\mathrm{MF}} = -\frac{1}{\beta|\Lambda_0|}\log Z_{\mathrm{MF}} = \frac{zJ}{2}\sum_{a=1}^{q} m_a^2 - \frac{1}{\beta}\log\left(\sum_{a=1}^{q}\exp\left[z\beta J m_a\right]\right) \tag{5.53}$$

be the mean-field "free energy" per site (cf. Eq. (5.9) for the Ising model). The "self-consistence" condition—i.e. the first part of the variational characterization of the physical free energy which requires to look for the extrema of F_{MF} in the family (5.53) parametrized by the m_a's,—yields

$$\frac{\partial F_{\mathrm{MF}}}{\partial m_a} \equiv zJ\left(m_a - \frac{e^{z\beta J m_a}}{\sum_b \exp[z\beta J m_b]}\right) = 0. \tag{5.54}$$

We stress that the self-consistent values of m_a automatically satisfy the constraint (5.48). Convexity (i.e. the second part of the variational characterization) requires the matrix

$$\frac{1}{zJ}\frac{\partial^2 F_{\mathrm{MF}}}{\partial m_a\, \partial m_b} \equiv (1 - z\beta J\, m_a)\delta_{ab} + z\beta J\, m_a m_b \tag{5.55}$$

to be positive semi-definite. From Eq. (5.54) we see that, for all $a, b = 1, \ldots, q$,

$$\alpha\, m_a - \log m_a = \alpha\, m_b - \log m_b \tag{5.56}$$

where we set $\alpha \equiv z\beta J$.

Lemma 5.2 *Depending on the two parameters α and y, the solutions of the equation*

$$f_\alpha(m) \equiv \alpha m - \log m = y \qquad (5.57)$$

in the interval $0 \le m \le 1$ are

	$y < 1 + \log \alpha$	$1 + \log \alpha \le y < \alpha$	$y > \alpha$
$\alpha \le 1$	No solution	No solution	1 solution: $0 \le m \le 1$
$\alpha > 1$	No solution	2 solutions: $\begin{cases} 0 \le m_- \le 1/\alpha \\ 1/\alpha \le m_+ \le 1 \end{cases}$	1 solution: $0 \le m \le 1/\alpha$

In particular the m_a's can take at most two distinct values, *so the only possible symmetry breaking patterns are $\mathfrak{S}_q \to \mathfrak{S}_{q_1} \times \mathfrak{S}_{q-q_1}$ where $0 \le q_1 \le q$ is the number of m_a's equal to $m_+ > 1/\alpha$.*

Proof The critical point of the function $f_\alpha(m)$ is $m = \alpha^{-1}$ with critical value $1 + \log \alpha$. If $\alpha \le 1$ the function is monotonically decreasing in the interval $[0, 1]$ from ∞ to the minimal value $f_\alpha(1) = \alpha$. So when $\alpha \le 1$ we have no solution for $y < \alpha$ and one solution otherwise. When $\alpha > 1$, the function is decreasing in the interval $[0, \alpha^{-1}]$ from ∞ down to $1 + \log \alpha$, and increasing in the interval $[\alpha^{-1}, 1]$ from the value $1 + \log \alpha$ to the maximal value α. Therefore we have no solution for $y < 1 + \log \alpha$, 2 solutions for $1 + \log \alpha < y \le \alpha$, and 1 solution for $y > \alpha$. For $y = 1 + \log \alpha$ the two solutions coincide. See Fig. 5.4 for the graphical solution.

Lemma 5.3 *When the m_a's take two distinct values $m_\pm$, the largest one m_+ has multiplicity 1, that is, in the disordered phase the unbroken symmetry is the full symmetry group $\mathfrak{S}_q$ while in the ordered phase the unbroken symmetry is $\mathfrak{S}_{q-1}$.*

Remark 5.2 The **Lemma** agrees, of course, with the physical intuition that in the ordered phase most spins are aligned in one direction since this is the case for the ground states where all "spin" ξ_i have the same value a_0, so that $m_{a_0} = 1$, $m_{b \neq a_0} = 0$.

Proof Suppose the statement is false. By a permutation we can arrange so that $m_1 = m_2 = m_+ > 1/\alpha$. The first 2×2 diagonal block of the matrix (5.55) (with $z\beta J = \alpha$) is a linear combination of a *negative* multiple of the identity and a matrix with a zero eigenvalue

$$- |1 - \alpha m_+| \begin{bmatrix} 1 & 0 \\ 0 & 1 \end{bmatrix} + \alpha m_+^2 \begin{bmatrix} 1 & 1 \\ 1 & 1 \end{bmatrix} \qquad (5.58)$$

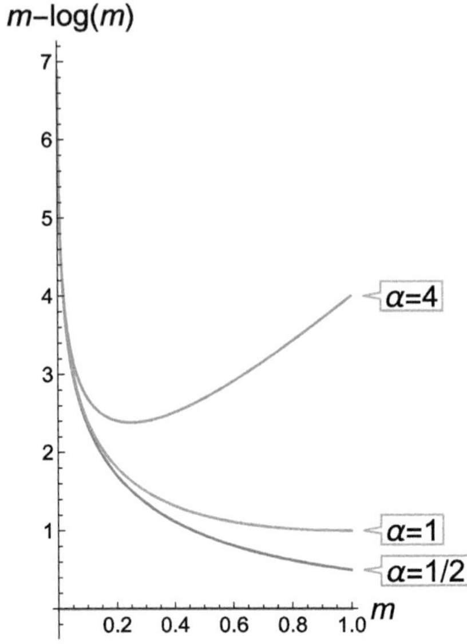

Fig. 5.4 Graphical solutions to Eq. (5.57)

so the diagonal block has a negative eigenvalue $-|1-\alpha m_+|$. Hence a matrix (5.55), associated to a configuration with at least two m_a's equal to m_+, is *never* positive definite, and these configurations are not solutions to the variational problem.

Then, by a suitable permutation of the a's, we may assume

$$m_1 \geq m_2 = m_3 = \cdots = m_q, \tag{5.59}$$

and write $\Delta = m_1 - m_2$, so that

$$m_1(\Delta) = \frac{1+(q-1)\Delta}{q}, \qquad m_a(\Delta) = \frac{1-\Delta}{q} \quad a = 2, 3, \ldots \tag{5.60}$$

and Eq. (5.56) reduces to

$$\alpha\,\Delta - \log\left(1 + \frac{q\,\Delta}{1-\Delta}\right) = 0. \tag{5.61}$$

It is easy to check that this equation is equivalent to

$$\frac{\partial}{\partial\Delta}\big(\beta F_{\mathrm{MF}}(\Delta)\big) = 0, \tag{5.62}$$

where $\beta F_{\mathrm{MF}}(\Delta) \equiv \beta F_{\mathrm{MF}}(m_a(\Delta))$ is the free energy computed on a configuration of the form (5.60):

$$
\begin{aligned}
\beta F_{\mathrm{MF}}(\Delta) &= \beta F_{\mathrm{MF}}(0) + \alpha \frac{\Delta}{q} + \alpha \frac{(q-1)\Delta^2}{2q} - \log\left(1 + \frac{e^{\alpha\Delta} - 1}{q}\right) \\
&= \beta F_{\mathrm{MF}}(0) + \frac{\alpha(q-1)(q-\alpha)}{2q^2}\Delta^2 - \frac{\alpha^3(q-1)(q-2)}{6q^3}\Delta^3 + O(\Delta^4)
\end{aligned}
\tag{5.63}
$$

Indeed,

$$
\frac{\partial}{\partial\Delta}\beta F_{\mathrm{MF}} = \frac{\alpha(q-1)}{q(q-1+e^{\alpha\Delta})}\left(1 + (q-1)\Delta - (1-\Delta)e^{\alpha\Delta}\right).
\tag{5.64}
$$

$\Delta = 0$ is a solution to (5.61) for all α. A non-zero order parameter, $\Delta \neq 0$, may be a solution for at most one value of α, namely for

$$
\alpha = \alpha_q(\Delta) \equiv \frac{1}{\Delta}\log\left(1 + \frac{q\,\Delta}{1-\Delta}\right).
\tag{5.65}
$$

We distinguish two cases: $q = 2$ (which is the Abelian Ising model), and $q \geq 3$ where the symmetry $\mathfrak{S}_q$ is non-Abelian.

Fact 5.2 *For $q = 2$ the function $\alpha_2(\Delta)$ is convex and monotonically increasing in $(0, 1]$ and its infimum is reached as $\Delta \to 0^+$*

$$
\inf_{\Delta\in(0,1]} \alpha_2(\Delta) = \lim_{\Delta\to 0^+} \alpha_2(\Delta) = 2,
\tag{5.66}
$$

see Fig. 5.5. Hence

(1) *for $\alpha < \alpha_c \equiv 2$ Eq. (5.61) has the single solution $\Delta = 0$ which is a minimum of βF_{MF} (the coefficient of Δ^2 in the second line of (5.63) is positive), that is, for $\alpha < \alpha_c$ we have $\Delta = 0$ and the system is in the* disordered *phase;*

(2) *for $\alpha > \alpha_c$ Eq. (5.61) has two solutions $\Delta = 0$ and $\Delta_+ > 0$. $\Delta = 0$ is now a local maximum of the free energy, while $\Delta = \Delta_+$ yields the absolute minimum,*

Fig. 5.5 The function $\alpha_2(\Delta)$

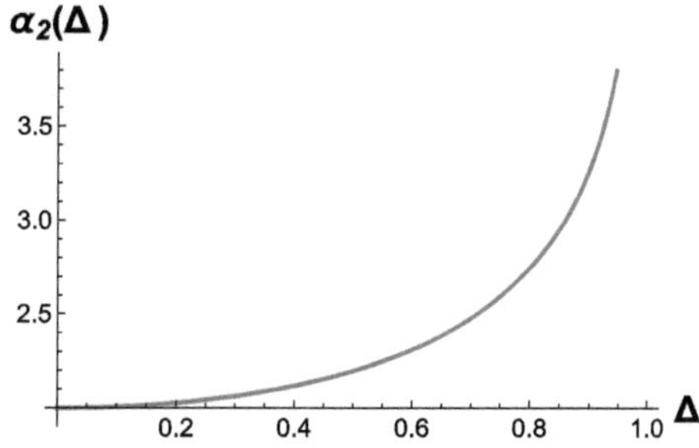

Fig. 5.6 Free energy for
$q = 2$ for α above and below
the critical value $\alpha_c = 2$

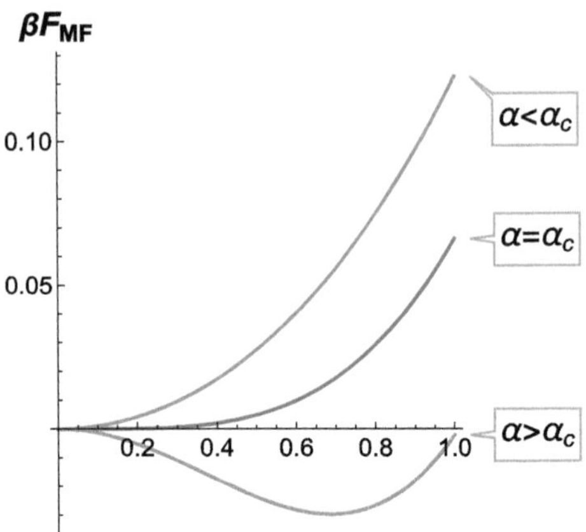

see Fig. 5.6. Therefore when $\alpha > \alpha_c$ the order parameter $\Delta > 0$ and the system is in an ordered phase.

(3) *as $\alpha \to \alpha_c^+$ we have $\Delta_+ \to 0$, so that the order parameter Δ and the energy*

$$U_{MF}(\Delta) \equiv \frac{\partial}{\partial \beta}(\beta F_{MF}) \equiv zJ \, \frac{\partial}{\partial \alpha}(\beta F_{MF}) = U_{MF}(0) + O(\Delta^2) \qquad (5.67)$$

are continuous across the phase transition, i.e. the order/disorder phase transition is of second order with no latent heat.

See Figs. 5.5 and 5.6. Of course these are just the results we got in Sect. 5.1 for the Ising model.[3] We reviewed them here for the sake of comparison with the situation for $q \geq 3$ which is summarized in the following **Fact**.

Fact 5.3 *For $q \geq 3$:*

(1) *$\alpha_q(\Delta)$ has a unique local minimum in $(0, 1]$, α_*, at the point $\Delta = \Delta_*$ where*

$$\alpha_q(\Delta_*) \equiv \alpha_* < \alpha_0 \equiv \lim_{\Delta \to 0^+} \alpha(\Delta) = q. \qquad (5.68)$$

See Fig. 5.7. One has

$$\alpha_* = \frac{q}{1 + (q-2)\Delta_* - (q-1)\Delta_*^2} < q, \qquad (5.69)$$

---

[3] The critical temperature $kT_c = zJ/2$ agrees with the one we computed in Sect. 5.1 when one keeps into account the different normalization of the coupling J.

Fig. 5.7 The function $\alpha_6(\Delta)$

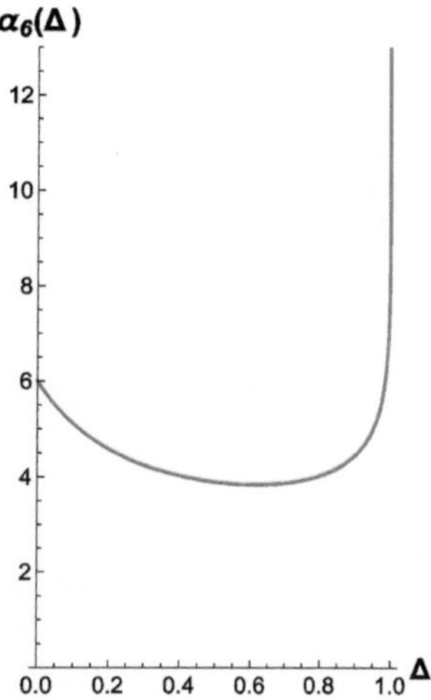

which implies

$$\Delta_* < \frac{q-2}{q-1}. \tag{5.70}$$

For $\alpha < \alpha_*$ the only solution to Eq. (5.61) is $\Delta = 0$ and the system is disordered;

(2) *if $\alpha > \alpha_0 \equiv q$ Eq. (5.61) has two solutions $\Delta = 0$ and $\Delta_+(\alpha) > 0$. The coefficient of Δ^2 in the second line of (5.63) is negative, and function $\beta F_{MF}(\Delta)$ has a local maximum at $\Delta = 0$: hence the disordered phase is ruled out in this regime. $\Delta_+(\alpha)$ gives the absolute minimum of the free energy, hence $\Delta = \Delta_+(\alpha)$;*

(3) *in the region $\alpha_* < \alpha < \alpha_0$, there are three solutions to Eq. (5.61) The first one at $\Delta = 0$ is a local minimum, the second one at some $0 < \Delta < \Delta_*$ is a local maximum, and the third one at some $\Delta = \Delta_+ > \Delta_*$ is a local minimum. One has to determine which one of the two local minima is the* absolute *minimum. The interval $\alpha_* < \alpha < \alpha_0$ splits in two parts*

$$\alpha_* < \alpha < \alpha_c \quad and \quad \alpha_c < \alpha < \alpha_0. \tag{5.71}$$

In the first interval the solution $\Delta = 0$ is the absolute minimum and the phase is disordered, while in the second one the absolute minimum is at $\Delta_+ > \Delta_c$ where Δ_c is the largest solution of

$$\alpha_c \Delta_c = \log\left(1 + \frac{q\,\Delta_c}{1-\Delta_c}\right). \tag{5.72}$$

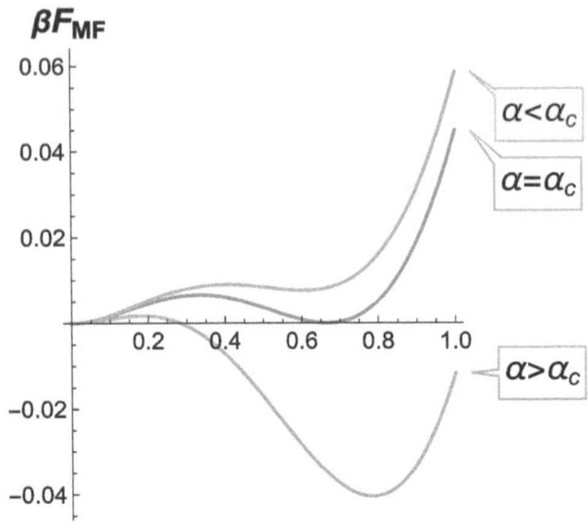

Fig. 5.8 Mean field free energy of the $q = 4$ Potts as a function of Δ for α below and above the critical value $\alpha_c = 3 \log 3 \approx 3.2958$

The critical point (α_c, Δ_c) is determined by this equation together with the condition that the free energy has the same value at the two local minima

$$\beta F_{MF}(\alpha_c, \Delta_c) - \beta F_{MF}(\alpha_c, 0) = 0. \tag{5.73}$$

The solution of the last two equations is

$$\Delta_c = \frac{q - 2}{q - 1}, \qquad \alpha_c = \frac{2(q - 1)}{q - 2} \log(q - 1). \tag{5.74}$$

See Fig. 5.8.

The proof of these statements is elementary, see Appendix 1. We summarize the conclusions of the mean field approximation for the q-state Potts model:

Fact 5.4 *When $q \geq 3$ we have a disordered phase for $\alpha < \alpha_c$ and an order phase for $\alpha > \alpha_c$. At the phase transition for $\alpha = \alpha_c$ i.e. at the critical temperature*

$$kT_c = \frac{zJ(q - 2)}{2(q - 1)\log(q - 1)} \tag{5.75}$$

The order parameter Δ jumps discontinuously

$$0 = \lim_{\alpha \to \alpha_c^-} \Delta \neq \lim_{\alpha \to \alpha_c^+} = \Delta_c \equiv \frac{q - 2}{q - 1}. \tag{5.76}$$

The internal energy is also discontinuous

$$U_{MF}(\alpha_c, 0) - U_{MF}(\alpha_c, \Delta_c) = -zJ\frac{\partial}{\partial\alpha}(\beta F_{MF}(\Delta) - \beta F_{MF}(0))\Big|_{\substack{\alpha=\alpha_c \\ \Delta=\Delta_c}} =$$

$$= zJ\frac{(q-1)\Delta^2}{2q}\Big|_{\Delta=\frac{(q-2)}{q-1}} = zJ\frac{(q-2)^2}{2q(q-1)} > 0$$

$$(5.77)$$

Therefore in the mean field approximation in the $q \geq 3$ Potts model the order/disorder phase transition is of first order with a positive latent heat given by Eq. (5.77).

Supercooling and Superheating

The following figure represents schematically the values of Δ which realize a minimum of the free energy at any given temperature T: solid blue curves correspond to values of Δ which yield global minima of the mean field free energy while dashed blue curves correspond to mere *local* minima:

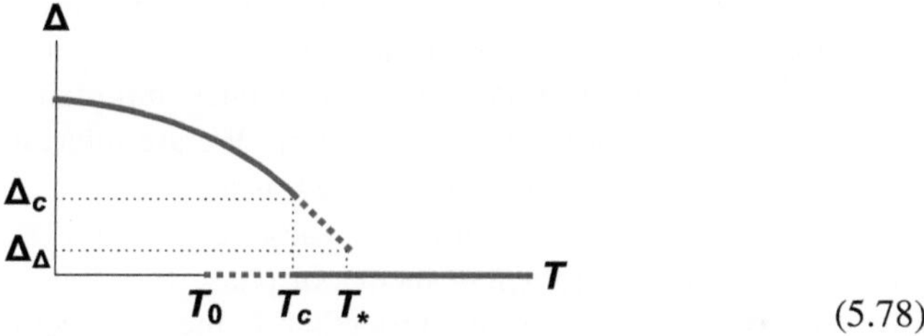

$$(5.78)$$

The equilibrium states are the points on the solid lines, and the transition between ordered and disordered *equilibrium* phases happens at $T = T_c$. *What is the meaning of the dashed curves?* Suppose we start from some large temperature, where the system is disordered, and cool it down. What happens when we reach $T = T_c$? If we allow time for the system to thermalize, it will end up in the ordered phase. But there is an energy gap (latent heat) between the two phases, and if we decrease the temperature rapidly enough so that the system has not enough time to exchange heat with the surroundings in order to adjust its internal energy to the equilibrium value below T_c, the system will remain in a disordered state until we reach the temperature T_0 below which no disordered state is possible since $\Delta = 0$ is not even a local minimum. At this point our system settles in the equilibrium ordered state.

The disordered states for $T_0 < T < T_c$ described by the dashed portion of the T-axis in figure (5.78) are *meta-stable* states, i.e. states which are not at equilibrium, but may persist for a long time because of a barrier in internal energy (latent heat). When we decrease the temperature rapidly without giving time to our system to thermalize in a proper equilibrium state, it will typically "freeze" itself into a meta-stable state. The phenomenon of disordered states persisting *below the critical temperature* is called *supercooling.*

The converse phenomenon, called *superheating* is the persistence of ordered (meta-stable) states at temperatures above T_c (but below T_*) along the upper dashed curve in the figure. Its existence is also a consequence of the gap in energy between the two equilibrium phases.

Supercooling and superheating are instances of the general phenomenon called *hysteresis.* The hysteresis phenomena represented in figure (5.78) are quite typical of first order phase transitions while they are absent in higher order phase transitions. The main point is that the meta-stable state (local minimum) is separated from the equilibrium state (global minimum) by an energy barrier, and the time needed to cross the barrier (the time to thermalize) is exponentially large in the height of the barrier. Consequently the meta-stable state is long-lived.

5.3   Landau (Pheno) Theory of Phase Transitions

We have seen (in the mean field approximation[4]) that the order of the phase transition in the q-state Potts model varies from second to first as we increase q from 2 to 3. The order of the transition encodes important qualitative aspects of the critical phenomena at the phase transition. We are interested in an easy diagnostic technique to determine the order of the transition and to get a qualitative picture of the physics around the critical point. These are the purposes of the crudest form of the Landau *phenomenological* theory of phase transitions which (like mean field) disregards thermal fluctuations altogether. In the next section we shall present a less crude version of the theory which takes in some account the effects of fluctuations and allows to estimate the error implicit in neglecting them.

The basic idea of Landau theory is to study the (trial) free energy near the critical point by expanding it in powers of the order parameter *assumed to be small.* For instance, in the mean field Potts model one has

$$\beta F = \frac{\alpha(q-1)(q-\alpha)}{2q^2}\Delta^2 - \frac{\alpha^3(q-1)(q-2)}{6q^3}\Delta^3 - $$
$$- \frac{\alpha^4(q-1)(6-6q+q^2)}{24q^4}\Delta^4 + O(\Delta^5). \tag{5.79}$$

We now present a few prototypical situations described by Landau's theory of phase transitions to illustrate the methods and the typical results. The first two instances will be introduced through the example of the Potts model in large

---

[4] Recall that the approximation is reliable only if the dimension is large enough. For instance, in the exact solution of the two-dimensional Potts model one finds that the phase transition is second order for $q \leq 4$ and first order for $q \geq 5$, while the mean field answer was first order for ≥ 3.

dimension $d \gg 1$ that we have already solved in the mean field approach. We use this model as our motivating example for the theory.

Case I: Ising Model

We know that for $q = 2$ (the Ising model) the order/disorder phase transition is of *second order:* at the critical point $\alpha_c = 2$ the order parameter is continuous (but certainly not analytic!) as a function of $\alpha \propto T^{-1}$ and there is no latent heat. However the heat capacity C_V is discontinuous at α_c. Let us see how this behavior around the critical point is reflected in the properties of the "free energy"[5] F as a function of the order parameter $\sigma \equiv \Delta$ at temperatures near the critical one

$$kT_c = \frac{1}{2}zJ. \tag{5.80}$$

To discuss the behavior of the system near the critical point, where σ is small, we can expand $F(\sigma)$ in powers of σ and truncate the expansions after the first few terms. For $|T - T_c|$ small we get

$$F(\sigma) \approx F(0) + \frac{1}{2}a(T - T_c)\sigma^2 + c\,\sigma^4, \tag{5.81}$$

$$F'(\sigma) \approx a(T - T_c)\sigma + 4c\,\sigma^3 \tag{5.82}$$

where a, c are positive constants which can be read from Eq. (5.79) with $q = 2$, and all terms of odd degree in σ vanish because of the Ising symmetry $\sigma \leftrightarrow -\sigma$.

When $T > T_c$ the term quadratic in σ in $F(\sigma)$ is *positive,* and there is a unique real solution to $F'(\sigma) = 0$, i.e. $\sigma = 0$, which is the minimum of the function. Hence at high temperature $\sigma = 0$ and the phase is disordered.

When $T < T_c$ the equation $F'(\sigma) = 0$ has three real solutions

$$\sigma_0 = 0, \qquad \sigma_\pm = \sqrt{\frac{a(T_c - T)}{4c}} \tag{5.83}$$

(the two solutions $\sigma_\pm$ are related by the $\mathbb{Z}_2$ symmetry). Now the solution $\sigma_0 = 0$ is a local maximum and should be disregarded. The other two correspond to the two ordered phases at low temperature. As $T \to T_c^-$, $\sigma_\pm \to 0$, and the phase transition is continuous without latent heat: see Fig. 5.9.

Since there is no latent heat, the transition is of order > 1 in the temperature. To check that it is of second order in T, we have to compute the heat capacity and show

---

[5] We write "free energy" between quotes because this function is the "off shell" free energy; the actual free energy is the value of this function at its global minimum.

Fig. 5.9 The typical form of the "free energy" as a function of the order parameter σ near a second order transition point $T \approx T_c$

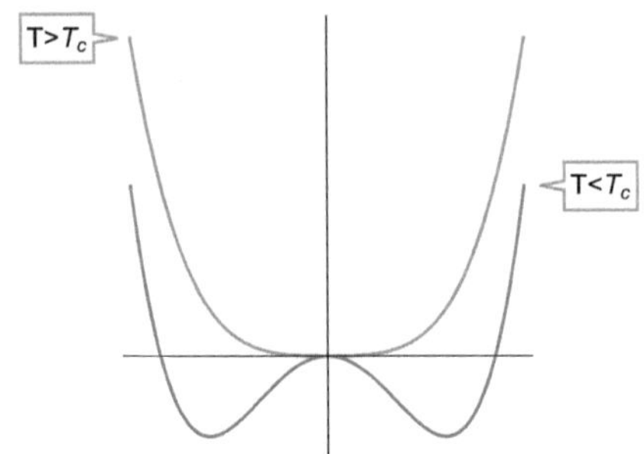

that it is discontinuous. The free energy is

$$F = \begin{cases} F(0) & T > T_c \\ F(0) - \frac{a^2(T_c-T)^2}{8c} & T < T_c \end{cases} \tag{5.84}$$

the entropy is

$$S = -\frac{\partial F}{\partial T} = S(0) + \begin{cases} 0 & T > T_c \\ \frac{a^2(T-T_c)}{4c} & T < T_c \end{cases} \tag{5.85}$$

and the heat capacity

$$C_V \approx T_c \frac{\partial S}{\partial T} = C_V(0) + \begin{cases} 0 & T > T_c \\ \frac{a^2}{4c} & T < T_c \end{cases} \tag{5.86}$$

which is indeed discontinuous at $T = T_c$ (as expected).

The *Landau theory* of critical phenomena states that *this is the typical behavior of the "free energy" as a function of the order parameter σ for temperatures T near the critical temperature T_c of a second order phase transition in a system which is described by a single order parameter σ.* The characteristic features of a second order phase transition then are:

2ndO1   in the expansion of F in powers of σ the cubic term vanishes;
2ndO2   the coefficient of σ^2 is linear in $T - T_c$ with a positive coefficient;
2ndO3   the coefficient of σ^4 is positive (so σ remains small and the use of a series expansion is justified).

We shall see in **Case IV** below that condition **2ndO3** does not hold when the second order phase transition happens at higher critical points (which belong to a higher codimension locus in parameter space). For a *generic* second order phase transition (i.e. in codimension 1 in control parameter space) Landau theory predicts

$$\sigma \propto (T_c - T)^{1/2} \quad \text{as } T \to T_c^- \tag{5.87}$$

so that the critical exponent β is $1/2$ in the Landau approximation (which gets exact only as $d \to \infty$, when the fluctuation can be safely neglected).

Case II: $q = 3, 4$ **Potts Models**

As a second instance of Landau's method, let us look to the characterization of *first order* phase transitions with positive latent heat $L > 0$. We may use the $q = 3, 4$ Potts model (in high dimension d!) as the leading examples. We have an expansion

$$F \approx \frac{1}{2}a(T - T_0)\sigma^2 - b\sigma^3 + c\sigma^4 \tag{5.88}$$

$$F' \approx a(T - T_0)\sigma - 3b\sigma^2 + 4c\sigma^3 \tag{5.89}$$

where a, b, c are positive constants and T_0 is merely the temperature at which the quadratic term flips sign (cf. figure (5.78)). As before, we neglect all higher terms. The solutions to $F'(\sigma) = 0$ are $\sigma = 0$ and the roots of

$$4c\sigma^2 - 3b\sigma + a(T - T_0) = 0 \quad \Rightarrow \quad \sigma_\pm = \frac{3b \pm \sqrt{9b^2 - 16a(T - T_0)c}}{8c} \tag{5.90}$$

The roots $\sigma_\pm$ are real iff the discriminant is positive, that is, iff

$$T < T_0 + \frac{9b^2}{16ac} \equiv T_*. \tag{5.91}$$

For $T > T_*$ the only real solution is $\sigma = 0$ and the phase is disordered (cf. figure (5.78)). For $T < T_*$ we have three solutions $0 = \sigma_0 < \sigma_- < \sigma_+$, and we have to see which one gives the absolute minimum. σ_+ is a local minimum while in the interval $T_0 < T < T_*$ the origin $\sigma = 0$ is a local minimum and σ_- a local maximum. Next we have to determine which one between the two local minima $\sigma = 0$ and $\sigma = \sigma_+$ has the lesser free energy. Since the free energy of the first configuration is zero for all T, the phase transition between $\sigma = 0$ and $\sigma = \sigma_+ > 0$ happens at the crossover between the two local minima, that is, when the free energy $F(\sigma_+)$ reaches zero. Before that point $F(\sigma_+) < F(0) = 0$. Hence the critical temperature T_c of the phase transition is given by the solution to the coupled system of equations

$$\begin{cases} F(\sigma_c) = \dfrac{1}{2}a(T_c - T_0)\sigma_c^2 - b\sigma_c^3 + c\sigma_c^4 = 0 \\[2mm] F'(\sigma_c) = a(T_c - T_0)\sigma_c^2 - 3b\sigma_c^2 + 4c\sigma_c^3 = 0. \end{cases} \tag{5.92}$$

whose solution is

$$T_c = T_0 + \frac{b^2}{2ac} > T_0, \qquad \sigma_c = \frac{b}{2c}. \tag{5.93}$$

Note that the first-order phase transition happens *above* the temperature at which the quadratic term flips sign for both signs of b, while $\sigma_c \to 0$ and $T_c \to T_0$ as $b \to 0$.

The *latent heat* is given by $L = T_c \, \Delta S$ where ΔS is the discontinuity in entropy

$$\Delta S = -\frac{\partial F}{\partial T}\bigg|_{\sigma=0} + \frac{\partial F}{\partial T}\bigg|_{\sigma=\sigma_c} = \frac{1}{2}a\sigma_c^2 + \frac{\partial F}{\partial \sigma}\frac{\partial \sigma}{\partial T} = \frac{a}{2}\sigma_c^2 \tag{5.94}$$

($\partial F/\partial \sigma$ vanishes at $\sigma = \sigma_c$, Eq. (5.92)). Then the latent heat is

$$L = \frac{ab^2}{8c^2}\, T_c. \tag{5.95}$$

The *Landau theory* of critical phenomena states that all first-order phase transitions with no symmetry breaking and a single order parameter "look like this" near the critical point (assuming the discontinuity in σ is small). The characteristic features of this **Case II** (first order) of phase transitions are:

1ˢᵗ**O1**   the existence of a non-zero cubic term in σ (no $\sigma \leftrightarrow -\sigma$ symmetry);
1ˢᵗ**O2**   the coefficient of σ^2 is linear in $T - T_0$ with a positive coefficient;
1ˢᵗ**O3**   a higher term which keeps σ small (generically a term $c\sigma^4$ with $c > 0$).

The order parameter/temperature relation of any system with these characteristics is given (schematically) by figure (5.78). In particular *superheating* and *supercooling* phenomena will be present.

Case III: First-Order Transition with Symmetry Breaking
We now consider a situation where we have the symmetry $\sigma \leftrightarrow -\sigma$, so that the cubic term vanishes, but the coefficient of σ^4 is negative. Since the free energy should be bounded below as a function of σ, we assume that the coefficient of σ^6 is positive:

$$F(\sigma) = \frac{1}{2}a(T - T_0)\sigma^2 - b\sigma^4 + c\sigma^6 + \cdots \tag{5.96}$$

where $b, c > 0$. Now the extrema are given by the equation

$$F'(\sigma) = a(T - T_0)\sigma - 4b\sigma^3 + 6c\sigma^5 = 0 \tag{5.97}$$

whose solutions are $\sigma = 0$ and

$$\sigma_{\pm}^2 = \frac{b \pm \sqrt{b^2 - 3ca(T - T_0)/2}}{3c}. \tag{5.98}$$

The discriminant is positive for

$$T < T_* = T_0 + \frac{2b^2}{3ac}. \tag{5.99}$$

From Eq. (5.98) we see that:

(I)   when $T < T_0$ the squared root σ_-^2 is negative and so Eq. (5.97) has only three *real* solutions $\sigma_0 = 0$ and $\pm\sigma_+$. σ_0 is a local maximum, and $\pm\sigma_+$ are absolute minima interchanged by the $\mathbb{Z}_2$ symmetry. In this regime the phase is *ordered* and the $\mathbb{Z}_2$ symmetry is *broken;*

(II) for $T_0 < T < T_*$ we have five *real* solutions. $\sigma_0 = 0$ is a local minimum, $\pm\sigma_-$ are local maxima, and $\pm\sigma_+$ are (equivalent) local minima. As in **Case II** there is an intermediate *critical temperature* $T_0 < T_c < T_*$ such that:

   (i)   for $T_0 < T < T_c$ we have $F(\sigma_+) < F(0)$ so that $\sigma = \pm\sigma_+ \neq 0$ and the phase is *ordered*;
   (ii) for $T_c < T < T_*$ we have $F(\sigma_+) > F(0)$ so that $\sigma = 0$ and the phase is *disordered*;

   T_c is the *critical temperature* for the order/disorder transition;

(III) when $T > T_*$ Eq. (5.97) has only one *real* solution, $\sigma = 0$, and the phase is *disordered.*

As in **Case II** the critical temperature T_c and the critical order parameter

$$\sigma_c = \lim_{T \to T_c^-} \sigma \tag{5.100}$$

are the solutions to the system of equations

$$\begin{cases} F(\sigma_c) = \dfrac{1}{2}a(T_c - T_0)\sigma_c^2 - b\sigma_c^4 + c\sigma_c^6 = 0 \\[2mm] F'(\sigma_c) = a(T_c - T_0)\sigma_c - 4b\sigma_c^3 + 6c\sigma_c^5 = 0 \end{cases} \tag{5.101}$$

whose solution is

$$T_c = T_0 + \frac{b^2}{2ac} \qquad \sigma_c = \sqrt{\frac{b}{2c}}. \tag{5.102}$$

The shape of the free energy in the various regimes is plotted in Figs. 5.10 and 5.11.

We leave to the reader to compute the latent heat of the transition at T_c and check that indeed this is a first order phase transition which is also a transition between a $\mathbb{Z}_2$ symmetric phase at high temperature to a phase with $\mathbb{Z}_2$ broken at lower temperature. As typical for first order phase transitions, we have hysteresis phenomena. The figure (5.78) of **Case II** applies to the present situation too, except

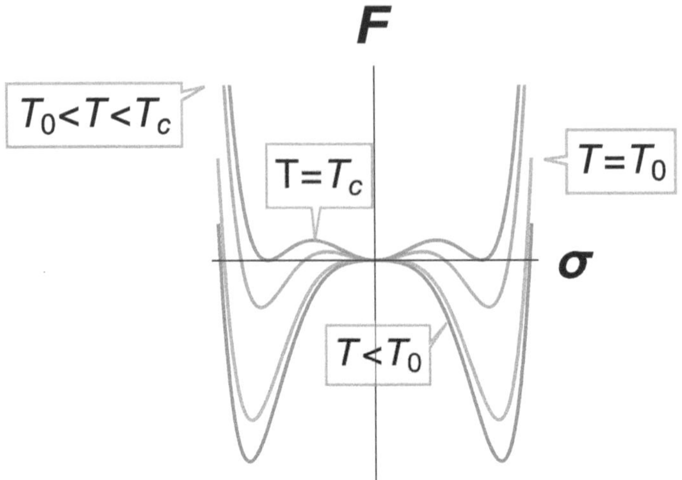

Fig. 5.10 The free energy as a function of the temperature around a "case III" phase transition for $T \leq T_c$

Fig. 5.11 The free energy as a function of the temperature around a "case III" phase transition for $T \geq T_c$

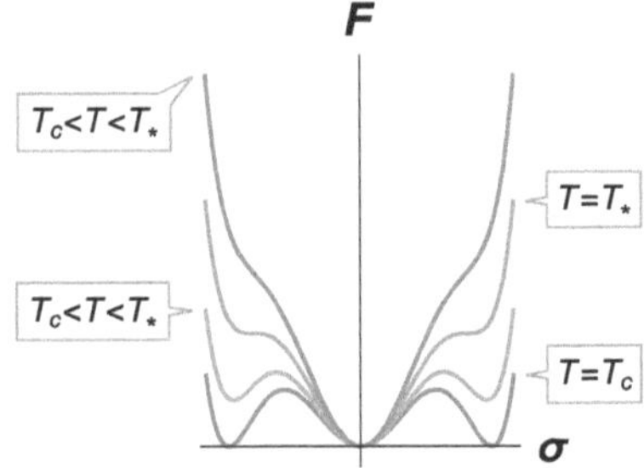

that now we have *two* ordered phases (interchanged by the $\mathbb{Z}_2$ symmetry) instead of one:

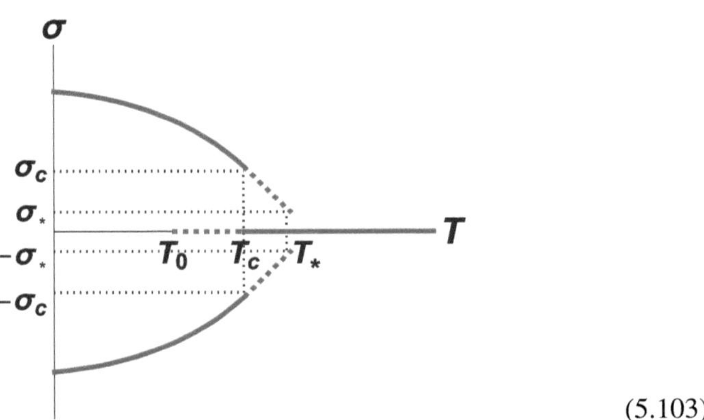

$$\tag{5.103}$$

Fig. 5.12 Phases of $\mathbb{Z}_2$-symmetric system with free energy $F(\sigma) = \frac{1}{2}a(T, P)\sigma^2 + b(T, P)\sigma^4 + c\sigma^6 + \cdots$ ($c > 0$) as a function of the coordinates a, b in control parameter space. Along the solid blue curve there is a first-order phase transition, while in the region between the solid and dashed blue curves there are meta-stable ordered states. The red line represents a second-order/disorder phase transition. The purple point marks the end of the first-order curve and is the tricritical point. The shape of the "free energy" in each region is represented by an inserted plot. The dotted red curve is the crossover between the second-order behavior and the tricritical one

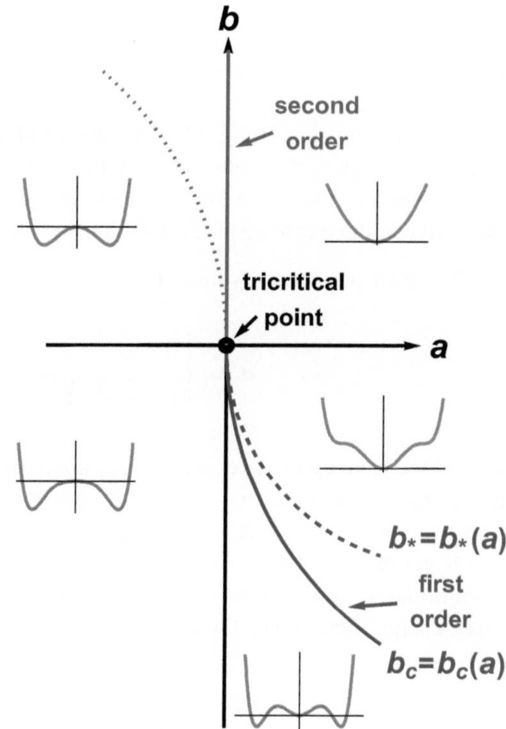

Case IV: Tricritical Point

Lastly we consider the case with only even powers of σ, where the coefficient of σ^4 is identically zero, i.e.

$$F(\sigma) = \frac{1}{2}a(T - T_0)\sigma^2 + c\sigma^6 + \cdots \tag{5.104}$$

with $a, c > 0$. The physical situation we have in mind is represented in Fig. 5.12: our system has a $\mathbb{Z}_2$ symmetry $\sigma \leftrightarrow -\sigma$ and it depends on more control parameters than just the temperature T. For instance, we may have the two control parameters T and P, so that the "free energy" takes the form

$$F(\sigma) = \frac{1}{2}a(T, P)\sigma^2 + b(T, P)\sigma^4 + c\sigma^6 + \cdots \tag{5.105}$$

with $c > 0$. Tuning the two control parameters, we can make both the coefficients of σ^2 and σ^4 to vanish. This cannot be obtained by just adjusting T since the vanishing of both coefficients is a codimension-2 condition. Hence we are discussing a phenomenon akin to the triple point in the water phase diagram (see Fig. 4.1) which happens in codimension 2 in $\mathcal{E}$, that is, in codimension 1 inside the critical locus itself.[6] In this sense we are discussing a "higher" critical phenomenon. A point where this happens is called a *tricritical point*. The various phases in the two control parameters space are represented in Fig. 5.12; using our previous results we can describe the physics along all codimension-1 loci, but we still have to understand what happens in codimension 2 i.e. at the tricritical point itself. To do this, we restrict our attention to the curve $b(T, P) = 0$ and use T as the control parameter along it. We arrive at the expression (5.104) that we now analyze *á la* Landau.

The order parameter is now

$$\sigma = \begin{cases} 0 & T > T_c \equiv T_0 \\ \left(\frac{a(T_c-T)}{6c}\right)^{1/4} & T < T_c, \end{cases} \tag{5.106}$$

so that we have again a *second order phase transition* but now the critical exponent of the order parameter is $\beta = 1/4$ instead of $1/2$, which implies that the phase transition belongs to a *different universality class* (cf. Sect. 4.10).

Of course we can consider even higher *n-critical phenomena* which appear in even higher codimension (when our manifold $\mathcal{E}$ has large enough dimension). The "free energy" then takes the form

$$F(\sigma) = \frac{1}{2}a(T - T_0)\sigma^2 + c\,\sigma^{2n} + \cdots \tag{5.107}$$

with $a, c > 0$. In this case the critical exponent becomes $\beta = 1/(2n - 2)$.

Landau Theory with Several Order Parameters
Up to now we made the simplifying assumption that the system has just *one* order parameter σ. In general we may have more than one. The expansion of the "free energy" then takes the form

$$F(\sigma_i) = a(T)_{ij}\,\sigma_i\sigma_j + O(\sigma^4) \tag{5.108}$$

where we assumed that no cubic term is present and that the quartic terms stabilize all σ_i to "small" values.

At high temperature the system is in a disordered phase. As we lower the temperature, we find a phase transition at the first ($\equiv$ highest) temperature T_c where

$$\det a(T_c) = 0. \tag{5.109}$$

Let $\eta = \sum_i c_i\,\sigma_i$ be the linear combination of the order parameters which corresponds to the zero eigenvector of the matrix $a(T_c)_{ij}$. Below the critical point

---

[6] Landau called the tricritical point the "critical point of the critical transition".

T_c the order parameter η becomes non-zero

$$\begin{cases} \eta = 0 & T > T_c \\ \eta \neq 0 & T < T_c \end{cases} \tag{5.110}$$

The linear combination η is called the *critical order parameter.* The linear combinations of the σ_i's orthogonal to η remain zero just below T_c: they are *non-critical* order parameters at T_c.

5.4 Landau-Ginzburg Theory I: Generalities

The Landau treatment of critical phenomena given in the previous section was rather crude, neglecting all thermal fluctuations in the system which, of course, are important. They are especially crucial in low dimension d where we got significant discrepancies with respect to the exact result. In particular Landau's rough method gives us *no estimate* of the error we make by replacing the actual ensemble with the mean field one. For these reasons, Landau (working with Ginzburg) soon after the formulation of his "theory" introduced a refinement of it which allows to take fluctuations in some account. The treatment, while substantially improved, is still a phenomenological one, and valid for small values of the order parameter(s) and space dimension d not too small.

For the sake of clarity we introduce the Landau-Ginzburg theory of phase transitions from the very scratch. In doing this we shall repeat once again several observations we already made a number of times. We apologize for the repetitions which may be annoying but help to make the argument straight.

We wish to study critical phemomena, i.e. the thermal physics in the vicinity of a phase transition. We know that the critical points correspond to loci in control parameter space where the thermodynamic potentials are not analytic,[7] and hence the physical quantities on one side of the transition cannot be obtained from the ones on the other side by analytic continuation across the transition.

The transition may be described in terms of a set of order parameters. For easy of exposition we first assume that one order parameter suffices to capture the essence of the critical behavior of our system. The generalization to several (finitely many!) order parameters is straightforward and will be summarized below. Usually the order parameter is an extensive quantity, and we consider its value m per "particle" (per *site* in lattice systems). For instance, for a magnetic system m is the spontaneous magnetization, for a gas-liquid transition m is the difference in specific volume between the gas and the liquid, etc. The order parameter m is the most important quantity in the description of the system at criticality. As discussed in Sect. 1.10,

---

[7] The "off-shell" free energy $F(\sigma)$ is still analytic, but the physical free energy (equal to the critical value of $F(\sigma)$ at its global minimum) is *not* analytic.

there is a dual intensive variable B such that the First/Second Law contact form κ contains the term $NB\,dm$, that is, the quantity of (virtual) work per "particle" done in a process where the order parameter m varies by dm is

$$\frac{\delta W}{N} = B\,dm. \tag{5.111}$$

When m is a magnetization, B is the external magnetic field, in the gas-liquid transition B is the pressure P, etc. By abuse of language we call B the "magnetic field" independently of its physical interpretation in the particular system at hand.

We work with an ensemble in which T and B are fixed. The corresponding thermodynamical potential is the Gibbs free energy (since we keep fixed two *intensive* parameters). However we shall write it as $F(T, B)$ because this (abusive) notation is more common in the literature. The partition function is

$$Z(T, B) = e^{-\beta F} = \mathrm{Tr}[e^{-\beta H(B)}] \tag{5.112}$$

where $H(B)$ is the Hamiltonian in presence of a magnetic field B, and we used the notation valid in the quantum set-up with our usual understanding that the trace over the Hilbert space should be replaced by an integral over the phase space $\mathfrak{W}$ in the classical case and by the obvious discrete state sums for discrete systems.

Our goal is to replace the exact partition function (5.112), which is typically hard to compute, with an effective "phenomenological" version of it. To this end we refine the procedure used in the mean field approach: there we worked with a *constant* magnetization m, equal to the average over all spins (5.3), thus killing all local fluctuations. In the actual system the fluctuations create spatial *bubbles* inside which the spins have diverse orientations, so that the mean magnetization *inside the bubble* is different from its overall average m, see Fig. 4.8. We then refine the mean field procedure by taking a *coarse-grained mean,* where we average the spins over some small (but still "macroscopic") domain of size a (a typical bubble size), producing a magnetization $m(x)$ which depends on the position x of the domain in $\mathbb{R}^d$, thus keeping track of the fluctuation of the magnetization $m(x)$ from region to region. By construction, $m(x)$ has no fluctuation on scales $\lesssim a$, that is, all its Fourier modes with frequency $\gtrsim 1/a$ vanish.[8] Then we compute the (free) energy of such a coarse-grained configuration $m(x)$ in presence of an external "magnetic field" B

$$E[m(x), B] \tag{5.113}$$

which is formally a functional depending on the function $m(x)$ defined in $\mathbb{R}^d$. However, restricting the system in a finite box $K \subset \mathbb{R}^d$ with the truncation to the Fourier modes $\lesssim 1/a$, the coarse-grained order parameter $m(x)$ depends on only finitely-many Fourier coefficients, so that $E[m(x), B]$ is actually a function in a space of huge *but finite* dimension (before taking the thermodynamic limit).

---

[8] Mathematically: $m(x) = \sum_s c_s \exp(i\boldsymbol{k}_s \cdot \boldsymbol{x})$ is a Fourier polynomial.

The effective partition *á la* Landau-Ginzburg is the canonical partition function of the effective statistical system whose degree of freedom is the function $m(x)$ and whose Hamiltonian is the functional (5.113):

$$Z(T, B)_{\text{LG}} \equiv e^{-\beta F_{\text{LG}}} = \mathcal{N} \int [Dm] \, e^{-\beta E[m, B]}, \tag{5.114}$$

where $[Dm]$ denotes the functional integral measure over the space of all functions $m \colon K \to \mathbb{R}$ from our finite d-dimensional box K to $\mathbb{R}$, or, in the general case of several order parameters, over the space of maps

$$m \colon K \to \mathcal{M} \tag{5.115}$$

where $\mathcal{M}$ is the manifold in which the order parameters take their values. In Eq. (5.114) $\mathcal{N}$ is a convenient normalization constant which cancels out in the computation of correlation functions: we use it to absorb all annoying overall factors in the following formulae.

Formally (5.114) is a *path integral* in d dimensions of the kind one encounters in Quantum Field Theory (QFT) [2–6]. Despite physicists use path integrals all the time, mathematically they are well defined only in very special cases.[9] In a sense, the fact that they are not well defined *mathematically* is a blessing, since making sense of them requires Renormalization [7] and this leads to the application of the Renormalization Group techniques in Statistical Mechanics, which produces the ultimate theory of critical phenomena.[10] We shall discuss examples of "nice" path integrals in the next section. For the moment we limit to stress that in our set-up the "space of functions" over which we have to integrate is actually a finite dimensional space parametrized by finitely-many Fourier coefficients, so our Landau-Ginzburg path integrals (5.114) are just ordinary integrals in finite dimension hence well-defined (but hard to compute since the dimension is typically huge).

What is the relation between the Landau-Ginzburg effective partition function $Z(T, B)_{\text{LG}}$ and the exact partition function $Z(T, B)$?

Here we sketch a formal answer. More precise arguments will be given below in the context of concrete models. Let O_x be the *local operator* such that $m = \langle O_x \rangle$: for instance in the Ising case $O_x = \sigma_x$ ($x \in \Lambda$). Let $\rho(y)$ be a smooth positive *test function* with support in a small region of size a and total mass 1, i.e. $\int d^d y \, \rho(y) = 1$. We identify the coarse-grained order parameter $m(x)$ with the smeared operator

$$m(x) = \int d^d y \, \rho(x - y) \, O_y \tag{5.116}$$

---

[9] For instance, they are well defined for all reasonable one-dimensional systems.

[10] Renormalization Group methods are beyond an introduction to Statistical Mechanics, and will not be covered in this book. See [8–10] for an introduction to Renormalization Group techniques.

(the integral is replaced by a sum for a discrete system). Let us define

$$\exp\left(-\beta\,E[m(x),B]_{\text{ex}}\right) \overset{\text{def}}{=} \text{Tr}\left[e^{-\beta H}\,\delta\left(m(x) - \int d^d y\,\rho(x-y)\,O_y\right)\right]$$

$$(5.117)$$

Clearly the exact partition function is

$$Z(T,B) = \mathcal{N}\int [Dm]\,\exp\left(-\beta\,E[m(x),B]_{\text{ex}}\right). \qquad (5.118)$$

Equation (5.118) is just a tautological rewriting not a simplification. Now the idea is to guess a physically reasonable (approximate) expression $E[m(x),B]$ which replaces the exact (but very complicate) functional $E[m(x),B]_{\text{ex}}$.

From (5.117) we see that $E[m(x),B]_{\text{ex}}$ is nothing else than the free energy computed in the generalized canonical ensemble where the finitely-many "almost macroscopic" variables (Fourier coefficients) which parametrize the function $m(x)$ have definite values. Hence

$$E[m(x),B]_{\text{ex}} = U[m(x)] - T\,S[m(x)] \qquad (5.119)$$

where $U[m(x)]$ and $S[m(x)]$ are, respectively, the internal energy and the entropy computed in this peculiar ensemble. Since both sides are extensive quantities we have

$$E[m(x),B]_{\text{ex}} = \int d^d x\,\mathcal{E}(m(x),B)_{\text{ex}} \qquad (5.120)$$

for some density $\mathcal{E}(m(x),B)_{\text{ex}}$ which is a local expression in the function $m(x)$ and its derivatives of arbitrary high order. The basic idea is to replace $\mathcal{E}(m(x),B)_{\text{ex}}$ by an approximate expression $\mathcal{E}(m(x),B)$ obtained by truncating its series expansion in powers of the "field" $m(x)$ and its derivatives. In Sect. 5.6 we shall present examples of *exact* functionals $\mathcal{E}(m(x),B)_{\text{ex}}$ and of their approximate Landau-Ginzburg versions $\mathcal{E}(m(x),B)$. But first we have to address a more urgent question: *Is the ensemble* (5.117) *physically sound?*

We expect that in a typical system the order parameter is slow to come to the thermal equilibrium (think of the phenomenon of hysteresis), so the microscopic degrees of freedom will thermalize much before the profile $m(x)$ of the order parameter settles to its equilibrium shape. Thus the profile $m(x)$ may *effectively* be treated as an "almost" macroscopic control parameter which we may keep fixed in defining an ensemble. Therefore the proposed procedure—while not exact—is physically reasonable.

The main improvement of the Landau-Ginzburg refinement is that now we can get a *qualitative* understanding of the fluctuations of the order parameter $m(x)$ in the x-space, and check whether or not the assumption that they are small is justified in each particular regime of our system. Replacing $\mathcal{E}(m(x), B)_{\mathrm{ex}}$ with the simpler energy density $\mathcal{E}(m(x), B)$ is fully justified when the exact and the approximate systems belong to the same universality class. Recall from the discussion in Sect. 4.10 that we are entitled to replace our system with the "simplest one" in its universality class, so whenever the Landau-Ginzburg functional $E[m(x), B]$ belongs the right class the Landau-Ginzburg phenomenological approach becomes exact. As anticipated in Sect. 4.10 this will happen when the space dimension d is larger than the second critic dimension d_2. We shall resume the discussion of the Landau-Ginzburg theory after a short digression on the computation of simple functional integrals.

5.5 DIGRESSION: Elementary Functional Integrals

While it is hard to define (let alone to compute) functional integrals[11] in general, some basic examples may be evaluated explicitly. To focus the ideas we suppose that our space is a periodic box K in d dimensions (so $K \simeq (S^1)^d$ is a d-torus) of size L, and that the several order parameters $m_i(x)$ (called *fields* in this context) take value in $\mathbb{R}^m$

$$m : K \to \mathbb{R}^m. \tag{5.121}$$

The periodic boundary conditions allow to expand the field $m_i(x)$ in Fourier modes

$$m_i(x) = \sum_{k \in \mathbb{Z}^n} \mu_i(k)\, e^{2\pi i k \cdot x / L}. \tag{5.122}$$

The functional integral is then concretely an integral over the Fourier coefficients $\mu_i(k)$ of the fields $m_i(x)$.

Generalities on Correlations

We start from some general consideration about functional integrals of the form

$$Z = \int [Dm] e^{-S[m(x)]} \quad \text{with} \quad S[m(x)] \equiv \int d^d x\, L\big(m(x)\big) \tag{5.123}$$

where $L(m(x))$ is a local expression[12] in the order parameters $m_i(x)$ ($i = 1, \ldots, s$) and their derivatives. The path integrals of the Landau-Ginzburg theory are of this form for $L(m(x)) = \beta \mathcal{E}(m(x); B)$.

---

[11] The terms "functional integral" and "path integral" are used interchangeably.

[12] In Field Theory $L(m(x))$ is known as the *Lagrangian* while the functional $S[m(x)]$ is called the *action*.

We wish to compute the ℓ-point correlation functions of the order parameters

$$\big\langle m_{i_1}(x_1)\cdots m_{i_\ell}(x_\ell)\big\rangle \overset{\text{def}}{=} \frac{\int [Dm]\, e^{-S[m(x)]}\, m_{i_1}(x_1)\cdots m_{i_\ell}(x_\ell)}{\int [Dm]\, e^{-S[m(x)]}} \tag{5.124}$$

as well as the *connected* correlation functions which often are more useful. See the off-text box for the combinatorics of connected vs. ordinary correlation functions.

Connected Correlation Functions

Let O_i $(i = 1,\ldots,s)$ be random variables (physical observables) and z_i indeterminates. The generating functions of the ordinary and connected correlation functions are, respectively,

$$\sum_{n_1\ldots n_s} \frac{z_1^{n_1}\ldots z_s^{n_s}}{n_1!\ldots n_s!}\langle O_1^{n_1}\ldots O_s^{n_s}\rangle = \Big\langle \exp\Big(\sum_i z_i O_i\Big)\Big\rangle$$

$$\sum_{n_1\ldots n_s} \frac{z_1^{n_1}\ldots z_s^{n_s}}{n_1!\ldots n_s!}\langle O_1^{n_1}\ldots O_s^{n_s}\rangle_{\text{conn.}} = \log\Big\langle \exp\Big(\sum_i z_i O_i\Big)\Big\rangle$$

The second equation may be taken as the definition of the connected correlation functions. The ordinary ℓ-point functions are polynomials in the connected k-point correlations with $k \leq \ell$ and *viceversa*. These polynomial relations between the ordinary and connected correlations give the combinatorial definition of the later ones. The polynomial relations are obtained by equating the coefficients of $z_1^{n_1}\cdots z_s^{n_s}$ in the identity

$$\Big\langle \exp\Big(\sum_i z_i O_i\Big)\Big\rangle = \exp\Big\langle \exp\Big(\sum_i z_i O_i\Big)\Big\rangle_{\text{conn.}}$$

In particular

$$\langle O\rangle_{\text{conn.}} = \langle O\rangle$$

$$\langle O_1 O_2\rangle_{\text{conn.}} = \langle O_1 O_2\rangle - \langle O_1\rangle\langle O_2\rangle$$

The terminology *connected* arises from the fact that in the graphical representation of the correlation function (widely used in QFT), the connected correlations are given by connected graphs, see e.g. [2, 4, 6, 11, 12]. Connected correlation functions are also known as *Ursell functions* [13]. The *physical* interpretation of the connected correlation functions is explained in the main text.

It is convenient to construct the *generating **functional*** of the correlation functions. We define the functional

$$Z[J(x)] \stackrel{\text{def}}{=} \int [Dm]\, e^{-S[m(x);J(x)]} \tag{5.125}$$

$$\text{with}\quad S[m(x); J(x)] \stackrel{\text{def}}{=} \int d^d x \Big(L\big(m(x)\big) + J_i(x)\, m_i(x) \Big). \tag{5.126}$$

The functional $Z[J(x)]$ is the partition function for the system modified by the introduction of the (background[13]) *sources* $J_i(x)$ for the fields $m_i(x)$ in the action $S[m(x)] \rightsquigarrow S[m(x); J(x)]$. The external magnetic field $B(x)$ in the Ising model, which may be position-dependent, cf. Eq. (4.51), is the prototypical example of such a source when the field (order parameter) is the magnetization $m(x)$.

The correlation functions (resp. the connected correlation functions) in presence of the (background) sources $J_i(x)$ are given by

$$\big\langle m_{i_1}(x_1) \cdots m_{i_\ell}(x_\ell) \big\rangle_J = \frac{1}{Z[J(x)]} \frac{\delta^\ell Z[J(x)]}{\delta J_{i_1}(x_1) \cdots \delta J_{i_\ell}(x_\ell)} \tag{5.127}$$

$$\big\langle m_{i_1}(x_1) \cdots m_{i_\ell}(x_\ell) \big\rangle_J^{\text{conn.}} = \frac{\delta^\ell \log Z[J(x)]}{\delta J_{i_1}(x_1) \cdots \delta J_{i_\ell}(x_\ell)} \tag{5.128}$$

where $\delta/\delta J_i(x)$ is the *functional derivative* defined in the obvious way, by the rule

$$\frac{\delta J_j(y)}{\delta J_i(x)} = \delta_{ij}\, \delta(x - y). \tag{5.129}$$

To get the correlation functions of the original system one sets the sources to zero, $J_i(x) = 0$, in the RHS of Eqs. (5.127) and (5.128).

The generating functional of the connected correlations

$$W[J(x)] = \log Z[J(x)] \tag{5.130}$$

is the *logarithm* of the partition function $Z[J(x)]$ in presence of the sources ($\equiv$ generalized magnetic fields) $J_i(x)$, thus $W[J(x)]$ is the thermodynamic potential for the corresponding ensemble. From their very definition (5.128), the connected

---

[13] A *background* is a field configuration which is kept fixed, i.e. we do not integrated over it in the path integral.

correlation functions are the derivatives of the thermodynamic potential $W[J(x)]$. Comparing with the thermodynamic formalism of Chap. 1 we learn that

> The connected correlation functions have the physical interpretation of *generalized susceptibilities* for the order parameters and their dual sources

This statement may be seen as a version of the fluctuation-dissipation theorem (see Chap. 6). This result explains why the connected correlations have a more direct physical significance than the ordinary ones. Since $W[J(x)]$ is a thermodynamical potential, it is natural to consider the dual potential $\Gamma[m(x)]$ obtained by a Legendre transform with respect to the sources $J(x)$ (cf. Chap. 1). The functional $\Gamma[m(x)]$ is known in QFT as the *effective action* or the generating functional of the *one-particle irreducible correlation functions,* see e.g. [4, 6, 11].

Functional δ-Function

The first example of an easy functional integral is when the Lagrangian $L(m(x))$ is linear in the fields $m_i(x)$ and *purely imaginary,*[14] that is,

$$L\big(m(x)\big) = \mathrm{i} \int \frac{\mathrm{d}^d x}{(2\pi)^d} \sum_i A_i(x)\, m_i(x) \tag{5.131}$$

for certain fixed (real) functions $A_i(x)$ on $K \equiv (S^1)^d$ which are independent of the fields $m_i(x)$. The path integral takes the form

$$
\begin{aligned}
\mathcal{N} \int [Dm] \exp&\left[-i \int_K \frac{\mathrm{d}^d x}{(2\pi)^d} \sum_i A_i(x)\, m_i(x) \right] = \\
&= \mathcal{N} \int \exp\left[-i \sum_i \sum_k \alpha_i(-k)\, \mu_i(k) \right] \prod_{i,k} \mathrm{d}\mu_i(k) = \\
&= \mathcal{N} \prod_{i,k} \int \mathrm{d}\mu_i(k)\, \exp\Big(-i\alpha_i(-k)\, \mu_i(k) \Big) = \\
&= \mathcal{N} \prod_{i,k} 2\pi\, \delta\big(\alpha_i(-k)\big) = \mathcal{N}' \prod_i \delta[A_i(x)]
\end{aligned}
\tag{5.132}
$$

---

[14] Depending on the application, $L(m(x))$ may be imaginary from the very beginning, or otherwise we replace it by $sL(m(x))$, evaluate the integral for $s \in \mathrm{i}\mathbb{R}$, then analytically continue to complex s, and finally set $s = 1$ in the resulting expression. In physics path integrals are quite often computed by such *analytic continuations.*

where $\alpha_i(\mathbf{k}) \equiv \alpha_i(-\mathbf{k})^*$ are the Fourier coefficients of the sources $A_i(x)$ and $\mathbf{k} \in \mathbb{Z}^d$. The expression $\delta[A_i(x)]$ in the RHS is the functional δ-function, i.e. the distribution in field-space defined by the self-reproducing property for smooth functionals $\mathcal{F}[\cdot]$ of the fields, that is, $\delta[\cdot]$ satisfies the functional identity

$$\int [DA(x)]\, \mathcal{F}[A_i(x)]\, \delta[A_i(x) - B_i(x)] = \mathcal{F}[B_i(x)] \qquad (5.133)$$

for all functionals $\mathcal{F}[\cdot]$ and all background fields $B_i(x)$ in $\mathbb{R}^d$. As always, $\mathcal{N}, \mathcal{N}'$ are inessential normalization factors.

Gaussian Integrals

The next simple instance is when $L(m(x))$ is quadratic in the fields $m_i(x)$ and their derivatives

$$S[m(x)] = \frac{1}{2} \int_K \mathrm{d}^d x \, m_i(x) D_{ij}\, m_j(x). \qquad (5.134)$$

D_{ij} is typically a symmetric matrix of (real) Hermitian linear differential operators acting on functions[15] defined in the finite box K. When the system is invariant under spatial translations in the coordinates x^a, the differential operator D_{ij} has constant coefficients. In most physical applications the operator D_{ij} is *elliptic* with a discrete spectrum (when working in a finite box K) made of positive eigenvalues with no accumulation point in $\mathbb{R}_{>0}$ and finite-dimensional eigenspaces. For, say, periodic boundary conditions, the prototypical example is

$$D_{ij} = -\delta_{ij}\,\Delta + M_{ij}^2 \qquad \Delta \equiv \sum_a \frac{\partial^2}{\partial x^a\, \partial x^a} \quad \text{Laplacian in } (S_1)^d \qquad (5.135)$$

with M_{ij}^2 a positive semi-definite (real) symmetric constant $n \times n$ matrix where n is the number of fields $m_i(x)$ $(i = 1, \ldots, n)$.[16] The spectrum of (5.135) in $(S_1)^d$ with circles of length L, is

$$\mathsf{Spec}(-\delta_{ij}\,\Delta + M_{ij}^2) = \left\{ \frac{4\pi^2}{L^2} k^2 + m_i^2 : \mathbf{k} \in \mathbb{Z}^d,\ m_i^2 \text{ eigenvalue of } M_{ij}^2 \right\}$$
$$(5.136)$$

Going back to the general case, we write $\{\lambda_a\}$ for the eigenvalues of the operator $\mathbf{D} = (D_{ij})$, ordered in a non-decreasing order $\lambda_1 \leq \lambda_2 \leq \cdots$, and $\psi_a(x)$ for the

---

[15] More generally on sections of bundles over our physical space.

[16] The square-root M_{ij} of M_{ij}^2 is called the *mass matrix* following the QFT interpretation.

corresponding normalized eigenfunction in $\mathbb{C}^n \otimes L^2((S^1)^d)$

$$\boldsymbol{D}\psi_a = \lambda_a \psi_a, \qquad \int d^d x \; \psi_a(x)^* \psi_b(x) = \delta_{ab}, \tag{5.137}$$

which we assume may be chosen to be real; this is the case for (5.135). The extension to the complex case is trivial. We expand the field $m(x) \equiv (m_i(x))$ in the orthonormal eigenfunction basis

$$m(x) = \sum_{a=0}^{\infty} c_a \psi_a(x) \quad c_a \in \mathbb{R}, \tag{5.138}$$

so that

$$S[m(x)] = \frac{1}{2} \sum_{a=0}^{\infty} \lambda_a \, c_a^2 \tag{5.139}$$

Now (choosing the overall normalization $\mathcal{N}$ in a convenient way)

$$\int [Dm] \, e^{-S[m(x)]} = \int [Dm] \, \exp\left(-\frac{1}{2} \int d^d x \; m(x)_i \, D_{ij} m(x)\right) =$$

$$= \prod_a \int \frac{dc_a}{\sqrt{2\pi}} e^{-\frac{1}{2}\lambda_a c_a^2} = \prod_a \lambda_a^{-1/2} = \left(\mathrm{Det}[\boldsymbol{D}]\right)^{-1/2}$$

$$\tag{5.140}$$

where $\mathrm{Det}[\boldsymbol{D}]$ is the *functional determinant* of the differential operator $\boldsymbol{D}$. There are various ways to define and compute the functional determinant: a convenient one is ζ*-regularization*, see the off-text box.

ζ-Regularization of Functional Determinants
One starts from the elementary identities

$$\log \mathrm{Det}[\boldsymbol{D}] = \mathrm{Tr} \log \boldsymbol{D} \equiv \sum_k \log \lambda_k$$

$$= -\frac{\partial}{\partial s} \sum_k \lambda_k^{-s}\bigg|_{s=0} \equiv -\frac{\partial}{\partial s} \mathrm{Tr}[\boldsymbol{D}^{-s}]\bigg|_{s=0} \tag{det}$$

(continued)

The function

$$\zeta_D(s) \stackrel{\text{def}}{=} \text{Tr}\left[\frac{1}{D^s}\right]$$

is the ζ-*function of the elliptic operator* D. The trace converges in a semiplane $\text{Re}\, s > s_0$ and for a typical elliptic operator D the ζ-function $\zeta_D(s)$ can be analytically continued to a meromorphic function in the full complex s-plane. For instance, when $D = -\Delta + V(x)$ where Δ is the Laplacian on any compact Riemannian d-manifold K and $V: K \to \mathbb{R}$ is continuous, the Weyl law [14] gives the following asymptotic estimate for the eigenvalues of D

$$\#\{\text{eigenvalues} \leq \lambda)\} \approx \frac{S_d}{(2\pi)^d d}\text{vol}(K)\lambda^{d/2}$$

where S_d is the volume of the unit sphere in $\mathbb{R}^d$, see Eq. (2.61), and we get simple poles along the real axis, the rightmost pole being at $s = n/2$ (for more details see [15]). Around $s = 0$ the ζ-function is analytic, so its derivative at $s = 0$ exists and defines the ζ-regularized determinant through the formula (det). See Appendix 2 for explicit examples

Gaussian Correlation Functions In the Gaussian case the generating functional $Z[J(x)]$ of the correlation functions is

$$
\begin{aligned}
Z[J(x)] &= \mathcal{N} \int [Dm] \exp\left[-\frac{1}{2}\int d^d x \, m^\dagger D m + \int d^n x \, m^\dagger J\right] = \\
&= \mathcal{N}\exp\left(\frac{1}{2}\int d^d x \, J^\dagger D^{-1} J\right) \\
&\quad \times \int [Dm] \exp\left[-\frac{1}{2}\int d^d x (m - D^{-1}J)^\dagger D(m - D^{-1}J)\right] \\
&= \exp\left(\frac{1}{2}\int d^d x \, J^\dagger D^{-1} J\right)\left[\text{Det}(D)\right]^{-1/2}
\end{aligned}
\tag{5.141}
$$

where in the second line we made the functional change of functional integration variables $m \rightsquigarrow m - D^{-1}J$ whose (functional) Jacobian is 1 and then used (5.140). As always, we absorbed all annoying normalizations in the definition of $\mathcal{N}$ which we choose so that the overall coefficient in the last line is 1. Thus, up to an inessential factor independent of $J(x)$, we have

$$Z[J(x)] = \exp\left(\frac{1}{2} \int d^d x \, d^d y \, J(x)^\dagger G(x, y) J(y)\right) \tag{5.142}$$

$$W[J(x)] \equiv \log Z[J(x)] = \frac{1}{2} \int d^d x \, d^d y \, J(x)^\dagger G(x, y) J(y), \tag{5.143}$$

where $G(x, y) \equiv \boldsymbol{D}^{-1}(x, y)$ is the (integral kernel of the) *inverse of the differential operator* $\boldsymbol{D}$, also called the *Green function* of the differential operator $\boldsymbol{D}$ (*propagator* in the QFT jargon), which satisfies the differential equation

$$\boldsymbol{D} \, G(x, y) = \delta(x - y). \tag{5.144}$$

Since the operator $\boldsymbol{D}$ is translation invariant, reinserting the indices, we have

$$G(x, y)_{ij} \equiv G(x - y)_{ij}, \tag{5.145}$$

while the condition that $\boldsymbol{D}$ is real Hermitian yields

$$G(x - y)_{ij} = G(y - x)_{ji}. \tag{5.146}$$

The correlation functions for the Gaussian path integral with sources (5.141) are

$$\langle m_i(x)\rangle = \int d^d y \, G(x - y)_{ij} \, J_j(y) \tag{5.147}$$

$$\langle m_i(x) \, m_j(y)\rangle_{\text{conn}} = G(x - y)_{ij}, \tag{5.148}$$

while the connected correlation functions with ≥ 3 points vanish. Setting the sources to zero, the ordinary correlation

$$\langle m_{i_1}(x_1) \cdots m_{i_\ell}(x_\ell)\rangle \tag{5.149}$$

vanishes for ℓ odd, while for ℓ even it is given by the *Wick theorem* [2, 4, 11]

$$\sum_{\text{pairings}} \prod_{\text{pairs } i_a, i_b} G(x_{i_a} - x_{i_b})_{i_a i_b} \tag{5.150}$$

where the sum is over the

$$(2\ell - 1)!! \equiv \frac{\ell!}{2^{\ell/2}(\ell/2)!} \equiv \frac{2^\ell}{\sqrt{\pi}} \Gamma(\ell + \tfrac{1}{2}) \tag{5.151}$$

distinct ways of dividing the indices $\{1, 2, \ldots, l\}$ into $\ell/2$ pairs. For later reference we rewrite Eq. (5.141) in the form

$$\exp\left(\frac{1}{2}\sum_{ij} J_i \, A_{ij} \, J_j\right) = \mathcal{N} \int \prod_i d\phi_i \, \exp\left(-\frac{1}{2}\sum_{ij}\phi_i \, A_{ij}^{-1}\,\phi_j + \sum_i \phi_i J_i\right) \tag{5.152}$$

where A may be a finite matrix or a linear operator acting on a Hilbert space, and A^{-1} is its inverse. $\mathcal{N}$ is a constant which is independent of the "sources" J_i and proportional to $1/\sqrt{\mathrm{Det}(A)}$. The source-field relation (5.147) is then

$$\langle \phi_i \rangle = A_{ij}\,\langle J_j \rangle. \tag{5.153}$$

Remark 5.3 The Gaussian statistical distributions are characterized by the fact that the connected correlations of $s \geq 3$ operators vanish, so that the distribution is fully determined by its the 1-point and 2-point correlations, that is, by its mean and variance. The ordinary s-point correlations are polynomials in the connected correlations with $r \leq s$ points. Specializing these polynomials to the case that all $r \geq 3$ connected correlation vanish, we get back Wick's theorem.

5.6 Landau-Ginzburg Theory II: Examples

We give some explicit examples of the Landau-Ginzburg strategy.

Ising Model
For the (ferromagnetic) Ising model on the lattice Λ of any dimension d, with nearest neighbor interactions, we define the $|\Lambda_0| \times |\Lambda_0|$ matrix

$$A_{ij} = \begin{cases} \beta J & \text{if } \langle i, j \rangle \\ 0 & \text{otherwise} \end{cases} \tag{5.154}$$

so that the Hamiltonian reads

$$\beta H = -\frac{1}{2}\sum_{i,j\in\Lambda_0} A_{ij}\,\sigma_i\,\sigma_j. \tag{5.155}$$

Then, using the Gaussian identity (5.152), we get (setting $N \equiv |\Lambda_0|$)

$$
\begin{aligned}
Z &\equiv \sum_{\{\sigma_i=\pm 1\}} \exp\left[\tfrac{1}{2}\sigma_i A_{ij}\sigma_j\right] = \\
&= \mathcal{N} \sum_{\{\sigma_i=\pm 1\}} \int \exp\left[-\tfrac{1}{2}\phi_i A_{ij}^{-1}\phi_i + \phi_i\sigma_i\right] \mathrm{d}^N\phi = \\
&= \mathcal{N} \int \mathrm{d}^N\phi\, \mathrm{e}^{-\phi_i A_{ij}^{-1}\phi_j/2} \prod_i [2\cosh(\phi_i)] = \mathcal{N}' \int \mathrm{d}^N\phi\, \mathrm{e}^{-E[\phi_i]}
\end{aligned}
\tag{5.156}
$$

where

$$
E[\phi_i] = \frac{1}{2}\sum_{ij} \phi_i A_{ij}^{-1}\phi_j - \sum_i \log\cosh(\phi_i)
\tag{5.157}
$$

while the magnetization is

$$
m = \langle\sigma_i\rangle = A_{ij}^{-1}\langle\phi_j\rangle.
\tag{5.158}
$$

We may use the real field ϕ_i as the (position-dependent) Landau-Ginzburg order parameter. All our manipulations up to now were *exact*. We rewrote the Ising model partition function Z as an integral (5.156) which resembles a Landau-Ginzburg path integral (5.114). In facts the functional (5.157) is precisely the exact energy functional $\beta\, E[m(x); B]_{\mathrm{ex}}$ defined in Eq. (5.117) (for $B = 0$: the extension to $B \neq 0$ is straightforward).

The final integral in the last line of (5.156), while exact, is not simpler to compute than the original partition function in the first line of the equation. However the bottom line of (5.156) is a good starting point to construct an *approximate* description *á la* Landau-Ginzburg by replacing $E[\phi_i]$ with a related but simpler functional.

To be concrete, we take Λ to be the infinite cubic lattice $\mathbb{Z}^d \subset \mathbb{R}^d$. The sites are labelled by $k = (k_1, \ldots, k_d) \in \mathbb{Z}^d$, so that for nearest neighbor interactions we have

$$
\begin{aligned}
A_{k,l} &= \beta J \sum_{r=1}^{d} (\delta_{k_r,l_r+1} + \delta_{k_r+1,l_r}) \prod_{s\neq r} \delta_{k_s,l_s} \equiv \\
&\equiv 2\beta J \int_{[-\pi,\pi]^d} \frac{\mathrm{d}^d p}{(2\pi)^d}\, \mathrm{e}^{\mathrm{i}\,p\cdot(k-l)} \sum_{r=1}^{d} \cos p_r
\end{aligned}
\tag{5.159}
$$

where $\boldsymbol{p} = (p_1, \ldots, p_d) \in [-\pi, \pi]^d$ and we used the Fourier integral representation of the Kronecker delta

$$\delta_{k,l} = \int_{-\pi}^{+\pi} \frac{\mathrm{d}p}{2\pi}\, \mathrm{e}^{\mathrm{i}p(k-l)}, \qquad k, l \in \mathbb{Z}. \tag{5.160}$$

Hence, we have the Fourier integral representation

$$\phi_k A_{k,l}^{-1} \phi_l = \int_{[-\pi,\pi]^d} \frac{\mathrm{d}^d \boldsymbol{p}}{(2\pi)^d} A(\boldsymbol{p})^{-1} |\phi(\boldsymbol{p})|^2, \tag{5.161}$$

where

$$A(\boldsymbol{p}) = 2\beta J \sum_r \cos p_r \quad \text{and} \quad \phi(\boldsymbol{p}) = \sum_k \mathrm{e}^{\mathrm{i}\boldsymbol{k}\cdot\boldsymbol{p}} \phi_k. \tag{5.162}$$

We stress that these formulae hold in units where the lattice spacing ($\equiv$ the length of an edge) is 1. In units where the spacing is a we must replace

$$\boldsymbol{p} \rightsquigarrow a\boldsymbol{p} \tag{5.163}$$

with $\boldsymbol{p}$ taking value in $[-\pi/a, \pi/a]^d$, so that as the lattice spacing $a \to 0$ the momentum $\boldsymbol{p}$ takes values in the full $\mathbb{R}^d$. Since we are interested in fluctuations over lengths $\gg a$, the relevant momenta satisfy $a|\boldsymbol{p}| \ll 1$, and it make sense to expand $A(\boldsymbol{p})$ in powers series keeping only the first non-trivial term

$$A(\boldsymbol{p}) = 2\beta J \sum_{r=1}^d \cos p_r = 2\beta J \sum_{r=1}^d \left(1 - \tfrac{1}{2}p_r^2\right) + O(p^4) = \tag{5.164}$$

$$= z\beta J - \beta J \boldsymbol{p}^2 + O(\boldsymbol{p}^4) = z\beta J(1 - \rho^2 \boldsymbol{p}^2) + O(\boldsymbol{p}^4)$$

where we wrote $z = 2d$ (the coordination number for the d-dimensional cubic lattice) and $\rho^2 = a^2/z$ (henceforth we use units such that $a = 1$). Therefore,

$$A^{-1}(\boldsymbol{p}) = \frac{1}{z\beta J}\left(1 + \rho^2 \boldsymbol{p}^2 + O(\boldsymbol{p}^4)\right). \tag{5.165}$$

Hence, neglecting terms $O(\boldsymbol{p}^4)$ in momenta, and taking the Fourier transform back to position space, we get the *approximate* Landau-Ginzburg free energy

$$E[\phi(x)]_{\mathrm{appr}} = \int \mathrm{d}^d x \left(\frac{1}{2}\frac{\rho^2}{z\beta J}\partial_i\phi(x)\,\partial_i\phi(x) + \frac{1}{2}\frac{\phi(x)^2}{z\beta J} - \log[\cosh\phi(x)]\right)$$

$$\tag{5.166}$$

which is valid as long as the length-wave of the fluctuations is large in lattice spacing units or, equivalently, iff $\|\partial_i\phi\| \ll 1$. The field redefinition

$$\varphi(x) = \frac{\rho}{\sqrt{z\beta J}}\phi(x) \tag{5.167}$$

sets the kinetic (derivative) term in its canonical form

$$E[\varphi(x)]_{\mathrm{appr}} =$$

$$= \int d^d x \left[\frac{1}{2}\partial_i\varphi(x)\,\partial_i\varphi(x) + \frac{1}{2\rho^2}\varphi(x)^2 - \log\cosh\left(\sqrt{\frac{z\beta J}{\rho^2}}\varphi(x)\right)\right] \tag{5.168}$$

Starting from this expression, we may proceed in different ways:

MF   we may neglect the fluctuations (i.e. the derivative term) altogether: we get back the mean field description of Sect. 5.1;

Lan   we may neglect the fluctuations and moreover assume that the order parameter φ is small, and replace the transcendent potential in Eq. (5.168) by its series expansion, dropping the terms of order higher than 4 as we did in Sect. 5.3. Since

$$\log\cosh\phi = \frac{1}{2}\phi^2 - \frac{1}{12}\phi^4 + O(\phi^6), \tag{5.169}$$

this gives us back the crudest Landau treatment;

LG   we may assume φ to be small and expand the potential up to φ^4 while assuming the fluctuations to be small *but non-zero* so keeping the terms with two derivatives in Eq. (5.168) (we already dropped terms with four or more derivatives). This is the Landau-Ginzburg procedure.

The mean field story **MF** is reviewed at the end of this section. Here we focus on **LG**. Using Eq. (5.169) we get

$$E[\varphi(x)]_{\mathrm{appr}} = \int d^d x \left[\frac{1}{2}\partial_i\varphi(x)\,\partial_i\varphi(x)+ \right.$$
$$\left. + \frac{1}{2\rho^2}(1 - z\beta J)\varphi(x)^2 + \frac{1}{12}\frac{(z\beta J)^2}{\rho^4}\varphi(x)^4 \right] \tag{5.170}$$

Note that $kT_c = zJ$ is the mean field critical temperature, so we set

$$\frac{1}{\rho^2}(1 - z\beta J) = \frac{1}{\rho^2 T}(T - T_c) \approx a(T - T_c) \qquad \text{as } T \approx T_c \tag{5.171}$$

$$z\beta J = 1 + O(T - T_c),$$

where $a \equiv (\rho^2 T_c)^{-1}$ is a positive constant. Near the critical point $T \approx T_c$ we get

$$E[\varphi(x)]_{\text{LG}} = \int d^d x \left[\frac{1}{2} \partial_i \varphi(x)\, \partial_i \varphi(x) + \frac{1}{2} a(T - T_c)\varphi(x)^2 + b\,\varphi(x)^4 \right]$$

$$(5.172)$$

with $b \equiv 1/(12\rho^4)$ a positive constant. This is the (approximate) Landau-Ginzburg free energy for the d-dimensional Ising model we were after. It differs from the crudest Landau free energy of Sect. 5.3 by a term proportional to the square of the gradient of the order parameter.

Example: The XY Model

To get the Landau-Ginzburg formulation of the XY *model* we replace the spins σ_i of the Ising model with unit vectors in the x-y plane: $s_i = (\cos\theta_i, \sin\theta_i)$. The Hamiltonian is

$$\beta H = -\frac{1}{2} \sum_{k,l} A_{kl}\, s_k \cdot s_l \tag{5.173}$$

where the matrix A_{kl} is as in Eq. (5.159). Now

$$Z = \int e^{A_{kl} s_k \cdot s_l / 2} \prod_i d\theta_i = \mathcal{N} \int d^N\boldsymbol{\phi}\, e^{-A_{kl}^{-1}\boldsymbol{\phi}_k \cdot \boldsymbol{\phi}_l / 2 + \boldsymbol{\phi}_i \cdot s_i} \prod_i d\theta_i =$$

$$= \mathcal{N} \int d^N\boldsymbol{\phi}\, e^{-A_{kl}^{-1}\boldsymbol{\phi}_k \cdot \boldsymbol{\phi}_l / 2} \prod_j \left(\int_0^{2\pi} d\theta_j\, e^{\phi_j\, \cos(\theta_j - \psi_j)} \right) =$$

$$= \mathcal{N}' \int d^N\boldsymbol{\phi}\, \exp\left[-\frac{1}{2} A_{kl}^{-1}\boldsymbol{\phi}_k \cdot \boldsymbol{\phi}_l + \sum_i \log I_0(\phi_i) \right] \tag{5.174}$$

where we wrote

$$\boldsymbol{\phi}_i = (\phi_i \cos\psi_i, \phi_i \sin\psi_i) \in \mathbb{R}^2, \tag{5.175}$$

and used

$$\int_0^{2\pi} e^{z\cos\theta}\, d\theta = 2\pi\, I_0(z), \tag{5.176}$$

where $I_0(z)$ is the modified Bessel function of the first kind with index 0 [16]. The exact partition function can be written in the Landau-Ginzburg form (5.114) with

$$E[\phi_i] = \frac{1}{2} \sum_{k,l} A_{kl}^{-1} \phi_k \cdot \phi_l - \sum_i \log I_0(|\phi_i|) \tag{5.177}$$

The rest of the analysis parallels the one for the Ising model.

Mean Field Revisited We may recover the mean field description as the crudest approximate evaluation of the path integral

$$\int [D\phi(x)] \, e^{-E[\phi(x)]} \tag{5.178}$$

in the RHS of Eq. (5.156) where we replace the integral by the maximal value of the integrand, that is, by the leading term in the saddle-point approximation as $N \to \infty$ (cf. Sect. 2.5). Note that the maximum is attained by a *constant* configuration, $\phi(x) = \phi$, since—when $|p| \ll 1$—the derivative term in $E[\phi(x)]$ is non-negative and zero only for constant configurations.

For instance, in the Ising case, setting in Eq. (5.166) the field $\phi(x)$ to be the constant configuration $z\beta J m$, we get:

$$\beta F[m] \stackrel{\text{def}}{=} E[z\beta J m] = \frac{1}{2} N z\beta J m^2 - N \log \cosh(z\beta J m) \tag{5.179}$$

whose minimization yields back Eq. (5.13) i.e. the mean field theory.

5.7   Fluctuations: The Gaussian Model

As stated at the beginning of Sect. 5.4, the improvement of the Landau-Ginzburg approach, based on the functional integral (5.114), with respect to the original Landau theory (Sect. 5.3) is that the refined method yields some control on the long-wave fluctuations, allowing us to estimate the error we made in forgetting them altogether, and hence giving a criterion to test the validity of mean field theory.

The Functional Integral We start from (5.166) where we set

$$\phi(x) = z\beta J \, m(x) \equiv \alpha \, m(x) \tag{5.180}$$

with $m(x)$ the local magnetization. In principle we have to compute the path integral

$$\mathcal{N} \int [Dm(x)] \exp\left(-\int d^d x \left(\frac{1}{2} K \, \partial_i m(x) \, \partial_i m(x) + \beta F[m(x)]_{\text{MF}}\right)\right) \tag{5.181}$$

where the function $F[m]_{\mathrm{MF}}$ is the mean field "free energy" in Eq. (5.9), that is,

$$\beta F[m]_{\mathrm{MF}} = \frac{1}{2}\alpha\, m^2 - \log[2\cosh(\alpha m)] \tag{5.182}$$

with $\alpha = z\beta J \equiv T_c/T$ and $K \equiv \alpha\rho^2$. The crudest mean field approach consists in replacing the functional integral by the value of the integrand at the saddle-point. To improve the approximation, we perform the detailed saddle-point evaluation of the path integral as in Sect. 2.5. From a more physical viewpoint we may describe the procedure as follows: we write the local magnetization $m(x)$ as its spatial mean m plus a fluctuation $\mu(x)$ which depends on the point $x \in \mathbb{R}^d$

$$m(x) = m + \mu(x), \tag{5.183}$$

where m is the solution to the mean field consistency equation

$$\frac{\partial F_{\mathrm{MF}}}{\partial m} = 0, \tag{5.184}$$

which yields the absolute minimum of the function $F[m]_{\mathrm{MF}}$, see Eqs. (5.13), so that the argument of the exponential in Eq. (5.181) becomes

$$E[m(x)] \equiv E[m + \mu(x)] = \beta V\, F_{\mathrm{MF}}|_{\min} +$$

$$+ \int \mathrm{d}^d x \left[\frac{K}{2}\partial_i\mu\,\partial_i\mu + \frac{\beta}{2}\frac{\partial^2 F_{\mathrm{MF}}}{\partial m^2}\bigg|_{\min}\mu^2 + \frac{\beta}{3!}\frac{\partial^3 F_{\mathrm{MF}}}{\partial m^3}\bigg|_{\min}\mu^3 + \cdots \right] \tag{5.185}$$

where $V \equiv |\Lambda_0|$ is the volume of the finite box, $F_{\mathrm{MF}}|_{\min}$ is the mean field free energy per site evaluated on the solution of Eq. (5.184) which realizes its absolute minimum. In the Taylor expansion of F_{MF} the term linear in μ vanishes by (5.184). The coefficient of μ^2 is positive by convexity of the mean field free energy; this term prevents the fluctuations μ to get too large. The Gaussian approximation consists in neglecting all higher order terms in μ, keeping only the quadratic terms. This is justified if the fluctuations are small enough, that is, when the coefficient of the quadratic term is large enough. Below we shall give a more precise criterion for the validity of this Gaussian approximation.

The coefficient of μ^2, evaluated on a solution of Eq. (5.13), is (cf. Eq. (5.15))

$$\frac{\partial^2(\beta F_{\mathrm{MF}})}{\partial m^2}\bigg|_{\min} = \frac{T_c}{T}\left(1 - \frac{T_c}{T} + \frac{T_c}{T}m^2\right) \approx \begin{cases} 2\frac{T_c - T}{T} + \cdots & T < T_c \\ \frac{T_c(T-T_c)}{T^2} & T > T_c \end{cases} \tag{5.186}$$

so is positive for all $T \neq T_c$ but goes to zero as $T \to T_c$ from both sides: we know from Landau theory that a vanishing quadratic term at the critical temperature is the trademark of second order phase transitions (cf. **Case I** in Sect. 5.3).

In the Gaussian approximation the free energy F is

$$\beta V F = \beta V F\Big|_{\text{min}} + \frac{1}{2}\log \text{Det}[-\Delta + M^2] \tag{5.187}$$

where we set[17]

$$M^2 = \frac{1}{K}\frac{\partial^2(\beta F_{\text{MF}})}{\partial m^2}\Big|_{\text{min}}. \tag{5.188}$$

M^2 goes to zero linearly as $T \to T_c$ cf. Eq. (5.186). The second term in Eq. (5.187),[18]

$$\log \text{Det}[-\Delta + M^2] = V\int \frac{\mathrm{d}^d p}{(2\pi)^d}\log(p^2 + M^2) =$$
$$= \frac{V}{(2\pi)^d}\frac{2\pi^{d/2}}{\Gamma(d/2)}\int_0^\infty p^{d-1}\,\mathrm{d}p\,\log(p^2 + M^2), \tag{5.189}$$

is formally divergent as $p \to \infty$ (UV divergence). However recall that in our problem the momenta take value in $[-\pi/a, \pi/a]^d$ (a is the lattice spacing) so in the second line the integration extends up to π/a and is actually finite. Alternatively, as it is usual in QFT, we define the determinant as its ζ-regularized version (Appendix 2).

Heat Capacity The second term in Eq. (5.187) yields corrections to the mean field values of thermal quantities. The internal energy becomes

$$U = \frac{\partial}{\partial\beta}(\beta F) = U_{\text{MF}} + \frac{1}{2}\frac{\partial}{\partial\beta}\log \text{Det}[-\Delta + M^2] \tag{5.190}$$

and the heat capacity

$$C_V = (C_V)_{\text{MF}} - \frac{1}{2kT^2}\frac{\partial^2}{\partial\beta^2}\log \text{Det}[-\Delta + M^2]. \tag{5.191}$$

The relation between the determinant defined by the integral expression in the second line of (5.189) (where we integrate only over moment up to a certain cut-off scale π/a) and its ζ-regularized version (see Appendix 2) becomes clear by

---

[17] The reason beyond the choice of notation is that in the QFT interpretation of the path integral the RHS is the square of the mass.

[18] We use that the volume of the sphere S^{d-1} is $\frac{2\pi^{d/2}}{\Gamma(d/2)}$, cf. Eq. (2.61).

comparing their k-fold derivatives: the formal equalities

$$\frac{V}{(2\pi)^d} \frac{2\pi^{d/2}}{\Gamma(d/2)} \lim_{a\to 0} \frac{\partial^k}{\partial M^{2k}} \int_0^{1/a} p^{d-1} \, dp \, \log(p^2 + M^2) =$$

$$= \frac{(-1)^k}{(k-1)!} \frac{V}{(2\pi)^d} \frac{2\pi^{d/2}}{\Gamma(d/2)} \int_0^{\infty} \frac{p^{d-1} \, dp}{(p^2 + M^2)^k} = \qquad (5.192)$$

$$= \frac{\partial^k}{\partial M^{2k}} \log \mathrm{Det}[-\Delta + M^2]\Big|_{\zeta\text{-reg.}}$$

are valid whenever the integral in the second line is convergent, that is, for $k > d/2$. Since $M^2 \approx \mathrm{const.}(\beta - \beta_c)$, the new term in Eq. (5.191) is proportional to the second derivative with respect to M^2, and both definitions give the same result for $d < 4$.

The logarithm of the ζ-determinant is proportional to $V M^d$ up to log corrections, see Appendix 2. Hence the second derivative with respect to M^2 scales as

$$C_V - (C_V)_{\mathrm{MF}} \approx \mathrm{const.} \, V \, |T - T_c|^{(d-4)/2} \qquad (5.193)$$

and the critical exponent α (defined in Sect. 4.10) is

$$\alpha = \frac{4-d}{2}. \qquad (5.194)$$

We conclude that the deviation from the mean field answer is large at criticality (as $T \to T_c$) in dimension $d \le 4$ (for $d = 4$ the divergence is only logarithmic), but small in dimension larger than 4. We conclude that in dimension larger than 4 the mean field approximation is reliable (as an approximation!) for all temperatures, even near the critical point. Below we shall make this point sharper by an analysis of the fluctuations around the mean field configuration.

Correlations of Fluctuations Next we consider the correlation functions of the fluctuation field $\mu(x)$. From these functions we can read how large are the fluctuations $m(x) - \langle m \rangle$, and hence understand under which conditions the mean field approximation is reliable.

Since the field $\mu(x)$ is Gaussian, the formulae in Sect. 5.5 apply: the correlation

$$\langle \mu(x_1) \, \mu(x_2) \cdots \mu(x_\ell) \rangle \qquad (5.195)$$

vanishes if ℓ is odd, otherwise it is given by the Wick theorem in terms of the two-point correlation function

$$\langle \mu(x) \, \mu(y) \rangle = \frac{1}{K} \, G(x - y) \qquad (5.196)$$

where $G(x - y) = G(y - x)$ is the scalar *Green function* (a.k.a. *propagator*) which is the (functional) inverse of the differential operator $-\Delta + M^2$ in $\mathbb{R}^d$, that is,

$$(-\Delta + M^2)\, G(x - y) = \delta^{(d)}(x - y) \tag{5.197}$$

where the RHS is the Dirac δ-function in d dimensions. From (5.196) we see that $G(x - y)$ is (up to normalization) the covariance of the (Gaussian) distribution of fluctuations around the mean field answer.

Taking the Fourier transform of both sides of (5.197) we get the momentum space form of the Green's function (called *propagator* in QFT)

$$\hat{G}(p) = \frac{1}{p^2 + M^2} \qquad p \equiv |p|. \tag{5.198}$$

To get the Green's function $G(x)$ in x-space, we Fourier transform it back ($r = |x|$):

$$
\begin{aligned}
G(x) &= \int \frac{\mathrm{d}^d p}{(2\pi)^d} \frac{e^{ix \cdot p}}{p^2 + M^2} && \textbf{(inverse Fourier transform)} \\[2mm]
&= \int \frac{\mathrm{d}^d p}{(2\pi)^d} \, e^{ix \cdot p} \int_0^\infty \mathrm{d}t \; e^{-(p^2 + M^2)t} && \textbf{(representation} \frac{1}{x} = \int_0^\infty \mathrm{d}t \, e^{-xt}\,) \\[2mm]
&= \frac{1}{(4\pi)^{d/2}} \int_0^\infty \mathrm{d}t \; t^{-d/2}\, e^{-M^2 t - r^2/4t} && \textbf{(Gaussian integral in } p\,) \\[2mm]
&= \frac{M^{d-2}}{(4\pi)^{d/2}} \int_0^\infty \mathrm{d}y \; y^{-d/2}\, e^{-y - M^2 r^2/4y}
\end{aligned}
$$

$$\tag{5.199}$$

Now we recall the integral representation of the modified Bessel function of the second kind with index ν [16]

$$K_\nu(z) = \frac{1}{2} \left(\frac{z}{2}\right)^\nu \int_0^\infty \frac{\mathrm{d}y}{y^{\nu+1}}\, e^{-y - z^2/4y}. \tag{5.200}$$

comparing with (5.199) we get

$$G(x) \equiv G(r) = \frac{1}{(2\pi)^{d/2}} \left(\frac{M}{r}\right)^{\frac{d}{2}-1} K_{\frac{d}{2}-1}(Mr) \tag{5.201}$$

The large argument asymptotics of the Bessel function is [16]

$$K_\nu(z) \approx \sqrt{\frac{\pi}{2z}}\, e^{-z} \quad \text{as } z \to \infty \tag{5.202}$$

while

$$\text{as } z \to 0 \quad K_\nu(z) \approx \begin{cases} -\log z & \text{for } \nu = 0 \\ \frac{1}{2}\Gamma(\nu)(2/z)^\nu & \text{for } \nu > 0. \end{cases} \tag{5.203}$$

When the space dimension d is *odd* the relevant Bessel function is in facts an elementary function; for instance when $d = 3$ (the physical dimension!)

$$K_{1/2}(z) = \sqrt{\frac{\pi}{2z}}\, e^{-z} \quad \Rightarrow \quad G(r) = \frac{1}{4\pi}\frac{e^{-Mr}}{r} \tag{5.204}$$

is the *Yukawa potential,* which reduces to the Coulomb potential $1/(4\pi r)$ for $M = 0$.

Putting everything together, in d space dimensions, we have

$$\langle \mu(x)\, \mu(0)\rangle = \frac{1}{(2\pi)^{d/2}\, K}\left(\frac{M}{r}\right)^{\frac{d}{2}-1} K_{\frac{d}{2}-1}(Mr) \tag{5.205}$$

which, for large values of the argument Mr goes to zero exponentially, see (5.202). This result says that the fluctuations of the magnetization $m(x)$ are correlated on a length scale $\xi \sim 1/M$. The scale ξ is called the *correlation length* of the system: it is given by the inverse of the "mass" M, that is, by the inverse of the square root of the coefficient of μ^2 in the Landau-Ginzburg free energy $E[m + \mu(x)]$. We know that M^2 has the form $a_\pm|T - T_c|$ near the critical temperature T_c (for $B = 0$) so that

$$\text{correlation length:} \quad \xi \approx \frac{(\text{const.})_\pm}{|T - T_c|^{1/2}} \quad \text{for } T \approx T_c, \tag{5.206}$$

That is: *the correlation length ξ blows up at criticality.* The divergence of the correlation length may be used as an alternative characterization of the critical point of a second order phase transition.

When we approach the critical temperature T_c the argument Mr of the Bessel function goes to zero. Using Eq. (5.203) the correlator takes the form

$$G(r) \approx \begin{cases} -\frac{1}{2\pi}\log r & d = 2 \\ \frac{1}{4(\pi)^{d/2}}\Gamma\left(\frac{d}{2}-1\right)\frac{1}{r^{d-2}} & d > 2 \end{cases} \tag{5.207}$$

Note that

$$\frac{1}{4(\pi)^{d/2}}\Gamma\left(\frac{d}{2}-1\right)\frac{1}{r^{d-2}} \equiv \frac{1}{\text{Vol}(S^{d-1})}\int_r^\infty \frac{dr}{r^{d-1}} \tag{5.208}$$

is just the Coulomb potential in d space dimensions. We conclude that *at criticality the correlations of the magnetization fluctuation fall-off at infinity only with a power-law*, and so the cluster limit holds weakly i.e.

$$\langle \mu(\boldsymbol{x}_1)\ldots\mu(\boldsymbol{x}_r)\,\mu(\boldsymbol{y}_1)\ldots\mu(\boldsymbol{y}_s)\rangle =$$
$$= \langle \mu(\boldsymbol{x}_1)\ldots\mu(\boldsymbol{x}_r)\rangle\langle\mu(\boldsymbol{y}_1)\ldots\mu(\boldsymbol{y}_s)\rangle + O\!\left(|\boldsymbol{x}_i - \boldsymbol{y}_j|^{-(d-2)}\right) \tag{5.209}$$

with corrections which are only polynomially suppressed for large distances $|\boldsymbol{x}_i - \boldsymbol{y}_j|$. When $d \leq 2$ the connected correlations *actually grow* with the distance: in $d = 2$ logarithmically, see (5.207). This reflects the fact that at criticality $\mu(x)$ is not a meaningful field when $d \leq 2$: this statement is equivalent to the Mermin-Wagner and Coleman theorems, see Sect. 4.9. In particular this means that near the critical temperature T_c the mean field approximation is *poor* for $d = 2$ and *extremely poor* for $d = 1$ (as we already know by comparing with the exact solutions).

As we approach the critical point the fluctuations become correlated over large (diverging) distances, and the naive idea that we can disregard them becomes quite questionable. We need a quantitative criterion to determine when a mean field treatment is meaningful even in proximity of the critical point (the regime we are most interested to understand). Such a criterion has been proposed by Ginzburg.

Fluctuations: Ginzburg Criterion
The order parameter $m(x)$ fluctuations are strongly correlated inside regions of size ξ. For the mean field approach to be sensible, the fluctuations of the order parameter around its mean, $\mu(\boldsymbol{x})$, averaged in a correlation volume, must be small compared to the average of the order parameter in this same volume. When this is the case we can neglect the order-parameter fluctuations, and estimate the free energy using the mean field m. The condition of validity of mean field is then the following inequality

$$\left\langle\left[\int_{R_\xi} \mathrm{d}^d x \left(m(x) - \langle m(x)\rangle\right)\right]^2\right\rangle \ll \left\langle\int_{R_\xi} \mathrm{d}^d x \; m(x)\right\rangle^2 \equiv \left(\mathsf{Vol}(R_\xi)\right)^2 m^2 \tag{5.210}$$

where $R_\xi \subset \mathbb{R}^d$ is a region of linear size ξ and we set $m \equiv \langle m(x)\rangle$ as always. Equation (5.210) is known as the *Ginzburg criterion*. The LHS is

$$\int_{R_\xi} \mathrm{d}^d x \int_{R_\xi} \mathrm{d}^d y \; \langle\mu(\boldsymbol{x})\,\mu(\boldsymbol{y})\rangle = \mathsf{Vol}(R_\xi)\int_{R_\xi} \mathrm{d}^d x \; \langle\mu(\boldsymbol{x})\,\mu(0)\rangle. \tag{5.211}$$

We computed the explicit form of the 2-point function $\langle\mu(\boldsymbol{x})\,\mu(0)\rangle$ in the saddle point approximation—see Eq. (5.205)—but we wish to be more general and write formulae which apply beyond the approximation and in full generality to *all* systems. The exact 2-point correlation must have the form

$$\langle\mu(\boldsymbol{x})\,\mu(0)\rangle = \xi^{-(d-2+\eta)}\,F(r/\xi), \qquad r \equiv |\boldsymbol{x}| \tag{5.212}$$

for some $\eta \in \mathbb{R}$ and function $F(\cdot)$. By regularity and the definition of the *correlation length* ξ, the function $F(\cdot)$ should have the asymptotic behaviors

$$F(z) \approx C\, z^{-a} e^{-z} \quad \text{as } z \to \infty, \qquad F(z) \;\approx\; \frac{C'}{z^{d-2+\eta}} \quad \text{as } z \to 0 \qquad (5.213)$$

where C, C', a are constants and d is the space dimension as always. From Eq. (5.212) one sees that

$$(d - 2 + \eta)/2 \qquad (5.214)$$

is the dimension (in inverse length units) of the field $m(x)$. In the Gaussian approximation (which corresponds to weak coupling in QFT) the dimension is

$$\text{dimension } m(x) = (d - 2)/2 \quad \Leftrightarrow \quad \eta = 0 \qquad (5.215)$$

which is the result we found in Eq. (5.205). $(d - 2)/2$ is called the *canonical dimension* of a scalar field in d dimensions; it is determined by the fact that the expression

$$E[m(x)] = \int d^d x \left(\frac{1}{2} \partial_i m(x)\, \partial_i m(x) + \cdots \right) \qquad (5.216)$$

must have dimension zero for the exponential $\exp(-E[m(x)])$ to make sense. However, the interactions/fluctuations modify the physical dimension from its Gaussian value $(d - 2)/2$ to the exact value $(d - 2 + \eta)/2$. The correction η to the dimension is known as the *anomalous dimension* (this terminology arises from QFT [7]).

For definiteness we take R_ξ to be a ball of radius ξ, the conclusions will be independent of this choice. The RHS of Eq. (5.210) becomes

$$\frac{\xi^d}{d} \text{Vol}(S^{d-1})^2 \xi^{-(d-2+\eta)} \int_0^\xi r^{d-1}\, dr\, F(r/\xi) =$$
$$= \frac{\xi^d}{d} \text{Vol}(S^{d-1})^2 \xi^{2-\eta} \int_0^1 y^{d-1}\, dy\, F(y) = A\, \xi^{d+2-\eta} \qquad (5.217)$$

where A is a constant (depending essentially only on the dimension d) which absorbs the several constants in the formula. Then Eq. (5.210) becomes

$$A\, \xi^{d+2-\eta} \ll \frac{\text{Vol}(S^{d-1})^2}{d^2} \xi^{2d}\, m^2. \qquad (5.218)$$

We conclude

Ginzburg Criterion *The fluctuations are negligible iff*

$$A' \ll \xi^{(d-2+\eta)/2} m \tag{5.219}$$

where A' is a constant which depends (essentially) only on the space dimension d.

We know that the fluctuations are small away from the critical point. We focus on the behavior below but near the critical temperature T_c. In this regime

$$\xi = a(T_c - T)^{-\nu}, \qquad m = b(T_c - T)^{\beta}, \tag{5.220}$$

where a, b are constants, and ν, β are the *critical exponents* defined in Sect. 4.10. In the Gaussian approximation to the Ising model the critical exponents are

$$\beta = \nu = \frac{1}{2} \quad \text{and} \quad \eta = 0. \tag{5.221}$$

Instead at the tricritical point (**Case IV** of Sect. 5.3)

$$\beta = \frac{1}{4}, \quad \nu = \frac{1}{2}, \quad \eta = 0. \tag{5.222}$$

The Ginzburg criterion as $T \to T_c^-$ becomes

$$A'' \ll (T_c - T)^{2\beta - (d-2+\eta)\nu} \quad \Rightarrow \quad d > 2 - \eta + \frac{2\beta}{\nu} > 0. \tag{5.223}$$

For the Gaussian model this gives

$$d > d_c = 4. \tag{5.224}$$

The dimension $d_c = 4$ is called the *critical upper dimension* for the critical Gaussian model. Above this dimension the mean field theory is accurate even as we approach the critical temperature. For the tri-critical model (5.222) the critical dimension becomes $d_c = 3$.

Appendix 1: Proof of Fact 5.3

The first statement in (1) is obvious from the graph in Fig. 5.7. To get Eq. (5.69),
take the derivative

$$
\begin{aligned}
\Delta \frac{\partial \alpha_q(\Delta)}{\partial \Delta} &= \frac{q}{(1-\Delta)(1+(q-1)\Delta)} - \frac{1}{\Delta}\log\left(1+\frac{q\Delta}{1-\Delta}\right) = \\
&= \frac{q}{(1-\Delta)(1+(q-1)\Delta)} - \alpha_q(\Delta)
\end{aligned}
\tag{5.225}
$$

hence at the minimum Δ_* of $\alpha_q(\Delta)$ Eq. (5.69) holds. (2) is obvious. For (3) it is
enough to check that the values (5.74) solve the two Eqs. (5.72) and (5.73).

Appendix 2: ζ-Regularized Determinants

We compute the ζ-regularized determinant of the differential operator $-\Delta + M^2$ in a
period box in n-dimensions of volume V asymptotically for large V (i.e. neglecting
$o(V)$ corrections). The ζ-regularized determinant is defined as

$$
\log \mathrm{Det}[-\Delta + M^2] = \mathrm{Tr}\log[-\Delta + M^2] = -\frac{\partial}{\partial s}\mathrm{Tr}\frac{1}{(-\Delta + M^2)^s}\bigg|_{s=0}
\tag{5.226}
$$

where

$$
\mathrm{Tr}\frac{1}{(-\Delta + M^2)^s}
\tag{5.227}
$$

is the meromorphic function in $\mathbb{C}$ obtained by analytic continuation of the function

$$
\frac{1}{\Gamma(s)}\int_0^\infty dt\; t^{s-1}\, \mathrm{Tr}\left(e^{-(-\Delta+M^2)t}\right)
\tag{5.228}
$$

which a priori is defined in the domain of convergence of the integral.

Up to now everything was exact. Now we evaluate the functional trace by the
Euler-Maclaurin summation formula (valid up to $o(V)$ corrections)

$$
\mathrm{Tr}\, e^{\Delta t} \rightsquigarrow V\int \frac{d^n p}{(2\pi)^n} e^{-p^2 t} = \frac{V}{(4\pi)^{n/2}} t^{-n/2}.
\tag{5.229}
$$

Then

$$\mathrm{Tr}\frac{1}{(-\Delta + M^2)^s} = \frac{V}{(4\pi)^{n/2}}\frac{1}{\Gamma(s)}\int_0^\infty dt\, t^{s-n/2-1}e^{-M^2 t} =$$

$$= \frac{V}{(4\pi)^{n/2}}\frac{1}{\Gamma(s)}M^{n-2s}\int_0^\infty dy\, y^{s-n/2-1}\,e^{-y} = \frac{V}{(4\pi)^{n/2}}\frac{\Gamma(s-n/2)}{\Gamma(s)}M^{n-2s}$$

$$(5.230)$$

a meromorphic function of $s \in \mathbb{C}$. There are two situations: $n = 2m - 1$ is *odd* and

$$\frac{\Gamma(s-n/2)}{\Gamma(s)} = \frac{(-2)^m\sqrt{\pi}}{(2m-1)!!}s + O(s^2)$$

$$(5.231)$$

so

$$-\frac{\partial}{\partial s}\mathrm{Tr}\frac{1}{(-\Delta + M^2)^s}\bigg|_{s=0} = -\frac{V}{(4\pi)^m}\frac{(-2)^m\sqrt{\pi}}{(2m-1)!!}M^n.$$

$$(5.232)$$

The second situation is when $n = 2m$ is *even:*

$$\frac{\Gamma(s-n/2)}{\Gamma(s)} = \frac{(-1)^m}{m!}\left(1 + H_m s + O(s^2)\right)$$

$$(5.233)$$

where $H_m \overset{\mathrm{def}}{=} \sum_{k=1}^m \frac{1}{k}$ is the m-th *harmonic number.* Now

$$-\frac{\partial}{\partial s}\mathrm{Tr}\frac{1}{(-\Delta + M^2)^s}\bigg|_{s=0} = -\frac{V}{(4\pi)^m}\frac{(-1)^m}{m!}H_m M^m + \frac{2\,V}{(4\pi)^m}\frac{(-1)^m}{m!}M^n\,\log M.$$

$$(5.234)$$

References

1. M. Bensimhoun, Historical account and ultra-simple proofs of Descartes's rule of signs, De Gua, Fourier, and Budan's rule (2013), arXiv:1309.6664
2. J. Glimm, A. Jaffe, *Quantum Physics. A Functional Integral Point of View* (Springer, 1981)
3. R.J. Rivers, *Path Integral Methods in Quantum Field Theory*. Cambridge Monographs on Mathematical Physics (Cambridge University Press, 1987)
4. G.W. Semenoff, *Quantum Field Theory. An introduction.* Graduate Texts in Physics (Springer, 2023)
5. Z. Haba, *Lectures on Quantum Field Theory and Functional Integration* (Springer, 2023)
6. V. P. Nair, *Quantum Field Theory. A Modern Perspective* (Springer, 2005)
7. J.C. Collins, *Renormalization* (Cambridge University Press, 2023)
8. D.J. Amit, *Field Theory, the Renormalization Group, and Critical Phenomena.* International Series in Pure and Applied Physics (McGraw-Hill, 1978)
9. C. Itzykson, J.-M. Drouffe, *Statistical Field Theory*, vols. 1 and 2. Cambridge Monographs in Mathematical Physics (Cambridge University Press, 1989)

10. G. Mussardo, *Statistical Field Theory. An Introduction to Exactly Solved Models in Statistical Physics*, 2nd edn. Oxford Graduate Texts (Oxford University Press, 2020)
11. S. Weinberg, *The Quantum Theory of Fields. Volume 1: Foundations* (Cambridge University Press, 1995)
12. M.E. Peskin, D.V. Schroeder, *An Introduction to Quantum Field Theory* (Perseus, 1995)
13. H.D. Ursell, The evaluation of Gibbs phase-integral for imperfect gases. Proc. Cambridge Philos. Soc. **23**, 685–697 (1927)
14. W.A. Strauss, *Partial Differential Equations* (John Wiley & Sons, 2008)
15. M. Berger, P. Gauduchon, E. Mazet, *Le spectre d'une variété riemannienne*. Lecture Notes in Mathematics, vol. 194 (Springer, 1971)
16. D.M. Lozier, R.F. Boisvert, C.W. Clark (eds.), *NIST Handbook of Mathematical Functions* (Cambridge University Press, 2010). On-line version: *NIST Digital Library of Mathematical Functions,* available on-line https://dlmf.nist.gov

Part II
Other Fundamental Topics

This second part of the book contains materials which—while fundamental in Statistical Mechanics—are less elementary than the ones in the first part and may be omitted in a short introductory course, especially for students in mathematical curricula. Chapters in this part have no exercises.

Chapter 6
Linear Response Theory. Brownian Motion

Up to now we considered only systems at thermal equilibrium. In this chapter we study important but relatively simple off-equilibrium situations. We start with a system at thermal equilibrium at some temperature $T = \beta^{-1}$ and then apply to it an external force which varies periodically on time. The external force perturbs the system *out of equilibrium:* a system cannot remain in a stationary state when subjected to a time-dependent external force. The energy supplied from the exterior will be *dissipated* into heat in the system. We want to compute how much energy is dissipated (and show that it is non-negative as predicted by the Second Law). We do this for very weak external force, i.e. at the linear level in the force. Understanding how the system will react to such weak external force is the purpose of *linear response theory*. Its main result is the *dissipation-fluctuation theorem* which yields a simple and universal relation between the amount of energy dissipation and the thermal fluctuations of the dual variable at equilibrium. The theorem has many important generalizations. Moreover the theorem sheds light on how the *irreversibility* of real physical processes arises from microscopic dynamical equations which are time-reversal symmetric.

In the second part of the chapter we consider the Brownian motion both for its own sake and as an illustration of the dissipation-fluctuation mechanism. We discuss the Langevin, Fokker-Planck, and Kramers equations, their solutions, and how thermal equilibrium is eventually established.

6.1 Linear Response Theory

We start with a quantum system with (unperturbed) Hamiltonian H at thermal equilibrium at temperature β^{-1}, and then switch on a time-dependent external field $h(t)$ which couples to an operator Q of the quantum system, thus producing an external force which acts on the system.[1] We focus on quantum systems. The classical case may be obtained by taking $\hbar \to 0$, i.e. by replacing traces over the Hilbert space $\mathcal{H}$ by suitably normalized integrals over the phase space $\mathfrak{W}$, cf. Chap. 2.

In presence of the time-dependent external force the system will not remain in equilibrium, but will evolve with a time-dependent Hamiltonian of the form

$$H(t) = H - Q\, h(t). \tag{6.1}$$

We assume the external field $h(t)$ to be very weak. In the math analysis $h(t)$ is considered to be *infinitesimal*.

Quantum Liouville Equation Revisited

The quantum Liouville equation for the density matrix ϱ reads (cf. Sect. 2.2)[2]

$$i\frac{d\varrho}{dt} = H(t)\varrho - \varrho H(t) \equiv \big[H, \varrho\big] - h(t)\big[Q, \varrho\big] \tag{6.2}$$

with the initial boundary condition that at time $t = -\infty$ the system was in thermal equilibrium as described by the canonical density matrix at the given temperature

$$\varrho(t)\Big|_{t \to -\infty} = \varrho_e \overset{\text{def}}{=} \frac{e^{-\beta H}}{Z}, \qquad Z \equiv \mathrm{Tr}_{\mathcal{H}}\, e^{-\beta H}. \tag{6.3}$$

Setting

$$\phi(t) = e^{iHt}\varrho(t)\, e^{-iHt} \tag{6.4}$$

we get from (6.2)

$$
\begin{aligned}
i\dot{\phi} &= e^{iHt}\big(i\dot{\varrho} - [H, \varrho]\big)e^{-iHt} = \\
&= -h(t)\, e^{iHt}\big(Q\, e^{-iHt}\, \phi\, e^{iHt} - e^{-iHt}\, \phi\, e^{iHt}\, Q\big)e^{-iHt} = -h(t)\big[Q(t), \phi\big]
\end{aligned}
\tag{6.5}
$$

where

$$Q(t) \equiv e^{iHt}\, Q\, e^{-iHt}. \tag{6.6}$$

---

[1] Examples: $h(t)$ may be an electric potential and Q the electric charge, or $h(t)$ may be a magnetic field and Q the magnetic dipole of the system, etc.

[2] Through this chapter we set $\hbar = 1$.

We consider the unitary operator[3]

$$U(t) = T \exp\left(\int_{-\infty}^{t} i\, h(s)\, Q(s)\, ds\right), \tag{6.7}$$

defined by the ODE and boundary condition

$$\dot{U}(t) = i\, h(t)\, Q(t)\, U(t), \qquad U(-\infty) = 1. \tag{6.8}$$

$U(t)$ is given by the (convergent) series of reiterated integrals

$$U(t) = 1 + \int_{-\infty}^{t} i\, h(s)\, Q(s)\, ds + \int_{-\infty}^{t} i\, h(s_1)\, Q(s_1)\, ds_1 \int_{-\infty}^{s_1} i\, h(s_2)\, Q(s_2)\, ds_2 +$$
$$+ \int_{-\infty}^{t} i\, h(s_1)\, Q(s_1)\, ds_1 \int_{-\infty}^{s_1} i\, h(s_2)\, Q(s_2)\, ds_2 \int_{-\infty}^{s_2} i\, h(s_3)\, Q(s_3)\, ds_3 + \cdots \tag{6.9}$$

The solution to Eq. (6.5) is then

$$\phi(t) = U(t)\, \phi(-\infty)\, U(t)^{-1} = U(t)\, \varrho_e\, U(t)^{-1}, \tag{6.10}$$

that is, the time evolution of the density matrix is

$$\varrho(t) = \mathrm{e}^{-iHt}\, U(t)\, \varrho_e\, U(t)^{-1}\, \mathrm{e}^{iHt}. \tag{6.11}$$

The field $h(t)$ is very small (infinitesimal). At the linear order in $h(t)$ we have

$$Z\, \varrho(t) \approx \mathrm{e}^{-iHt} \left(1 + \int_{-\infty}^{t} i\, h(s)Q(s)ds\right) \mathrm{e}^{-\beta H} \left(1 - \int_{-\infty}^{t} i\, h(s)Q(s)ds\right) \mathrm{e}^{iHt}$$

$$\approx \mathrm{e}^{-\beta H} + i \int_{-\infty}^{t} ds\, h(s)\, \mathrm{e}^{-iHt}\left(\mathrm{e}^{iHs}\, Q\mathrm{e}^{-iHs}\mathrm{e}^{-\beta H} - \mathrm{e}^{-\beta H}\mathrm{e}^{iHs}\, Q\mathrm{e}^{-iHs}\right)\mathrm{e}^{iHt} =$$

$$= \mathrm{e}^{-\beta H} - i \int_{-\infty}^{t} ds\, h(s)\left[\mathrm{e}^{-\beta H}, Q(s-t)\right] \tag{6.12}$$

i.e. at the linear level the solution of the quantum Liouville equation (6.2) is

$$\varrho(t) \approx \varrho_e - i\left[\varrho_e, \int_{-\infty}^{t} ds\, h(s)\, Q(s-t)\right] \tag{6.13}$$

---

[3] The RHS of (6.7) is the time-ordered exponential defined by the differential equation (6.8) or, equivalently, by the convergent expansion (6.9). See e.g. Chap. 12 of [1].

Linear Response

Now consider a quantum observable O of our system. We study how its expectation value evolves with time in the process described in the previous paragraph:

$$\langle O \rangle_t \equiv \mathrm{Tr}(O\varrho(t)) \approx \mathrm{Tr}(O\varrho_e) - \mathrm{i} \int_{-\infty}^{t} \mathrm{d}s\, h(s)\, \mathrm{Tr}\big(O[\varrho_e,\, Q(s-t)]\big) =$$

$$= \mathrm{Tr}(\varrho_e O) - \mathrm{i} \int_{-\infty}^{t} \mathrm{d}s\, h(s)\, \mathrm{Tr}\big(\varrho_e[Q(s-t), O]\big) = \tag{6.14}$$

$$= \mathrm{Tr}(\varrho_e O) - \mathrm{i} \int_{-\infty}^{t} \mathrm{d}s\, h(s)\, \mathrm{Tr}\big(\varrho_e[Q, O(t-s)]\big)$$

where

$$O(t) \equiv \mathrm{e}^{\mathrm{i}Ht} O\, \mathrm{e}^{-\mathrm{i}Ht}. \tag{6.15}$$

In other words, the variation of the expectation value $\langle O \rangle_t$, with respect to the equilibrium one $\mathrm{Tr}(O\varrho_e)$, for effect of the infinitesimal external field $h(t)$ is

$$\delta\langle O \rangle_t = -\mathrm{i} \int_{-\infty}^{t} \mathrm{d}s\, h(s) \big\langle [Q(0), O(t-s)] \big\rangle, \tag{6.16}$$

where $\langle \cdots \rangle$ is the thermal expectation value at equilibrium. Formula (6.16) yields the *linear response* of the system to a small external perturbation $\delta H = -h(t)Q$.

The *response function* for the observable O is

$$\Phi(t)_O \overset{\text{def}}{=} -\mathrm{i}\big\langle [Q(0), O(t)] \big\rangle \equiv -\mathrm{i}\,\mathrm{Tr}(\varrho_e[Q(0), O(t)]) =$$
$$= -\mathrm{i}\,\mathrm{Tr}([\varrho_e, Q(0)]O(t)) \tag{6.17}$$

and

$$\delta\langle O \rangle_t = \int_{-\infty}^{t} \mathrm{d}s\, \Phi(t-s)_O\, h(s) \tag{6.18}$$

is the *linear response* of the system as measured by the quantity $\langle O \rangle$. We set

$$\Phi(t)_O = 0 \qquad \text{for } t < 0, \tag{6.19}$$

and rewrite (6.18) in the form

$$\delta\langle O \rangle_t = \int_{-\infty}^{+\infty} \mathrm{d}s\, \Phi(t-s)_O\, h(s). \tag{6.20}$$

Equation (6.19) is just *causality:* the value of $\langle O \rangle_t$ depends (in general) on the values of the field $h(s)$ at all past times $s \leq t$ but cannot depend on the values the external field will take at times $s > t$ *in the future.* Since the answer to a finite external perturbation acting for a short time is finite, the function $\Phi(t)_O$ must be *bounded* (up to, possibly, a "contact term" of the form $a_O \, \delta(t)$).

Our next task is to compute the operator entering in the definition (6.17) of Φ_O

$$-\frac{\mathrm{i}}{Z} \, [e^{-\beta H}, Q] = \frac{\mathrm{i}}{Z} \int_0^\beta \mathrm{d}\lambda \, \frac{\partial}{\partial \lambda} \left(e^{(\lambda - \beta)H} Q \, e^{-\lambda H} \right) =$$

$$= -\mathrm{i} \int_0^\beta \mathrm{d}\lambda \, \varrho_e \, e^{\lambda H} [Q, H] e^{-\lambda H} = \int_0^\beta \mathrm{d}\lambda \, \varrho_e \, e^{\lambda H} \, \dot{Q} \, e^{-\lambda H}$$

$$(6.21)$$

Canonical Correlations

Following Kubo [2], we introduce the *canonical correlation* of two quantum operators A, B

$$\langle\!\langle A, B \rangle\!\rangle \overset{\text{def}}{=} \frac{1}{\beta Z} \int_0^\beta \mathrm{d}\lambda \, \mathrm{Tr}\big[e^{-(\beta - \lambda)H} A \, e^{-\lambda H} B \big]. \tag{6.22}$$

The canonical correlation is related to the ordinary thermal correlation as follows

$$\langle [A(0), B(t)] \rangle \equiv \mathrm{Tr}\big(\varrho_e [A(0), B(t)] \big) = \mathrm{Tr}\big([\varrho_e, A(0)] B(t) \big) =$$

$$= \mathrm{i} \int_0^\beta \mathrm{d}\lambda \, \mathrm{Tr}\big(\varrho_e \, e^{\lambda H} \dot{A}(0) \, e^{-\lambda H} B(t) \big) = \mathrm{i}\,\beta \, \langle\!\langle \dot{A}(0), B(t) \rangle\!\rangle$$

$$(6.23)$$

where we used Eq. (6.21) with $Q \rightsquigarrow A$.

The canonical correlation has the following nice properties:

CC1 $\langle\!\langle A, B \rangle\!\rangle$ is symmetric in A, B. Indeed:

$$\mathrm{Tr}\left(e^{-(\beta - \lambda)H} A \, e^{\lambda H} B \right) = \mathrm{Tr}\left(e^{\lambda H} B \, e^{-(\beta - \lambda)H} A \right) = \mathrm{Tr}\left(e^{-(\beta - \lambda')H} B \, e^{\lambda' H} A \right)$$

$$(6.24)$$

CC2   the canonical expectation is invariant under time translations

$$\langle\!\langle A(t), B(s) \rangle\!\rangle = \langle\!\langle A(t + a), B(s + a) \rangle\!\rangle \tag{6.25}$$

CC3   When A, B are Hermitian, $\langle\!\langle A, B \rangle\!\rangle \in \mathbb{R}$. Indeed, since the trace is cyclic

$$\mathrm{Tr}\left(e^{-(\beta - \lambda)H} A \, e^{\lambda H} B \right)^{\dagger} = \mathrm{Tr}\left(e^{\lambda H} A \, e^{-(\beta - \lambda)H} B \right) \tag{6.26}$$

CC4   When A is Hermitian $\langle\!\langle A, A \rangle\!\rangle \geq 0$. Indeed

$$\mathrm{Tr}\left[e^{-(\beta-\lambda)H} A\, e^{\lambda H} A \right] = \mathrm{Tr}\left[\left(e^{-(\beta-\lambda)H/2} A\, e^{\lambda H/2} \right)\left(e^{-(\beta-\lambda)H/2} A\, e^{\lambda H/2} \right)^{\dagger} \right]$$

(6.27)

CC5   When the operators A, B transform with a definite sign under time-reversal (we can always reduce to this case), $\langle\!\langle A(0), B(t) \rangle\!\rangle$ is *even* or respectively *odd* as a function of t depending on the time-reversal properties of A, B:

$$\langle\!\langle A(0), B(-t) \rangle\!\rangle = \begin{cases} \langle\!\langle A(0), B(t) \rangle\!\rangle & \text{if } A, B \text{ have \textbf{same} parity for time reversal} \\[2mm] -\langle\!\langle A(0), B(t) \rangle\!\rangle & \text{if } A, B \text{ \textbf{opposite} parity for time reversal} \end{cases}$$

(6.28)

In view of Eqs. (6.17), (6.21), and (6.22) we can rewrite the response function as

$$\Phi(t)_O = \begin{cases} \beta\, \langle\!\langle \dot{Q}(0), O(t) \rangle\!\rangle & t \geq 0 \\ 0 & t < 0. \end{cases}$$

(6.29)

Susceptibility

Now we suppose that the external field $h(t)$ has a definite frequency[4] ω

$$h(t) = h_0\, e^{-i\omega t} + \text{c.c.}$$

(6.30)

Then the response is

$$\delta\langle O \rangle_t = \chi(\omega)_O\, h_0\, e^{-i\omega t} + \text{c.c.,}$$

(6.31)

where the *susceptibility* (also called *admittance*[5]) $\chi(\omega)_O$ of the observable O is

$$\chi(\omega)_O \overset{\text{def}}{=} \int_{-\infty}^{+\infty} \Phi(t)_O\, e^{i\omega t}\, dt = \beta \int_0^{\infty} \langle\!\langle \dot{Q}(0), O(t) \rangle\!\rangle\, e^{i\omega t}\, dt.$$

(6.32)

The condition that $\delta\langle O \rangle_t$ is real yields

$$\chi(-\omega) = \chi(\omega)^*,$$

(6.33)

---

[4] Here and below "+ c.c." stands for *plus complex conjugate*.

[5] The *admittance* is the inverse of the *impedance* which is the quantity more commonly used in engineering.

that is,

$$\operatorname{Re}\chi(-\omega) = \operatorname{Re}\chi(\omega), \qquad \operatorname{Im}\chi(-\omega) = -\operatorname{Im}\chi(\omega). \tag{6.34}$$

The KMS Condition

We return to the "ordinary" thermal expectation values in the canonical ensemble

$$\langle A(0)\,B(t)\rangle \equiv \frac{1}{Z}\operatorname{Tr}\!\left(e^{-\beta H} A(0)\,B(t)\right) \tag{6.35}$$

analytically continued to complex values of t. They satisfy the functional equation

$$\langle A(0)\,B(t)\rangle = \langle B(t - i\beta)\,A(0)\rangle \tag{6.36}$$

called the *KMS* (Kubo-Martin-Schwinger) *boundary condition*. Equation (6.36) is the basic functional equation which defines the correlation functions in a *bona fide* equilibrium state at definite temperature β^{-1}. In axiomatic treatments of quantum Statistical Physics the KMS condition is taken as the *definition* of an equilibrium state at temperature β^{-1}.

Taking the Fourier transform of both sides of (6.36), we rewrite the KMS condition in the form

$$\textbf{KMS} \quad \int_{-\infty}^{+\infty} \langle A(0)\,B(t)\rangle\, e^{i\omega t}\, dt = e^{-\beta\omega} \int_{-\infty}^{+\infty} \langle B(t)\,A(0)\rangle\, e^{i\omega t}\, dt \tag{6.37}$$

We write a few important consequences of the KMS condition:

$$\int_{-\infty}^{+\infty} \langle [A(0), B(t)]\rangle\, e^{i\omega t}\, dt = (1 - e^{\beta\omega}) \int_{-\infty}^{+\infty} \langle A(0)\,B(t)\rangle\, e^{i\omega t}\, dt \tag{6.38}$$

$$\int_{-\infty}^{+\infty} \langle \{A(0), B(t)\}\rangle\, e^{i\omega t}\, dt = (1 + e^{\beta\omega}) \int_{-\infty}^{+\infty} \langle A(0)\,B(t)\rangle\, e^{i\omega t}\, dt \tag{6.39}$$

where, as always, $\{A(0),\, B(t)\} \stackrel{\text{def}}{=} A(0)B(t) + B(t)A(0)$ is the *anti*-commutator. We combine the last two equations in the formula

$$\int_{-\infty}^{+\infty} \langle [A(0), B(t)]\rangle\, e^{i\omega t}\, dt = -\frac{1}{2}\frac{\omega}{E_\beta(\omega)} \int_{-\infty}^{+\infty} \langle \{A(0), B(t)\}\rangle\, e^{i\omega t}\, dt \tag{6.40}$$

where

$$E_\beta(\omega) = \frac{1}{2} \frac{\omega}{\tanh(\beta\omega/2)} > 0 \tag{6.41}$$

is the mean energy of a harmonic oscillator of frequency ω at temperature β^{-1}, cf. Sect. 3.6. Using Eq. (6.23)

$$\frac{d}{dt} \langle\!\langle A(0), B(t) \rangle\!\rangle = \frac{i}{\beta} \langle [A(0), B(t)] \rangle, \tag{6.42}$$

we get

$$\int\limits_{-\infty}^{+\infty} \langle\!\langle A(0), B(t) \rangle\!\rangle \, e^{i\omega t} \, dt = \frac{1}{\beta E_\beta(\omega)} \int\limits_{-\infty}^{+\infty} \frac{1}{2} \langle \{A(0), B(t)\} \rangle \, e^{i\omega t} \, dt. \tag{6.43}$$

Properties of $\chi(\omega)$. Kramers-Kronig Dispersion Relations

The generalized susceptibility $\chi(\omega) \equiv \chi(\omega)_O$ is constrained by two properties:

$\chi 1$ *reality:* the response to a real external force is a real variation $\delta\langle O \rangle_t$ in the expectation value of the physical observable O;

$\chi 2$ *causality:* the response cannot come before the perturbing force.

Properties $\chi 1$, $\chi 2$ have strong consequences which hold in full generality.

Proposition 6.1 *For all Hermitian operator O, the generalized susceptibility $\chi(\omega) \equiv \chi(\omega)_O$ has the following properties:*

$\chi 3$ *$\chi(\omega)$ extends to a holomorphic function on the upper complex half-plane*

$$\mathbb{H} \equiv \{\omega \in \mathbb{C}: \operatorname{Im}\omega > 0\} \tag{6.44}$$

$\chi 4$ *$\chi(\omega)$ is regular along the real axis except possibly at $\omega = 0$;*

$\chi 5$ *$\chi(-\omega^*) = \chi(\omega)^*$. In particular $\chi(\omega)$ is real along the positive imaginary axis $\operatorname{Re}\omega = 0$, $\operatorname{Im}\omega > 0$;*

$\chi 6$ *the limit of $\chi(\omega)$ as $\omega \to \infty$ is a real number (arising from the "contact term") which vanishes for a regular response function $\Phi(t)_O$. In this case, and assuming that $\chi(\omega)$ has no pole in the origin, $\chi(\omega)$ satisfies the Kramers-Kronig dispersion relations in the form*

$$\operatorname{Re}\chi(\omega) = \frac{1}{\pi} P\!\int\limits_{-\infty}^{+\infty} \frac{\operatorname{Im}\chi(\xi)}{\xi - \omega} \, d\xi, \qquad \operatorname{Im}\chi(\omega) = -\frac{1}{\pi} P\!\int\limits_{-\infty}^{+\infty} \frac{\operatorname{Re}\chi(\xi)}{\xi - \omega} \, d\xi$$

$$\tag{6.45}$$

where P denotes the principal prescription *for the singular integral*

$$P\int_{-\infty}^{+\infty} \frac{A(x)\,dx}{x-a} \overset{\text{def}}{=} \lim_{\epsilon \to 0} \left(\int_{-\infty}^{a-\epsilon} \frac{A(x)\,dx}{x-a} + \int_{a+\epsilon}^{+\infty} \frac{A(x)\,dx}{x-a} \right). \tag{6.46}$$

In the general case (with a pole at 0 and/or a "contact term") replace in (6.45)

$$\chi(\omega) \rightsquigarrow \chi(\omega) - \chi(\infty) - \frac{\lim_{\eta \to 0} \eta\, \chi(\eta)}{\omega}. \tag{6.47}$$

Using Eqs. (6.45) we can recover the full susceptibility $\chi(\omega)$ from its imaginary (or real) part. As we shall show in the next section, the dissipation-fluctuation theorem allows to recover either $\operatorname{Im}\chi(\omega)$ or $\operatorname{Re}\chi(\omega)$ directly from the thermal correlations *at equilibrium* in the canonical ensemble. Hence the susceptibilities can be fully computed using the equilibrium physics.

Proposition 6.2 *Moreover, in the particular case $O = Q$:*

$\chi 7$ *In the upper half-plane $\mathbb{H}$ the holomorphic function $\chi(\omega)$ cannot take real values except along the imaginary axis;*

$\chi 8$ *Along the imaginary axis the function $\chi(\omega)$ is monotonically decreasing from a positive value (possibly $+\infty$) at $\omega = 0$ to 0 at $\omega = i\infty$;*

$\chi 9$ *The function $\chi(\omega)$ has no zero in the upper half-plane $\mathbb{H}$.*

Proof $\chi 3$. From Eq. (6.32) and $\Phi(t) = 0$ for $t < 0$

$$\chi(-iy) = \int_{0}^{+\infty} \Phi(t)\, e^{-yt}\, dt \tag{6.48}$$

is the Laplace transform of the response function $\Phi(t)$. It is well known that the Laplace transform is analytic in the open semi-plane of convergence. The convergence semi-plane contains $\operatorname{Re} y \equiv \operatorname{Im}\omega > 0$ since $\Phi(t)$ is bounded up to a contact term whose only effect is to add a constant to $\chi(\omega)$. $\chi 4$ follows from regularity of the response. $\chi 5$ is obvious from Eq. (6.32).

χ**6.** Consider the integral

$$\int_C \frac{\chi(\xi)\, d\xi}{2\pi i(\xi - \omega)} \tag{6.49}$$

where C is the following contour

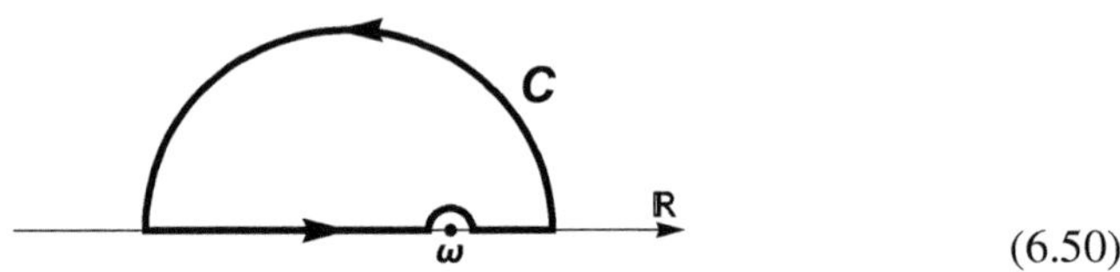

$$\tag{6.50}$$

The radius of the large semi-circle is sent to infinity, while the radius of the small semi-circle centered at the point $\omega \in \mathbb{R}$ is sent to zero. The integral along C vanishes since the integrand is holomorphic in its interior. The integral along the big circle goes to zero because of the exponentially small factor $e^{i\omega t}$ (the Jordan lemma [3]). The integrals along the segments on the real axis produce the principal integral (6.46), while the integral on the small semicircle gives minus one-half the residue of the integrand at $\xi = \omega$. Thus

$$0 = \frac{1}{2\pi i} P \int_{-\infty}^{+\infty} \frac{\chi(\xi)\, d\xi}{\xi - \omega} - \frac{1}{2}\chi(\omega) = 0 \tag{6.51}$$

For Proposition 6.2 consider the integral

$$\int_C \frac{d\chi(\omega)}{2\pi i(\chi(\omega) - a)} \qquad a \in \mathbb{R}. \tag{6.52}$$

The integral gives the number of zeros minus the number of poles of the function $\chi(\omega) - a$ in the upper half-plane, hence the number of zeros since the function is holomorphic in $\mathbb{H}$. The full large semicircle is mapped by $\chi : \mathbb{H} \to \mathbb{C}$ in the origin of the complex plane $\mathbb{C}$. The origin of $\mathbb{H}$ is mapped to some real value $\chi_0 \in \mathbb{R}$. By Lemma 6.1 (proven in the next section) the negative real axis $\mathbb{R}_{\leq 0}$ is mapped into some curve γ laying completely in the lower half-plane $\mathbb{H}^*$ except for the two endpoints $\omega = -\infty$ and $\omega = 0$ which are mapped on the real axis at 0 and χ_0 respectively. The positive real axis $\mathbb{R}_{\geq 0}$ is mapped in the reflected curve $\gamma^* \subset \mathbb{H}$ in the upper half plane. Hence $\chi(C)$ is a closed curve in $\mathbb{C}$, symmetric by reflection with respect to the real axis, which contains in its interior the segment $[0, \chi_0] \subset \mathbb{R}$ of the real axis.[6] Thus for $a \in \mathbb{R}$

$$\int_C \frac{\mathrm{d}\chi(\omega)}{2\pi\mathrm{i}(\chi(\omega) - a)} = \begin{cases} 1 & \text{iff } a \in [0, \chi_0] \\ 0 & \text{otherwise} \end{cases} \tag{6.53}$$

and each real value in the interval $[0, \chi_0]$ is taken exactly once in the upper half plane and no other real values are possible. Since all these real value are taken on the positive imaginary axis, there are no other points where $\chi(\omega)$ is real. On the positive real axis the function should be monotonic decreasing since otherwise it would take some real values more than once. $\qquad\square$

6.2 Dissipation-Fluctuation Theorem

Energy Dissipation

In our time-dependent system (6.1) the energy is *not* conserved[7]

$$\frac{\mathrm{d}H(t)}{\mathrm{d}t} = \frac{\partial H(t)}{\partial t} = -\dot{h}(t)\,Q(t), \tag{6.54}$$

and the variation with time of the internal energy of the system is[8]

$$\frac{\mathrm{d}}{\mathrm{d}t}\mathrm{Tr}\big(H(t)\varrho(t)\big) = -\dot{h}(t)\,\mathrm{Tr}\big(Q(t)\varrho(t)\big) = -\dot{h}(t)\langle Q\rangle - \dot{h}(t)\,\delta\langle Q\rangle_t. \tag{6.55}$$

We are interested in the time-average of the *dissipation of energy* when the system is perturbed by a weak field $h(t)$ of non-zero frequency ω (6.30). The first term in the RHS of (6.55) has zero mean, and the second one is

$$i\omega\big(h_0\mathrm{e}^{-i\omega t} - h_0^*\mathrm{e}^{i\omega t}\big)\big(\chi(\omega)\,h_0\mathrm{e}^{-i\omega t} + \chi(\omega)^*\,h_0^*\mathrm{e}^{i\omega t}\big), \tag{6.56}$$

where $\chi(\omega) \equiv \chi(\omega)_Q$. Neglecting terms proportional to $\exp(\pm 2i\omega t)$, which average to zero, we get for the mean energy dissipation

$$\overline{\frac{\mathrm{d}E}{\mathrm{d}t}} = -i\omega\big(\chi(\omega) - \chi(\omega)^*\big)|h_0|^2 = -i\omega\big(\chi(\omega) - \chi(-\omega)\big)|h_0|^2. \tag{6.57}$$

---

[6] We stress that $\chi_0 > 0$: indeed a constant field h leads to a time-independent Hamiltonian $H' = H - h\,Q$ to which we can apply standard equilibrium methods: at equilibrium $\langle Q\rangle$ and h must have the same sign in order to minimize the energy.

[7] We use Eq. (6.41) of [4] with $F \equiv H$.

[8] Recall that the quantum Liouville equation (6.2) implies $\mathrm{Tr}(H(t)\dot{\varrho}(t)) = 0$.

We see that the dissipation is given by the *imaginary part of the susceptibility*. Now

$$
\chi(\omega) - \chi(-\omega) =
$$

$$
= i\omega\beta \left(\int_0^\infty dt\, \langle\!\langle Q(0), Q(t) \rangle\!\rangle\, e^{i\omega t} + \int_0^\infty dt\, \langle\!\langle Q(0), Q(t) \rangle\!\rangle\, e^{-i\omega t} \right) =
$$

$$
= i\omega\beta \left(\int_0^\infty dt\, \langle\!\langle Q(0), Q(t) \rangle\!\rangle\, e^{i\omega t} + \int_0^\infty dt\, \langle\!\langle Q(0), Q(-t) \rangle\!\rangle\, e^{-i\omega t} \right) =
$$

$$
= i\omega\beta \left(\int_0^\infty dt\, \langle\!\langle Q(0), Q(t) \rangle\!\rangle\, e^{i\omega t} + \int_{-\infty}^0 dt\, \langle\!\langle Q(0), Q(t) \rangle\!\rangle\, e^{i\omega t} \right) =
$$

$$
= i\omega\beta \int_{-\infty}^{+\infty} dt\, \langle\!\langle Q(0), Q(t) \rangle\!\rangle\, e^{i\omega t} = \frac{i\omega}{E_\beta(\omega)} \int_{\infty}^{+\infty} dt\, \langle Q(0)\, Q(t) \rangle\, e^{i\omega t}.
$$

$$\tag{6.58}$$

Using the identity

$$
\omega \int_{-\infty}^{+\infty} \langle Q \rangle^2 e^{i\omega t}\, dt = 2\pi \langle Q \rangle^2 \omega\, \delta(\omega) = 0.
\tag{6.59}
$$

and (6.41), Eq. (6.58) can be rewritten as the basic formula

$$
\mathrm{Im}(\chi(\omega)) = \tanh\left(\frac{\beta\omega}{2}\right) \int_{-\infty}^{+\infty} \left(\langle Q(0)\, Q(t) \rangle - \langle Q \rangle^2 \right) e^{i\omega t}\, dt
\tag{6.60}
$$

which relates the dissipation of energy by effect of a weak periodic external force, Eq. (6.57), to the thermodynamical fluctuations *at equilibrium* of the coupling operator Q. The fundamental result (6.60) is called the *dissipation-fluctuation theorem:* the energy (supplied by the work produced by the external force acting on the system) which gets *dissipated into heat* is

$$
\frac{\overline{dE}}{dt} = \frac{\omega^2}{E_\beta(\omega)} |h_0|^2 \int_{-\infty}^{+\infty} \left(\langle Q(0)\, Q(t) \rangle - \langle Q \rangle^2 \right) e^{i\omega t}\, dt,
\tag{6.61}
$$

that is, proportional to the *thermal fluctuations at equilibrium* of the observable Q which couples to the external field $h(t)$.

The Second Law implies that the expression (6.61) must non-negative: *friction dissipates energy, does not produce it!* Hence the following statement should be true:

Lemma 6.1 *For $\omega \in \mathbb{R}$ one has*

$$
\mathrm{Im}(\chi(\omega)) = \begin{cases} < 0 & for\ -\infty < \omega < 0 \\ = 0 & for\ \omega = 0 \\ > 0 & for\ 0 < \omega < +\infty \\ = 0 & for\ \omega = \pm\infty. \end{cases} \tag{6.62}
$$

This statement is consistent with the reality condition $\chi(-\omega) = \chi(\omega)^*$.

Proof Let us show that the physical expectation (6.62) is mathematically correct. We write $\tilde{Q}(t)$ for the Hermitian operator $Q(t) - \langle Q \rangle$. Then

$$
\langle \tilde{Q}(0)\,\tilde{Q}(t) \rangle = \mathrm{Tr}\!\left(e^{-\beta H}\,\tilde{Q}(0)\,\tilde{Q}(t)\right) =
$$

$$
= \sum_{m,n} e^{-\beta E_m}\,\langle m|\tilde{Q}|n\rangle\langle\tilde{Q}|m\rangle e^{i(E_n - E_m)t} = \sum_{m,n} e^{-\beta E_m}\left|\langle m|\tilde{Q}|n\rangle\right|^2 e^{i(E_n - E_m)t}
$$

$$
\tag{6.63}
$$

since $\tilde{Q}$ is Hermitian. Now

$$
\int_{-\infty}^{+\infty} \langle \tilde{Q}(0)\,\tilde{Q}(t) \rangle\, e^{i\omega t}\, dt = 2\pi \sum_{m,n} e^{-\beta E_m}\left|\langle m|\tilde{Q}|n\rangle\right|^2 \delta(E_m - E_n - \omega) \tag{6.64}
$$

which is a non-negative measure. Comparing with (6.60) yields the **Lemma**. □

Remark 6.1 That (generalized) friction is a non-reversible process is, of course, the essence of the statement that real processes are irreversible and that the entropy increases until it reaches its maximum. Here we deduced this basic result from the first principles of quantum physics with the additional ingredient of the KMS boundary condition as definition of what we mean by *equilibrium* at fixed temperature.

Remark 6.2 We have deduced the dissipation-fluctuation theorem in the context of quantum physics. The theorem holds in the classical set-up as well. Just take the $\hbar \to 0$ limit of the previous formulae replacing traces in Hilbert space by integrals over the phase space (the order of operators now becomes irrelevant).

The General Theorem
We return to the susceptibility of a general observable O, Eq. (6.32)

$$
\chi(\omega)_O = i\omega\beta \int_0^\infty \langle\!\langle Q(0), O(t) \rangle\!\rangle\, e^{i\omega t}\, dt. \tag{6.65}
$$

There are two possible cases (cf. Eq. (6.28)): either Q and O have the same time-reversal parity, or they have the opposite one. In the first case

$$2\,\mathrm{i}\,\mathrm{Im}\chi(\omega)_O = \mathrm{i}\omega\beta\left(\int_0^\infty \langle\!\langle Q(0), O(t)\rangle\!\rangle\, e^{\mathrm{i}\omega t}\, dt - \int_0^\infty \langle\!\langle Q(0), O(t)\rangle\!\rangle\, e^{-\mathrm{i}\omega t}\, dt\right) =$$

$$= \mathrm{i}\omega\beta\left(\int_0^\infty \langle\!\langle Q(0), O(t)\rangle\!\rangle\, e^{\mathrm{i}\omega t}\, dt + \int_{-\infty}^0 \langle\!\langle Q(0), O(-t)\rangle\!\rangle\, e^{\mathrm{i}\omega t}\, dt\right) =$$

$$= \mathrm{i}\omega\beta\int_{-\infty}^{+\infty} \langle\!\langle Q(0), O(t)\rangle\!\rangle\, e^{\mathrm{i}\omega t}\, dt = \frac{\mathrm{i}\,\omega}{E_\beta(\omega)}\int_{-\infty}^{+\infty} \frac{1}{2}\langle\{Q(0), O(t)\}\rangle\, e^{\mathrm{i}\omega t}\, dt \qquad (6.66)$$

where in the last step we used Eq. (6.43). In the second case we have

$$2\,\mathrm{Re}\chi(\omega)_O = \mathrm{i}\omega\beta\left(\int_0^\infty \langle\!\langle Q(0), O(t)\rangle\!\rangle\, e^{\mathrm{i}\omega t}\, dt - \int_{-\infty}^0 \langle\!\langle Q(0), O(-t)\rangle\!\rangle\, e^{\mathrm{i}\omega t}\, dt\right) =$$

$$= \mathrm{i}\omega\beta\int_{-\infty}^{+\infty} \langle\!\langle Q(0), O(t)\rangle\!\rangle\, e^{\mathrm{i}\omega t}\, dt = \frac{\mathrm{i}\,\omega}{E_\beta(\omega)}\int_{-\infty}^{+\infty} \frac{1}{2}\langle\{Q(0), O(t)\}\rangle\, e^{\mathrm{i}\omega t}\, dt \qquad (6.67)$$

The two expressions for the imaginary (resp. real) part of the susceptibility $\chi(\omega)_O$ for Q and O of the same (resp. opposite) time-reversal parity

$$\mathrm{Im}\,\chi(\omega)_O = \frac{\omega}{4\,E_\beta(\omega)}\int_{-\infty}^{+\infty} \langle\{Q(0), O(t)\}\rangle\, e^{\mathrm{i}\omega t}\, dt \qquad \textbf{same}$$

$$(6.68)$$

$$\mathrm{Re}\,\chi(\omega)_O = \frac{\mathrm{i}\,\omega}{4\,E_\beta(\omega)}\int_{-\infty}^{+\infty} \langle\{Q(0), O(t)\}\rangle\, e^{\mathrm{i}\omega t}\, dt \qquad \textbf{opposite}$$

are called the *(generalized) dissipation-fluctuation theorem.*

6.3 Brownian Motion

As an interesting example of the dissipation-fluctuation theorem we consider the *Brownian motion*. This is a very important topic with countless applications in Physics first understood by Einstein in 1905.

The experimental set-up is as follows. We have extremely small yet "macroscopic" bodies of mass m moving in a medium composed by a huge number of molecules with masses $\mu \ll m$. By the equipartition theorem (Chap. 2) the molecules have classically a kinetic energy $d\,T/2$ where T is the temperature and

d the number of spatial dimensions of the medium. The molecules move randomly at velocities of order $\approx \sqrt{Td/\mu}$ in every possible direction. The "macroscopic" bodies will move at a much smaller velocity since, while very light, have masses much larger than the molecules. The molecules will collide with each body quite frequently, coming from random directions. The collisions produce impulsive forces which make the body to move in an irregular way, changing direction swiftly. These erratic zigzag movements were actually observed by Brown in colloid particles and called the *Brownian motion* henceforth. However, until Einstein's breakthrough paper of 1905, no one had the slightest clue of the physical origin of this peculiar phenomenon. At the time most scientists refused the idea that matter is made of molecules and atoms. The accurate explanation of the experimental data by Einstein using the hypothesis that the Brownian motion is caused by collisions with much lighter molecules (too small to be detected at the time) was then considered the first positive evidence of the so-called "atomic hypothesis" about the microscopic structure of matter.

The Simplest Stochastic Differential Equation

In this paragraph **bold** symbols stand for d-dimensional vectors; their components are denoted by the non-bold version of the same symbol with an upper index taking the values $1, \ldots, d$.

We model the Brownian motion as follows. Let $\boldsymbol{v}(t)$ be the velocity at time t of a body moving in a d-dimensional medium. The body's effective Newton law reads

$$m\,\dot{\boldsymbol{v}}(t) = -m\gamma\,\boldsymbol{v}(t) + \boldsymbol{\xi}(t) \qquad (6.69)$$

where the *phenomenological* force in the RHS has two components:

PhF1 $-m\gamma\,\boldsymbol{v}(t)$ is the resistance of the medium (friction) and $m\gamma$ is the *friction parameter* $(\gamma > 0)$;

PhF2 $\boldsymbol{\xi}(t)$ is the random force on the body produced by the several molecular collisions. We call $\boldsymbol{\xi}(t)$ the (background) *noise*.

Being the result of a large number of independent casual events, the noise force $\boldsymbol{\xi}(t)$ is not a deterministic function of the time t, but rather a *random variable* described by a certain probability distribution. This distribution will eventually thermalize into an equilibrium one. In view of the central limit theorem and the isotropy of the problem, it is natural to assume that the distribution at equilibrium is *Gaussian with zero mean*

$$\langle \xi^i(t) \rangle = 0 \qquad i = 1, \ldots, d. \qquad (6.70)$$

The noise probability distribution is then fully determined by its 2-point correlation

$$\langle \xi^i(t)\,\xi^j(t')\rangle \qquad i,j = 1,\dots,d. \tag{6.71}$$

By its very nature there are *no* long-time correlations in the Brownian motion, so the 2-point function (6.71) vanishes for $|t - t'|$ larger than some small time ϵ equal (roughly) to the mean time between two successive collisions with the molecules (whose momenta have no correlation). To make the model analytically simple, while preserving all essential features, it is convenient to take the limit $\epsilon \to 0$, that is, to use a δ-function variance

$$\langle \xi^i(t)\,\xi^j(t')\rangle = \lambda\,\delta^{ij}\,\delta(t - t') \tag{6.72}$$

where λ is a positive constant whose physical meaning will be clarified later. A Gaussian noise satisfying (6.70) and (6.72) is called *white noise* because the Fourier transform of the correlation (6.72) is constant as a function of the frequency ω, which is the defining property of the white light. The noise is then distributed according to the functional Gaussian probability measure

$$\mathcal{N}\exp\left(-\frac{1}{2\lambda}\int \boldsymbol{\xi}(t)^2\,dt\right)\left[D\boldsymbol{\xi}(t)\right] \tag{6.73}$$

where $\mathcal{N}$ is the normalization constant such that the total probability is 1.

A differential equation of the form (6.69), where $\boldsymbol{\xi}(t)$ is a random "noise" is called a *stochastic differential equation* [5–7]. In our case the stochastic ODE (6.69) is linear of first order and hence can be solved quite easily:

$$\boldsymbol{v}(t) = \boldsymbol{v}_0\,e^{-\gamma t} + \frac{1}{m}\,e^{-\gamma t}\int_0^t e^{\gamma s}\,\boldsymbol{\xi}(s)\,ds \tag{6.74}$$

for initial condition $\boldsymbol{v}_0 \equiv \boldsymbol{v}(0)$. The variance of the velocity (say for $t' > t$) is then

$$\langle v^i(t)\,v^j(t')\rangle =$$

$$= v_0^i\,v_0^j\,e^{-\gamma(t+t')} + \delta^{ij}\,e^{-\gamma(t+t')}\frac{\lambda}{m^2}\int_0^t\int_0^{t'} ds\,ds'\,e^{\gamma(s+s')}\delta(s - s') =$$

$$= v_0^i\,v_0^j\,e^{-\gamma(t+t')} + \delta^{ij}\,e^{-\gamma(t+t')}\frac{\lambda}{m^2}\int_0^t ds\,e^{2s\gamma} = \tag{6.75}$$

$$= v_0^i\,v_0^j\,e^{-\gamma(t+t')} + \delta^{ij}\,e^{-\gamma(t+t')}\frac{\lambda}{2\gamma^2}\left(e^{2t\gamma} - 1\right)$$

For large $t, t' \gg \gamma^{-1}$, that is, when we reach equilibrium,

$$\langle v^i(t)\, v^j(t')\rangle \approx \delta^{ij}\, \frac{\lambda}{2\gamma m^2}\, \mathrm{e}^{-\gamma(t'-t)} \tag{6.76}$$

independently of initial condition $\boldsymbol{v}_0$. The mean square displacement for $t \gg \gamma^{-1}$ is

$$\langle x^i(t)^2\rangle = \frac{\lambda}{2\gamma m^2} \int_0^t \mathrm{d}s \int_0^t \mathrm{d}s'\, \mathrm{e}^{-\gamma|s-s'|} = \frac{\lambda}{\gamma^2 m^2}\, t + O(1) \tag{6.77}$$

which, after Einstein, is usually written in the form

$$\langle x^i(t)^2\rangle = 2Dt, \qquad D \equiv \frac{\lambda}{2\gamma^2 m^2}. \tag{6.78}$$

D is called the *diffusion constant* since it measures how fast the Brownian bodies diffuse in the medium: after a time t they will be spread out in a region of size

$$\approx \sqrt{\langle x^i(t)^2\rangle} = \sqrt{2Dt}. \tag{6.79}$$

The variance λ of the noise is thus a measure of the diffusion in the Brownian motion.

The Gaussian Ward Identity

Let $A[\boldsymbol{\xi}(t)]$ be any functional of the stochastic noise $\boldsymbol{\xi}(t)$. One has

$$\langle A[\boldsymbol{\xi}]\, \xi^i(t)\rangle = \mathcal{N}\int [D\boldsymbol{\xi}(s)]\, A[\boldsymbol{\xi}]\, \xi^i(t) \exp\left(-\frac{1}{2\lambda}\int \boldsymbol{\xi}(s)^2 \mathrm{d}s\right) =$$

$$= -\lambda\, \mathcal{N}\int [D\boldsymbol{\xi}(s)]\, A[\boldsymbol{\xi}]\, \frac{\delta}{\delta\xi^i(t)} \exp\left(-\frac{1}{2\lambda}\int \boldsymbol{\xi}(s)^2 \mathrm{d}s\right) = \tag{6.80}$$

$$= \lambda\, \mathcal{N}\int [D\boldsymbol{\xi}(s)]\, \frac{\delta A[\boldsymbol{\xi}]}{\delta\xi^i(t)}\, \exp\left(-\frac{1}{2\lambda}\int \boldsymbol{\xi}(s)^2 \mathrm{d}s\right)$$

where in the last line we integrated by parts in the path integral. The identity

$$\langle A[\boldsymbol{\xi}]\, \xi^i(t)\rangle = \lambda\left\langle \frac{\partial A[\boldsymbol{\xi}]}{\delta\xi^i(t)}\right\rangle \tag{6.81}$$

is a baby instance of *Ward identity* for correlation functions computed by path integrals. Its physical meaning will be further clarified below.

The Fokker-Planck Equation

We return to the stochastic linear ODE (6.69) whose solution $\boldsymbol{v}(t)$ is a random variable depending on the noise $\boldsymbol{\xi}(t)$, see Eq. (6.74). We are interested in computing the probability distribution of the velocity $\boldsymbol{v}(t)$ at time t defined as

$$\Psi(\boldsymbol{v}, t) \stackrel{\text{def}}{=} \langle \delta(\boldsymbol{v}(t) - \boldsymbol{v}) \rangle \tag{6.82}$$

where $\langle \cdots \rangle$ stands for the average over the statistic distribution (6.73) of the noise $\boldsymbol{\xi}(t)$. The function $\Psi(\boldsymbol{v}, t)$ yields a complete description of the simplest Brownian motion governed by the stochastic ODE (6.69).

We look for a partial differential equation (PDE) which characterizes the function $\Psi(\boldsymbol{v}, t)$. We first construct the PDE in an elementary ad hoc way, then in Sect. 6.4 we shall reproduce the result from a deeper perspective. To avoid making the notation too heavy, we work out explicitly the $d = 1$ case, the extension to general d being totally obvious.

We take the time derivative of the probability distribution $\Psi(v, t)$:

$$
\begin{aligned}
\frac{\partial}{\partial t} \Psi(v, t) &= \langle \delta'(v(t) - v)\, \dot{v}(t) \rangle = -\frac{\partial}{\partial v} \langle \delta(v(t) - v)\, \dot{v}(t) \rangle = \\
&= \gamma \frac{\partial}{\partial v} \langle \delta(v(t) - v)\, v(t) \rangle - \frac{1}{m} \frac{\partial}{\partial v} \langle \delta(v(t) - v)\, \xi(t) \rangle = \\
&= \gamma \frac{\partial}{\partial v} \left(v \langle \delta(v(t) - v) \rangle \right) - \frac{\lambda}{m} \frac{\partial}{\partial v} \left\langle \frac{\delta(v(t) - v)}{\delta \xi(t)} \right\rangle = \qquad \textbf{(use (6.81))} \\
&= \gamma \frac{\partial}{\partial v} v \langle \delta(v(t) - v) \rangle + \frac{\lambda}{m} \frac{\partial^2}{\partial v^2} \left\langle \delta(v(t) - v) \frac{\delta v(t)}{\delta \xi(t)} \right\rangle = \\
&= \gamma \frac{\partial}{\partial v} v \langle \delta(v(t) - v) \rangle + \frac{\lambda}{2m^2} \frac{\partial^2}{\partial v^2} \langle \delta(v(t) - v) \rangle = \\
&= \frac{\partial}{\partial v} \left(\frac{\lambda}{2m^2} \frac{\partial}{\partial v} + \gamma v \right) \Psi(v, t)
\end{aligned}
\tag{6.83}
$$

In d dimensions the differential equation takes the form[9]

$$\frac{\partial}{\partial t} \Psi(\boldsymbol{v}, t) = \frac{\partial}{\partial v^i} \left(\frac{\lambda}{2m^2} \frac{\partial}{\partial v^i} + \gamma\, v^i \right) \Psi(\boldsymbol{v}, t) \equiv L_f \Psi(\boldsymbol{v}, t) \tag{6.84}$$

where L_f is the second-order differential operator

$$L_f = \frac{\partial}{\partial v^i} \left(\frac{\lambda}{2m^2} \frac{\partial}{\partial v^i} + \gamma\, v^i \right) \equiv \frac{\lambda}{2m^2} \frac{\partial^2}{\partial v^i \partial v^i} + \gamma\, v^i \frac{\partial}{\partial v^i} + d\, \gamma. \tag{6.85}$$

Equation (6.84) is (a special instance of) the (forward) *Fokker-Planck equation* [8] giving the (forward) time evolution of the probability distribution $\Psi(\boldsymbol{v}, t)$. We stress

---

[9] Sum over the repeated index i implicit.

that the operator L_f is neither Hermitian nor anti-Hermitian: this reflects the *non-reversible* nature of the phenomena described by the Fokker-Planck equation (6.84).

The Heat Equation

For $\gamma = 0$ Eq. (6.84) reduces to the *heat equation*

$$\frac{\partial}{\partial t}\Psi(\boldsymbol{x},t) = \frac{\alpha}{2}\frac{\partial^2}{\partial x^i \partial x^i}\Psi(\boldsymbol{x},t) \equiv \frac{\alpha}{2}\Delta\Psi(\boldsymbol{x},t), \qquad \alpha \equiv \lambda/m^2 \tag{6.86}$$

which describes the diffusion of heat in a homogeneous medium (α being the diffusion constant). Now $\boldsymbol{x} \in \mathbb{R}^d$ have the physical interpretation of *spatial coordinates* in the medium not of velocities. The solution of the d-dimensional heat equation is

$$\Psi(\boldsymbol{x},t) = \int d^d\boldsymbol{y}\, K(\boldsymbol{x}-\boldsymbol{y};t)\,\Psi(\boldsymbol{y};0) \tag{6.87}$$

where the *heat kernel* $K(\boldsymbol{x};t)$ is

$$K(\boldsymbol{x};t) = \frac{1}{(2\pi\alpha t)^{d/2}}\exp\left(-\frac{\boldsymbol{x}^2}{2\alpha t}\right) \tag{6.88}$$

which satisfies the heat kernel PDE with initial condition

$$\frac{\partial}{\partial t}K(\boldsymbol{x};t) = \frac{\alpha}{2}\Delta K(\boldsymbol{x};t), \qquad K(\boldsymbol{x};0) = \delta(\boldsymbol{x}). \tag{6.89}$$

Assuming a δ-function initial condition at $t = 0$, the heat will be diffused at time t in a region of size $\sqrt{2\alpha t}$. This explains the meaning of the diffusion constant α.

Fokker-Planck vs. Schrödinger Equation

We return to our linear Fokker-Planck equation (6.84). We write

$$\Psi(\boldsymbol{v},t) = \exp\left(-\frac{\gamma m^2}{2\lambda}\boldsymbol{v}^2\right)\Phi(\boldsymbol{v},t). \tag{6.90}$$

The Fokker-Planck equation now becomes

$$\frac{\partial}{\partial t}\Phi = \frac{\lambda}{2m^2}\left(\frac{\partial}{\partial v^i} - \frac{\gamma m^2}{\lambda}v^i\right)\left(\frac{\partial}{\partial v^i} + \frac{\gamma m^2}{\lambda}v^i\right) \equiv -H\Phi \tag{6.91}$$

where H is the Hermitian operator

$$H = -\frac{\lambda}{2m^2}\frac{\partial^2}{\partial v^i \partial v^i} + V(v), \quad \text{with} \quad V(v) = \frac{\gamma^2 m^2}{2\lambda}v^i v^i + \frac{d\,\gamma}{2}, \tag{6.92}$$

that is, (setting $\hbar = 1$) the quantum Hamiltonian of d decoupled harmonic oscillators of mass $M = m^2/\lambda$ and frequency $\omega = \gamma$ shifted by a constant equal to the zero point energy (so that the energy levels of each oscillator are γn with $n \geq 1$). We conclude:

Fact 6.1 *The Fokker-Planck equation for the d-dimensional Brownian motion described by the simple stochastic ODE (6.69) is related to the Schrödinger equation for the d-dimensional harmonic oscillator at the imaginary time $\tau = -it$ by conjugation of evolution operators*

$$H \rightsquigarrow L_f \equiv e^{-\gamma m^2 v^2/2\lambda} \, H \, e^{\gamma m^2 v^2/2\lambda}. \tag{6.93}$$

Let

$$M(v, w; t) = \langle v|e^{-Ht}|w\rangle = \sum_n e^{-E_n t} \, \psi_n(v)^* \, \psi_n(w) \tag{6.94}$$

be the kernel for the imaginary-time evolution of the harmonic oscillator

$$\Phi(v, \tau) = \int M(v, \tau; v') \, \Phi(v', 0) \, dv' \tag{6.95}$$

whose explicit expression is given by the *Mehler formula* [9, 10]. To write the formula in a simpler way we first put the harmonic Hamiltonian in a normal form.

Fact 6.2 (Mehler's Formula) *Write the harmonic Hamiltonian (with $\hbar = 1$) in the normal form*

$$H = \frac{1}{2}(p^2 + q^2 - 1). \tag{6.96}$$

Then its imaginary-time evolution kernel is given by Mehler's formula

$$\langle \tilde{q}|e^{-tH}|q\rangle = \frac{1}{\sqrt{\pi(1 - e^{-2t})}} \exp\left(-\frac{\tilde{q}^2 - q^2}{2} - \frac{(e^{-t}\tilde{q} - q)^2}{1 - e^{-2t}}\right) \tag{6.97}$$

The formal solution to of our (baby) Fokker-Planck equation (6.84) is then

$$\Psi(v, t) = \int d^d w \, K(v, w; t) \, \Psi(w; 0), \tag{6.98}$$

where the Fokker-Planck kernel is

$$K(v, w; t) = e^{-\gamma m^2 v^2/2\lambda} \prod_i M(v^i, w^i; t)\, e^{\gamma m^2 w^2/2\lambda} =$$

$$= \sum_n e^{-E_n t}\, e^{-\gamma m^2 v^2/2\lambda}\, \psi_n(v)\, \psi_n(w)^*\, e^{\gamma m^2 w^2/2\lambda} \tag{6.99}$$

where $\{E_n\}$ are the energy levels of a system of d harmonic oscillators and $\psi_n(x)$ the corresponding eigenfunctions.

For large t, that is at equilibrium, only the ground state contribution to the sum in the RHS of (6.99) is relevant, all other terms being exponentially suppressed. Since the ground state wave-function of a harmonic oscillator is Gaussian, we conclude that at the equilibrium the velocity distribution is Gaussian. Indeed the stationary distribution

$$\Psi(v)_{\text{stat}} = \mathcal{N} \exp\left(-\frac{\gamma m^2}{\lambda} v^2\right) \tag{6.100}$$

satisfies

$$\left(\frac{\lambda}{2m^2}\frac{\partial}{\partial v^i} + \gamma v^i\right)\Psi(v)_{\text{stat}} = 0, \tag{6.101}$$

and so a fortiori solves the Fokker-Planck equation (6.84). For large t the probability distribution $\Psi(v, t)$ approaches $\Psi(v)_{\text{stat}}$ for almost all initial conditions.

Backward Fokker-Planck Equation

We have already mentioned that the Fokker-Planck equation describes irreversible phenomena, hence their time-evolution in the forward and backward directions are inherently different. Consider the Fokker-Planck kernel

$$K(v, t; w, s) \equiv K(v, w, t - s) \tag{6.102}$$

where the RHS is given by Eq. (6.99). It satisfies the equation

$$\frac{\partial}{\partial t}K(v, t; w, s) = L_{f,v}\, K(v, t; w, s) \tag{6.103}$$

where $L_{f,v}$ is the forward Fokker-Planck operator (6.85) acting as a differential operator on the variables v. The *backward* Fokker-Planck equation has the form

$$-\frac{\partial}{\partial s}K(v, t; w, s) = L_{b,w}\, K(v, t; w, s) \tag{6.104}$$

where now $L_{b,w}$ is the backward Fokker-Planck operator acting as a differential operator on the variables w. From Eq. (6.99)

$$L_{b,w}\left(e^{\gamma m^2 w^2/2\lambda}\,\psi_n(w)^*\right) = -e^{\gamma m^2 w^2/2\lambda}\,H^*\psi_n(w)^* \tag{6.105}$$

that is,

$$L_{b,w} = \frac{\lambda}{2m^2}\left(\frac{\partial}{\partial w^i} - \frac{2\gamma m^2}{\lambda}w^i\right)\frac{\partial}{\partial w^i}. \tag{6.106}$$

The Einstein Relation

From the stochastic differential description of the Brownian motion we got the expression (6.100) for the distribution of velocities at equilibrium. Einstein observed that—if the Brownian motion takes place at the constant temperature T—this expression should coincide with the thermal distribution of velocities at equilibrium given by the canonical ensemble (called the *Maxwell distribution of velocities*)

$$\mathcal{N}\exp\left(-\frac{E}{kT}\right) = \mathcal{N}\exp\left(-\frac{mv^2}{2kT}\right). \tag{6.107}$$

Comparing this formula with (6.100) we get a formula for the noise variance

$$\lambda = 2kT\gamma m \tag{6.108}$$

and for the diffusion constant

$$D \equiv \frac{\lambda}{2\gamma^2 m^2} = \frac{kT}{m\gamma} \tag{6.109}$$

which is called the *Einstein relation*.

Einstein Relation vs. Dissipation-Fluctuation Theorem

The Brownian motion allows us to see the dissipation-fluctuation theorem in action in a *classical* context and see the kind of information about the dissipative forces (such as mechanical friction or impedance in a resistor) arises from the thermal fluctuations. Again we focus on $d = 1$ to simplify the notation, the result being valid for all d with the obvious modifications.

The diffusion constant D was defined in Eq. (6.78)

$$D = \lim_{t\to\infty}\frac{1}{2t}\langle(x(t) - x(0))^2\rangle \tag{6.110}$$

where the average $\langle \cdots \rangle$ is taken over the canonical ensemble at thermal equilibrium. Since

$$x(t) - x(0) = \int_0^t v(s)\,\mathrm{d}s \tag{6.111}$$

Equation (6.110) can be rewritten as

$$
\begin{aligned}
D &= \lim_{t \to \infty} \frac{1}{2t} \int_0^t \mathrm{d}s \int_0^t \mathrm{d}u \,\langle v(s)\,v(u) \rangle = \lim_{t \to \infty} \frac{1}{t} \int_0^t \mathrm{d}s \int_0^{t-s} \mathrm{d}u \,\langle v(s)\,v(s+u) \rangle \\
&= -\lim_{t \to \infty} \frac{1}{t} \int_0^t \mathrm{d}s \, \frac{\partial(t-s)}{\partial s} \int_0^{t-s} \mathrm{d}u \,\langle v(s)\,v(s+u) \rangle = \\
&= \lim_{t \to \infty} \int_0^t \mathrm{d}s \,\langle v(0)\,v(s) \rangle - \lim_{t \to \infty} \int_0^t \mathrm{d}s \left(1 - \frac{s}{t}\right) \langle v(s)\,v(t) \rangle
\end{aligned}
\tag{6.112}
$$

Using the large t asymptotics (6.76) we see that the second limit in the RHS vanishes. Hence the Einstein relation (6.109) can be rewritten as

$$\mu = \frac{1}{m\gamma} = \frac{D}{kT} = \frac{1}{kT} \int_0^\infty \langle v(0)\,v(s) \rangle\,\mathrm{d}s \tag{6.113}$$

where μ is the *mobility* (the inverse of the friction constant). μ is given in terms of the fluctuations of the velocity in the Brownian motion. This is a manifestation of the (classical) dissipation-fluctuation theorem. Another manifestation is the formula

$$m\gamma = \frac{1}{kT} \int_0^\infty \langle \xi(t_0)\,\xi(t_0 + t) \rangle\,\mathrm{d}t \tag{6.114}$$

which says that the systematic part of the force which appears as friction is actually determined by the fluctuations of the random force.

Non-white Noise
When the noise is white the dissipation-fluctuation theorem in the form (6.114) implies that the friction coefficient is constant. A generalization of the statement, called the *Nyquist theorem*,[10] says that, more generally, the frequency spectrum of the friction force is determined by the fluctuation of the random component of the force. *Friction* (more generally the system's resistance) is the way the external work is dissipated into microscopic thermal energy and hence it is determined by

---

[10] Its main application is to make explicit the relation between the random electromotive force in a resistor and its impedance.

the dissipation-fluctuation theorem. In presence of a non-white noise the Langevin equation (6.69) should be generalized in the form

$$m\,\dot{v}(t) = -m \int_0^t \gamma(t-s)\,v(s)\,\mathrm{d}s + \xi(t) \quad t > 0, \tag{6.115}$$

We consider

$$\gamma(\omega) = \int_0^\infty \mathrm{e}^{\mathrm{i}\omega t}\,\gamma(t)\,\mathrm{d}t \tag{6.116}$$

while the mobility $\mu(\omega)$ (that is, the admittance of the velocity with respect to an exterior force periodic of frequency ω) is

$$\mu(\omega) = \frac{1}{m}\frac{1}{\gamma(\omega) - \mathrm{i}\omega}. \tag{6.117}$$

Using the previous methods one gets

$$\mu(\omega) = \frac{1}{kT} \int_0^\infty \big\langle v(t_0)\,v(t_0 + t)\big\rangle \mathrm{e}^{\mathrm{i}\omega t}\,\mathrm{d}t \tag{6.118}$$

$$m\gamma(\omega) = \frac{1}{kT} \int_0^\infty \big\langle \xi(t_0)\,\xi(t_0 + t)\big\rangle \mathrm{e}^{\mathrm{i}\omega t}\,\mathrm{d}t \tag{6.119}$$

where we assumed the equipartition law $m\langle v^2 \rangle = kT$. The two formulae (6.118) and (6.119) are called the *first* and the *second* dissipation-fluctuation theorem. The second formula is the general form of the Nyquist theorem which relates, say, the thermal noise in a resistor to its frequency-dependent impedance.

6.4   Langevin ODEs, Path Integrals and SUSY

The stochastic equation (6.69) is linear of first order. We want to discuss stochastic equations which, while of first order, are not necessarily linear. In doing this we exploit their direct relation with supersymmetry and its spontaneous breaking [11–13]. More precisely we consider *gradient flows* with a Gaussian *noisy* force

$$\dot{x}^i = \frac{\partial W(\boldsymbol{x})}{\partial x^i} + \xi^i, \tag{6.120}$$

where x^i $(i = 1, \ldots, n)$ are real functions of t, $W(\boldsymbol{x})$ is a smooth function of the x^i's, and $\xi^i(t)$ is a white noise with a Gaussian distribution with variance and mean

$$\big\langle \xi^i(t)\,\xi^j(s)\big\rangle = \delta_{ij}\,\delta(t-s) \quad \text{and} \quad \big\langle \xi^i(t)\big\rangle = 0. \tag{6.121}$$

Stochastic equations of the form (6.120) are called *Langevin equations*. Equation (6.69) corresponds to the case where $W(x)$ is a quadratic function of the x^i's. While formally it may look that the following equations are correct for all smooth functions $W(x)$, they are actually valid under more restricted conditions. The relation of (non-stochastic) gradient flow equations with Morse theory [14] clarifies the conditions. To make things easier, we make assumptions stronger than needed: we assume $W(x)$ to be a bounded below *convex* function with the property that for all $\Lambda \in \mathbb{R}$ the set

$$\{x \in \mathbb{R}^n : W(x) < \Lambda\} \subset \mathbb{R}^n \text{ is compact}. \tag{6.122}$$

Under these assumptions all our manipulations below can be fully justified.

We write $A[x(t)]$ for any functional of functions $x \colon \mathbb{R} \to \mathbb{R}^n$ evaluated on the solutions of the Langevin equation (6.120). Through this stochastic equation $A[x(t)]$ becomes an implicit functional of the noise $\xi(t)$. We are interested in its average $\langle A[x] \rangle$ over the Gaussian distribution of $\xi(t)$.

To make the problem well-defined we need to specify a boundary condition. We start from the simplest one: periodic boundary conditions with period β

$$\xi(t + \beta) = \xi(t), \qquad x(t + \beta) = x(t). \tag{6.123}$$

Then we have the equality

$$\langle A[x(t)] \rangle = \int [Dx][D\xi] \, A[x(t)] \, \delta\left[\dot{x} - \frac{\partial W}{\partial x} - \xi\right] \times$$
$$\times \operatorname{Det}\left[\delta_{ij}\frac{d}{dt} - \frac{\partial^2 W}{\partial x^i \partial x^j}\right] \exp\left(-\frac{1}{2}\int_0^\beta \xi(s)^2 \, ds\right) \tag{6.124}$$

as one checks by performing first the functional integral in $[Dx]$ and noticing that the combination

$$\delta\left[\dot{x} - \frac{\partial W}{\partial x} - \xi\right] \operatorname{Det}\left[\frac{\delta(\dot{x}(t)^i - \partial_{x^i(t)} W(x(t)))}{\delta x(t)^j}\right] \equiv$$
$$\equiv \delta\left[\dot{x} - \frac{\partial W}{\partial x} - \xi\right] \operatorname{Det}\left[\frac{\delta \xi^i(t)}{\delta x^j(s)}\right] \tag{6.125}$$

is locally in function space just the δ function

$$\delta\left[x(t) - x[t; \xi(s)]\right] \tag{6.126}$$

where $x[t; \xi(s)]$ is a solution of the stochastic equation (6.120) in presence of the given noise $\xi(s)$. Indeed, the functional determinant in Eq. (6.125) is just the functional Jacobian for the functional change of variables

$$\boldsymbol{\xi} \rightsquigarrow \boldsymbol{x} \tag{6.127}$$

given by solving the ODE (6.120) so the expression (6.125) coincides with (6.126) in the vicinity in function space of the particular periodic solution $\boldsymbol{x}[t; \boldsymbol{\xi}(s)]$. For instance, when W is quadratic, there is a unique solution $\boldsymbol{x}[t; \boldsymbol{\xi}(s)]$ given by the expression $\boldsymbol{v}(t)$ in the RHS of (6.74) with the initial condition $\boldsymbol{v}_0 \equiv \boldsymbol{v}_0[\boldsymbol{\xi}(s)]$ for produces a β-periodic solution i.e.

$$\boldsymbol{v}[\beta; \boldsymbol{\xi}(s)] = \boldsymbol{v}_0[\boldsymbol{\xi}(s)] \quad \Rightarrow \quad \boldsymbol{v}_0[\boldsymbol{\xi}(s)] = \frac{1}{m} \frac{e^{-\beta\gamma}}{1 - e^{-\beta\gamma}} \int_0^\beta e^{\gamma s} \boldsymbol{\xi}(d)\, \mathrm{d}s. \tag{6.128}$$

The global aspects of the functional measure are controlled using Morse theory: under our special assumptions on the function $W(\boldsymbol{x})$ no subtlety arises from the behavior in the large, and $\boldsymbol{x}[t; \boldsymbol{\xi}(s)]$ is well defined globally. Otherwise one should take into account other aspects (which are well known and somehow elementary, but will be omitted for brevity) such as the existence of multiple solutions (or no solution at all) for certain (or even all) profile $\boldsymbol{\xi}(t)$ of the noise.

We may represent the functional Jacobian in Eq. (6.125)—defined as the determinant of the first order differential operator

$$\delta_{ij} \frac{\mathrm{d}}{\mathrm{d}t} + \frac{\partial^2 W}{\partial x^i \partial x^j}\bigg|_{x^i \equiv x^i(t)} \tag{6.129}$$

acting on the vector space of periodic functions of period β—as a path integral over periodic Grassmannian fields (of period β) $\bar{\eta}, \eta$

$$\mathrm{Det}\left[\frac{\delta \xi^i(t)}{\delta x^j(s)}\right] = \int [D\bar{\eta}\, D\eta] \exp\left(-\int_0^\beta \mathrm{d}s\left[\bar{\eta}^i \dot{\eta}^i - \bar{\eta}^i \eta^j\, \partial_{x^i} \partial_{x^j} W\right]\right) \tag{6.130}$$

We may integrate away the noise $\boldsymbol{\xi}(t)$ with the effect that the argument of the exponential in Eq. (6.124) becomes

$$\frac{1}{2}\int_0^\beta \xi^i(s)\xi^i(s) \rightsquigarrow \frac{1}{2}\int_0^\beta \left[\frac{dx^i}{dt} - \frac{\partial W}{\partial x^i}\right]^2 =$$

$$= \frac{1}{2}\int_0^\beta ds \sum_i \left[(\dot{x}^i)^2 + \left(\frac{\partial W}{\partial x^i}\right)^2 - 2\frac{\partial W}{\partial x^i}\dot{x}^i\right]^2 = \tag{6.131}$$

$$= \frac{1}{2}\int_0^\beta ds \sum_i \left[(\dot{x}^i)^2 + \left(\frac{\partial W}{\partial x^i}\right)^2\right] + W(\beta) - W(0)$$

The boundary term cancels since $W(\beta) \equiv W(0)$ by the periodic boundary conditions. Putting everything together we get

$$\langle A[\xi(t)]\rangle = \int [Dx \, D\bar{\eta} \, D\eta] \, A[\xi(t)] \exp\left(-\int_0^\beta L_E \mathrm{d}s\right) \tag{6.132}$$

where

$$L_E = \frac{1}{2}\sum_i \left(\frac{dx^i}{dt}\right)^2 + \frac{1}{2}\sum_i \left(\frac{\partial W}{\partial x^i}\right)^2 + \sum_i \bar{\eta}_i \frac{d}{dt}\eta_i - \frac{\partial^2 W}{\partial x^i \partial x^j}\,\bar{\eta}_i \eta_j \tag{6.133}$$

is the *Euclidean* (i.e. imaginary time) Lagrangian of the quantum mechanical supersymmetric model with superpotential $W(x)$ [11]; we shall say a few words on supersymmetry (SUSY) at the end of this section. The Hamiltonian operator of this quantum model is

$$H = -\frac{1}{2}\frac{\partial^2}{\partial x_i \partial x_i} + \frac{1}{2}\sum_i \left(\frac{\partial W}{\partial x^i}\right)^2 + \frac{1}{2}\frac{\partial^2 W}{\partial x^i \partial x^j}(b_i^\dagger b_j - b_j b_i^\dagger) \tag{6.134}$$

where $b_i^\dagger$, b_j are fermionic creator/annihilator operators (cf. Sect. 3.7) which satisfy the Clifford algebra

$$b_i b_j + b_j b_i = b_i^\dagger b_j^\dagger + b_j^\dagger b_i^\dagger = 0, \quad b_i^\dagger b_j + b_j b_i^\dagger = \delta_{ij}. \tag{6.135}$$

In the Hilbert space of the fermionic degrees of freedom $b_i^\dagger$, b_j there is a special state $|0\rangle$, called the *Clifford vacuum*, such that

$$b_j|0\rangle = 0 \quad \text{for all } j = 1, \ldots, n. \tag{6.136}$$

Since the Hessian of W is assumed to be positive-definite everywhere, the fermion contribution to the energy (6.134) is minimized when the fermions are frozen in the Clifford vacuum $|0\rangle$. Restricting to this sector of the Hilbert space we remain with the effective Hamiltonian for the bosonic degrees of freedom

$$2H_{\mathrm{bos}} = -\frac{\partial^2}{\partial \partial x^i \partial x^i} + \frac{\partial W}{\partial x^i}\frac{\partial W}{\partial x^i} - \frac{\partial^2 W}{\partial x^i \partial x^i} = -\left(\frac{\partial}{\partial x^i} - \frac{\partial W}{\partial x^i}\right)\left(\frac{\partial}{\partial x^i} + \frac{\partial W}{\partial x^i}\right) \tag{6.137}$$

which we conjugate to the operator

$$-2L_f \equiv -\frac{\partial}{\partial x^i}\left(\frac{\partial}{\partial x^i} + 2\frac{\partial W}{\partial x^i}\right) = e^{-W}2H_{\text{bos}}\,e^{W}. \tag{6.138}$$

so that after the conjugation $\psi \to e^{W}\psi$ the imaginary time Schrödinger equation with the fermions frozen in their ground state becomes

$$\frac{\partial}{\partial t}\Psi = L_f\Psi. \tag{6.139}$$

We claim that this is the equation describing the probability distribution

$$\Psi(x, t) = \langle \delta(x(t) - x)\rangle \tag{6.140}$$

of the stochastic (Langevin) equation (6.120) with the white noise (6.121).

Let us prove our claim. The kernel for the propagation

$$K(x, y; t) = \langle \delta(x(t) - x)\,\delta(x(0) - y)\rangle \tag{6.141}$$

can be written as the path integral (6.124) over paths $x : [0, t] \to \mathbb{R}^n$ satisfying the boundary conditions $x(t) = x$ and $x(0) = y$ with the insertion

$$\delta(x(t) - x)\,\delta(x(0) - y). \tag{6.142}$$

In the manipulation (6.131) we get non-zero boundary terms from the new boundary conditions which produce extra factors

$$e^{-W(x)}\langle x|e^{-tH_{\text{bos}}}|y\rangle\,e^{W(y)}. \tag{6.143}$$

The factors from the boundary terms have precisely the effect of replacing H_{bos} by the conjugate operator L_f. We conclude

Fact 6.3 *For the Langevin equation* (6.120) *the Fokker-Planck equation is* (6.139) *with forward evolution operator* (6.138).

Remark 6.3 Without our special assumptions on the function $W(x)$ the fermionic degrees of freedom cannot be consistently frozen in a specific state, and we do not get an effective purely bosonic system. Nevertheless the interacting supersymmetric theory always gives the solutions to the stochastic equation.

Remark 6.4 In Eq. (6.130) the determinant over the Fermi fields is defined with periodic boundary conditions, while the trace on the Hilbert space is given by the path integral with *anti-periodic* boundary conditions (cf. Sect. 3.8). Thus the path integral (6.130) represents the trace of $e^{-\beta(H+\mu F)}$ on the Hilbert space where we introduce an imaginary chemical potential $\mu = i\pi/\beta$ for the Fermi number F which has the effect of weighting the states in the trace with an extra factor $e^{\beta\mu F} \equiv (-1)^F$ which is a sign -1 (resp. $+1$) for Grassmann odd Fermi states (resp. for Grassmann even Bose states). From the path integral viewpoint the insertion of $e^{\beta\mu F}$ with $\mu = i\pi/\beta$ flips between periodic and antiperiodic boundary conditions on the fermionic fields as one sees from Eqs. (3.226) and (3.227) [11]. Using Eq. (3.254) we see that for a quadratic (Gaussian) action the Hilbert space trace

$$\mathrm{Tr}\left[e^{-\beta H}(-1)^F\right] \tag{6.144}$$

is equal to the inverse of the Bose partition function (cf. Eq. (3.254)) so that now (under the assumption that the path integrals are Gaussian) the path integral in Eq. (6.124) with no insertion (i.e. $A[x] \equiv 1$) is just 1. See [11] for a discussion of the meaning of this result for supersymmetry as well as for the generalization of this result to path integrals which are *not* Gaussian.

SUSY in the Langevin/Fokker-Planck Equations

The imaginary-time ($\equiv$ Euclidean) Lagrangian L_E in Eq. (6.133) is invariant up to surface terms under two supersymmetries with Grassmann odd constant parameters $\bar\epsilon$ and ϵ, respectively. The first one is

$$\delta x^i = \bar\epsilon\eta^i \qquad\qquad \delta\eta^i = 0$$
$$\delta\bar\eta^i = \bar\epsilon\left(-\dot x^i + \frac{\partial W}{\partial x^i}\right) \qquad \delta L_E = \frac{\mathrm{d}}{\mathrm{d}t}\left(\bar\epsilon\frac{\partial W}{\partial x^i}\eta^i\right). \tag{6.145}$$

The second one is

$$\delta x^i = \bar\eta^i\epsilon \qquad\qquad \delta\bar\eta^i = 0$$
$$\delta\eta^i = \left(\dot x^i + \frac{\partial W}{\partial x^i}\right)\epsilon \qquad \delta L_E = \frac{\mathrm{d}}{\mathrm{d}t}\left(\bar\eta^i\dot x^i\epsilon\right). \tag{6.146}$$

In the supersymmetric context the function $W(x^i)$ is called the *superpotential*. Supersymmetry yields a reinterpretation of the Gaussian Ward identities of Sect. 6.3 as the SUSY Ward identities: with our assumptions on the superpotential W, SUSY is unbroken [11]; then for all operator O we have the identities

$$\langle\delta_\epsilon O\rangle = \langle\delta_{\bar\epsilon} O\rangle = 0, \tag{6.147}$$

where $\delta_\epsilon O$, $\delta_{\bar\epsilon} O$ are the variations of O with respect to the above supersymmetries. Let $O = A[x]\eta^i(t)$ where $A[x]$ is any functional of the $x^i(s)$'s; we have

$$
\begin{aligned}
0 =& \left\langle \delta_\epsilon \left(A[x]\eta^i(t) \right) \right\rangle = \left\langle \int ds\, \frac{\delta A[x]}{\delta x^j(s)} \delta_\epsilon x^j(s)\, \eta^i(t) \right\rangle + \left\langle A[x]\delta_\epsilon \eta^i(t) \right\rangle = \\
=& -\left\langle \int ds\, \frac{\delta A[x]}{\delta x^j(s)} \bar\eta^j(s)\eta^i(t) \right\rangle \epsilon + \left\langle A[x] \left(\dot x^i(t) + \frac{\partial W}{\partial x^i}(t) \right) \right\rangle \epsilon = \\
=& -\left\langle \int ds\, \frac{\delta A[x]}{\delta x^j(s)} \left\langle s, j \left| \left(\frac{d}{dt} + \frac{\partial^2 W}{\partial x \partial x} \right)^{-1} \right| t, i \right\rangle \right\rangle \epsilon \\
& + \left\langle A[x] \left(\dot x^i(t) + \frac{\partial W}{\partial x^i}(t) \right) \right\rangle \epsilon
\end{aligned}
\tag{6.148}
$$

where

$$
\left\langle s, j \left| \left(\frac{d}{dt} + \frac{\partial^2 W}{\partial x \partial x} \right)^{-1} \right| t, i \right\rangle
\tag{6.149}
$$

is the kernel of the inverse ($\equiv$ Green's function) of the differential operator

$$
\delta^{ij} \frac{d}{dt} + \frac{\partial^2 W}{\partial x^i \partial x^j} \equiv \frac{\delta \xi^i(t)}{\delta x^j(s)}
\tag{6.150}
$$

hence (6.149) coincide with the Jacobian operator

$$
\frac{\delta x^j(s)}{\delta \xi^i(t)}
\tag{6.151}
$$

so that the last line of Eq. (6.148) becomes

$$
\left\langle A[x]\xi^i(t) \right\rangle = \left\langle \int ds\, \frac{\delta A[x]}{\delta x^j(s)} \frac{\partial x^j(s)}{\partial \xi^i(t)} \right\rangle = \left\langle \frac{\delta A[x]}{\delta \xi^i(t)} \right\rangle
\tag{6.152}
$$

which is the Gaussian Ward identity (6.81).

6.5   Hamiltonian Stochastic Equations and Kramers' Equation

In the simplest stochastic equation (6.69) there is no force except for the resistance of the medium and the background noise, both arising from the thermal fluctuations of the surrounding molecules, in agreement with the dissipation-fluctuation theorem. More general Brownian motions are described by second-order, non-linear stochastic equations of the typical form

$$\begin{cases} \dot{x} = p \\ \dot{p} = F(x) - m\gamma\, p + \xi \end{cases} \tag{6.153}$$

where ξ is a (vector) white noise as in Eqs. (6.70) and (6.72) and $m\gamma$ is the friction constant. We assume that the *drift force* $F(x)$ has a Hamiltonian formulation, that is, we assume that the system is described by a (classical) Hamiltonian of the form

$$H = \frac{1}{2}p^2 + V(x) \tag{6.154}$$

perturbed by the two phenomenological forces arising at thermal equilibrium from the surrounding medium: a friction force $-m\gamma\, p$ and a random noise force ξ. The resulting stochastic equations of motion are

$$\frac{\mathrm{d}}{\mathrm{d}t}\begin{pmatrix} x \\ p \end{pmatrix} = \begin{pmatrix} \frac{\partial H}{\partial p} \\ -\frac{\partial H}{\partial x} \end{pmatrix} - \begin{pmatrix} 0 \\ \gamma\, p - \xi \end{pmatrix} \tag{6.155}$$

The equation $\dot{x} = p$ is just the definition of momentum and is not affected by the medium. We wish to get the PDE governing the time-evolution of the expectation value of the phase-space density

$$\varrho(x, p; t) \;\overset{\mathrm{def}}{=}\; \big\langle \delta(x(t) - x)\,\delta(p(x) - p)\big\rangle. \tag{6.156}$$

In absence of the medium the density would satisfy the Liouville equation (Sect. 2.2)

$$\frac{\partial}{\partial t}\varrho(x, p; t) - \big[H, \varrho(x, p; t)\big]_{\mathrm{PB}} = 0 \tag{6.157}$$

($[\cdot, \cdot]_{\mathrm{PB}}$ is the Poisson bracket). In presence of the two phenomenological forces produced by the thermal fluctuations the RHS is not zero; we can infer its form from the equation (6.84) which (setting $m = 1$) corresponds to the case $H = p^2/2$. We get

$$\frac{\partial}{\partial t}\varrho(x, p; t) - \big[H, \varrho(x, p; t)\big]_{\mathrm{PB}} = \frac{\partial}{\partial p}\left(\lambda\,\frac{\partial}{\partial p} + \gamma\, p\right)\varrho(x, p; t) \tag{6.158}$$

The reader is invited to prove this equation by the elementary technique used to show Eq. (6.84). This more elaborate version of the Fokker-Planck equation (6.158) is called the *Kramers equation* [8]. Now let us make the redefinition of variable

$$\varrho(\boldsymbol{x}, \boldsymbol{p}; t) = \exp\left[-\gamma\, H(\boldsymbol{x}, \boldsymbol{p})/(2\lambda)\right] \Psi(\boldsymbol{x}, \boldsymbol{p}; t) \tag{6.159}$$

which puts the Kramers equation in the form

$$\frac{\partial}{\partial t}\Psi(\boldsymbol{x}, \boldsymbol{p}; t) = -i\,\mathbb{H}\Psi$$
$$= \lambda\left(\frac{\partial}{\partial \boldsymbol{p}} - \frac{\gamma}{2\lambda}\boldsymbol{p}\right)\left(\frac{\partial}{\partial \boldsymbol{p}} + \frac{\gamma}{2\lambda}\boldsymbol{p}\right)\Psi(\boldsymbol{x}, \boldsymbol{p}; t) + \left[H, \Psi(\boldsymbol{x}, \boldsymbol{p}; t)\right]_{\mathrm{PB}} \tag{6.160}$$

where $\mathbb{H}$ is the sum of a self-adjoint operator given by the Poisson bracket and a anti-self-adjoint operator. This corresponds to the splitting of the time evolution into reversible and irreversible processes. The function

$$\Psi(\boldsymbol{x}, \boldsymbol{p}; t) = \exp\left[-\frac{\gamma}{2\lambda} H(\boldsymbol{x}, \boldsymbol{p})\right] \tag{6.161}$$

is an everywhere positive stationary solution of Eq. (6.160) which is a ground eigenstate of the irreversible part of the evolution operator. Hence this solution has the proper characteristics to be a probability distribution at equilibrium and, moreover, all other positive solutions will approach this one as $t \to \infty$ at an exponential rate. Therefore as $t \to \infty$ the system reaches the equilibrium phase-space distribution

$$\varrho(\boldsymbol{x}, \boldsymbol{p}; t) = \exp\left[-\frac{\gamma}{\lambda} H(\boldsymbol{x}, \boldsymbol{p})\right]. \tag{6.162}$$

Comparing with the canonical distribution in equilibrium at temperature T we get the generalized Einstein relation

$$\lambda = kT\gamma. \tag{6.163}$$

Again, this relation reflects the dissipation-fluctuation theorem which relates the macroscopic friction force to the fluctuations of the noise. As before we can extend the analysis to the case of non-white noises getting frequency dependent friction and mobility response functions.

References

1. K. Konishi, G. Paffuti, *Quantum Mechanics. A New Introduction* (Oxford, 2009)
2. R. Kubo, The fluctuation-dissipation theorem. Rep. Progr. Phys. **29**, 255–284 (1966)
3. J.W. Brown, R.V. Churchill, *Complex Variables and Applications*, 7th edn. (McGraw Hill, 2004)

4. S. Cecotti, *Analytic Mechanics. A Concise Textbook* (Springer, 2024)
5. L.C. Evans, *An Introduction to Stochastic Differential Equations* (American Mathematical Society, 2013)
6. C.W. Gardiner, *Handbook of Stochastic Methods: For Physics, Chemistry and the Natural Sciences* (Springer, 2004)
7. J. Cecconi, *Stochastic Differential Equations*. Lectures given at a Summer School of the Centro Internazionale Matematico Estivo (C.I.M.E.) held in Cortona (Arezzo), Italy, May 29-June 10, 1978 (Springer, 1979)
8. H. Risken, *The Fokker-Planck Equation. Methods of Solution and Applications*, 2nd edn. (Springer, 1996)
9. F.G. Mehler, Ueber die Entwicklung einer Function von beliebig vielen Variabeln nach Laplaceschen Functionen höherer Ordnung. J. Reine Angew. Math. **66**, 161–176 (1866)
10. J. Glimm, A. Jaffe, *Quantum Physics. A Functional Integral Point of View* (Springer, 1981)
11. S. Cecotti, L. Girardello, Functional measure, topology and dynamical supersymmetry breaking. Phys. Lett. B **110**, 39 (1982)
12. G. Parisi, N. Sourlas, Supersymmetric field theories and stochastic differential equations. Nucl. Phys. **B206**, 321 (1982)
13. S. Cecotti, L. Girardello, Stochastic and parastochastic aspects of supersymmetric functional measures: a new nonperturbative approach to supersymmetry. Ann. Phys. **145**, 81–99 (1983)
14. E. Witten, Supersymmetry and Morse theory. J. Differential Geom. **17**, 661–692 (1982)

Chapter 7
Frustrated and Quenched Systems

In this chapter we review some more advanced topic. We discuss frustrated and quenched systems and then focus on *spin glasses,* introducing the replica trick. We present the Sherrington-Kirkpatrick (SK) model, deduce the naive solution with unbroken replica symmetry, and then outline Parisi's exact solution of the SK model with broken replica symmetry and a functional order parameter.

7.1 Frustration

A simple example of *frustrated system* is an Ising-like model on a $d = 2$ regular triangular lattice Λ with *anti-ferromagnetic* nearest neighbor couplings

$$H = J \sum_{l \in \Lambda_1} \sigma_{h(l)} \sigma_{t(l)}, \quad J > 0. \tag{7.1}$$

The contribution of each link $l \in \Lambda_1$ to the energy H is now minimized when the two spins at its ends, $\sigma_{h(l)}$ and $\sigma_{t(l)}$, point in *opposite* directions. But if two sides of a triangular plaquette satisfy this condition, the two spins connected by its third side are automatically aligned

$$\tag{7.2}$$

and hence the third link (the rippled one in the figure) cannot minimize its energy: we say that this link is *frustrated.* The ground states are the configurations with a minimal number of frustrated links, i.e. just one per plaquette. Clearly there is a huge

© The Author(s), under exclusive license to Springer Nature Switzerland AG 2024
S. Cecotti, *Statistical Mechanics*, UNITEXT for Physics,
https://doi.org/10.1007/978-3-031-67874-5_7

number of spin configurations with this property. It is easy to see that the number
of ground states for a triangular lattice with N sites grows with N as $\exp(s\,N)$ for
some positive constant s. Hence the number of ground states gets infinite in the
thermodynamic limit, and the entropy per site at zero temperature, s, is *positive*. This
is due to the fact that there exist *local modifications* of the spin configuration which
keep fixed all spins outside a bounded region and transform a ground state into a
different ground state. As a simple example, consider a ground state configuration
containing a node which belongs to three frustrated links: locally at this node the
configuration looks as one of the two hexagons with non-frustrated side links

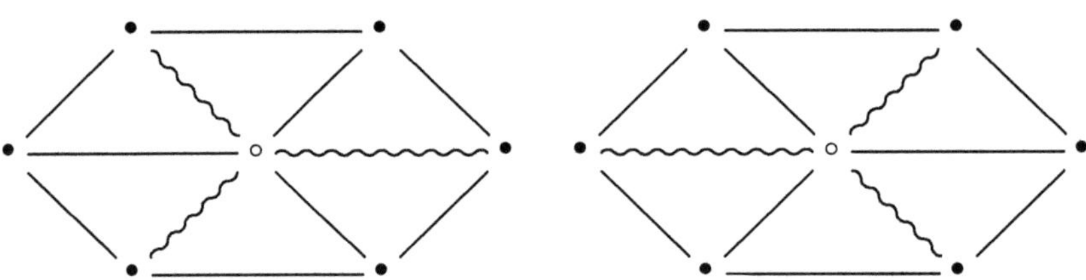

Flipping *only* the spin at the central white node o we interchange these two
hexagons, getting another valid ground state configuration. Seeing the triangular
lattice as the union of $N/3$ hexagons, we can replace each of them by one of the two
hexagons above, thus constructing $2^{N/3}$ ground state configurations, so the entropy
per site $s \geq \frac{1}{3}\log 2 > 0$. However there are many other ground states, as the reader
may easily check, and the exact entropy per site s at $T = 0$, computed by Wannier
[1], is

$$
\begin{aligned}
s &= \frac{2}{\pi}\int_0^{\pi/3}\log(2\cos\omega)\,d\omega = \frac{1}{\pi}\int_0^{\pi/3}\Big(\log(1+e^{2i\theta})+\log(1+e^{-2i\theta})\Big)d\theta \\
&= \frac{1}{2\pi i}\int_0^{-e^{2\pi i/3}}\log(1-z)\frac{dz}{z}+c.c. = \frac{1}{2\pi i}\Big(\mathrm{Li}_2(-e^{-2\pi i/3})-\mathrm{Li}_2(-e^{2\pi i/3})\Big) \\
&= \frac{3}{4\pi i}\Big(\mathrm{Li}_2(-e^{-\pi i/3})-\mathrm{Li}_2(-e^{\pi i/3})\Big) = \frac{3}{\pi}\int_0^{\pi/6}\log(2\cos\omega)\,d\omega \approx 0.323066
\end{aligned}
$$

$$(7.3)$$

where we wrote s in several convenient forms. $\mathrm{Li}_2(z)$ is the *dilogarithm function*, cf.
Sect. 3.3, and we used the identity $\mathrm{Li}_2(z^2) = 2\big(\mathrm{Li}_2(z)+\mathrm{Li}_2(-z)\big)$. The exact internal
energy of the 2d antiferromagnetic triangular Ising model is (see [1] ERRATA)

$$
\frac{U}{\frac{1}{2}NJ} = \frac{2}{1-\mu}\left(1-\frac{8}{\pi}\mu(3-\mu)\frac{K(k)}{4|\mu|^{1/2}+[(|\mu|+1)^3(3-|\mu|)]^{1/2}}\right) \qquad (7.4)
$$

where

$$\mu = 1 + 2\tanh(\beta J) \tag{7.5}$$

and $K(k)$ is the complete elliptic integral of the first kind with modulus k

$$K(k) = \int_0^{\pi/2} \frac{d\phi}{\sqrt{1 - k^2 \sin^2 \phi}} \tag{7.6}$$

with modulus k equal to

$$k = \frac{(|\mu| - 1)^3(3 - |\mu|)}{\left[4\sqrt{|\mu|} + \sqrt{(|\mu| + 1)^3(3 - |\mu|)}\,\right]^2}. \tag{7.7}$$

The complete elliptic integral is analytic in the complex k-plane except for two cuts along the real axis $(-\infty, -1]$ and $[1, +\infty)$. In the antiferromagnetic model $(J > 0)$ μ starts at 1 for $\beta = 0$ $(T = \infty)$ and ends at $\mu = 3$ for $\beta = \infty$ $(T = 0)$; one has

$$1 \le \mu \le 3 \quad \Rightarrow \quad 0 \le k \le \frac{71 - 17\sqrt{17}}{64} \approx 0.01417 \tag{7.8}$$

so the energy is analytic for the antiferromagnetic model at all temperatures and there is no phase transition. Flipping the sign of the coupling J we interchange antiferromagnetic and ferromagnetic models; thus in the ferromagnetic model μ starts at 1 for $\beta = 0$ and ends at -1 for $\beta = \infty$. $\mu = 0$ at the ferromagnetic critical temperature, $\tanh(\beta_c|J|) = 1/2$, which corresponds to the modulus $k = -1$ where the internal energy U gets non-analytic. In the antiferromagnetic case there is only a phase transition at zero temperature due to the presence of several ground states. In this respect the situation is similar to the $d = 1$ ferromagnetic case. For more on the antiferromagnetic Ising model on a triangular lattice, see e.g. [2].

An exponentially large number of ground states is typical of *frustrated systems* (i.e. systems with frustrated links). The ferromagnetic Ising model has two ground states and we found that in $d \ge 2$ at low temperature it has two distinct pure phases which correspond to the two ground states and are distinguished by the sign of the order parameter "magnetization". By analogy we may expect that a system with an exponentially large number of vacua has "infinitely many" pure states at low enough temperature and *large enough dimension* (this does *not* happen in $d = 2$ as we have seen). Then we need infinitely many order parameters to distinguish the states: the low-temperature phases of our frustrated system present very peculiar "orders" which superficially look as "disorder" since the spins point in different directions in a seemingly random way. It is obvious that frustrated systems will present new and subtler phenomena that were not present in the systems considered so far.

The frustration in an Ising-like model with general couplings J_l, which may depend on the link l,

$$H = -\sum_{l \in \Lambda_1} J_l \, \sigma_{t(l)} \sigma_{h(l)} \tag{7.9}$$

is measured by the *Toulouse frustration function* ($\equiv$ the $\mathbb{Z}_2$ holonomy element) defined on the closed paths[1] C in Λ

$$\Phi(C) \overset{\text{def}}{=} \prod_{l \in C} \text{sign} \, J_l. \tag{7.10}$$

A path C is *frustrated* (resp. unfrustrated) if $\Phi(C) = -1$ (resp. $\Phi(C) = 1$). One says that a plaquette is frustrated if its boundary is frustrated.

To provide some intuition on frustration, we study a baby model.

Ising Model with Long-Range Interactions

We consider an Ising-like model with long-range interactions where all spins couple to all other spins with the same strength. The total number of spins is N. We can think of this system as an Ising model in infinite dimension where the mean field approximation is exact. If we keep the coupling fixed as $N \to \infty$, the thermodynamic limit does not exist. To get a smooth thermodynamic limit we have to choose a coupling of order $1/N$. Adding a convenient constant, the Hamiltonian becomes

$$H = -\frac{J}{N} \sum_{i<j} \sigma_i \sigma_j - \frac{J}{2} = -\frac{J}{2N} \left(\sum_i \sigma_i \right)^2 + \frac{J}{2N} \sum_i (\sigma_i)^2 - \frac{J}{2} = -\frac{JN}{2} M^2 \tag{7.11}$$

where

$$M = \frac{N_+ - N_-}{N} \equiv 2\frac{N_+}{N} - 1 \qquad -1 \leq M \leq 1 \tag{7.12}$$

is the *magnetization*. Here N_+ (resp. N_-) are the numbers of spins pointing in the positive (resp. negative) direction. We first assume $J > 0$. In this case the system is ferromagnetic and there is no frustration. The entropy is (as for the Purcell-Pound

---

[1] A closed path C of length ℓ is a concatenated sequence of ℓ links $C \equiv \{l_a\}_{a=1}^{\ell} \subset \Lambda_1$ with $t(l_{a+1}) = h(l_a)$ and $h(l_\ell) = t(l_1)$.

spin model discussed at the end of Sect. 2.3)

$$S(M) = \log \binom{N}{N_+} = \log N! - \log N_+! - \log N_-! =$$

$$= \log \Gamma(N+1) - \log \Gamma(\tfrac{1}{2}N(1+M)+1) - \log \Gamma(\tfrac{1}{2}N(1-M)+1)$$

$$\approx N\left(\log 2 - \tfrac{1}{2}(1+M)\log(1+M) - \tfrac{1}{2}(1-M)\log(1-M)\right) \tag{7.13}$$

so that

$$\frac{\partial S(M)}{\partial M} = \operatorname{arctanh}(M) \tag{7.14}$$

and the minimization of the free energy $U - TS$ with respect to M gives back the mean-field equation (cf. Eq. (5.13))

$$M = \tanh(\beta J M) \tag{7.15}$$

which is *exact* for this infinite-range model. Recall from Remark 5.1 that this implies that in the high temperature phase for $T > T_c$ the entropy is constant

$$S = \log 2 \quad \text{while} \quad U = C_V = 0. \tag{7.16}$$

For the sake of comparison we present an alternative derivation of these results for the *ferromagnetic* model in the off-text box.

Ferromagnetic Ising Model with Long-Range Interactions: Alternative Derivation

The exact canonical partition function is

$$\sum_{N_+=0}^{N} \binom{N}{N_+} e^{\beta J (2N_+ - N)^2 / 2N}$$

$$= \sqrt{\frac{N}{2\pi\beta J}} \int_{-\infty}^{+\infty} dx \sum_{N_+=0}^{N} \binom{N}{N_+} \exp\left[-\frac{N}{2\beta J}x^2 + (2N_+ - N)x \right]$$

$$= \sqrt{\frac{N}{2\pi\beta J}} \int_{-\infty}^{+\infty} dx \, \exp\left[-N\left(\frac{x^2}{2\beta J} - \log(2\cosh x) \right) \right] \tag{7.17}$$

(continued)

In the thermodynamic limit $N \to \infty$, the saddle-point evaluation of the last integral becomes exact, and we get the same saddle-point equation (7.15) as in the mean field approach to the Ising model with magnetization $M \equiv x/(\beta J)$, cf. Sect. 5.1.

Now take $J < 0$ to make the long-range model *frustrated*. The ground states have $N_+ = N/2$ ($M = 0$) and the number of ground states is

$$\#(\text{ground states}) = \binom{N}{N/2} = \frac{2^N}{\sqrt{\pi N/2}}\left(1 + O(1/N)\right) \tag{7.18}$$

$$\frac{\#(\text{ground states})}{\#(\text{all states})} = \frac{1}{\sqrt{\pi N/2}}\left(1 + O(1/N)\right), \tag{7.19}$$

where we used the (precise) Stirling formula for the large N behavior of the factorial. We see that the number of ground states is exponentially large in the thermodynamic limit $N \to \infty$. The entropy is now maximal, $S = N \log 2$, at $M = 0$ (essentially equal to the log of the number of ground states) and decreases when we increase the energy. At the maximal value of the energy (for $M = \pm 1$) the entropy is *zero*. Hence, from the viewpoint of the *microcanonical ensemble* it looks as if the temperature is *negative* for all energies U (cf. Sect. 2.3). This is hardly a surprise: the thermodynamical potential βF depends only on the combination βJ, so the antiferromagnetic system "looks like" the ferromagnetic one at negative temperature.

Let us now compute the canonical partition function. We write $A = 1/(2\beta|J|) > 0$. The integral representation of the partition function in the off-text box is not convergent when $J < 0$ and we have to use a different one. We consider[2]

$$Z = \sum_{N_+=0}^{N} \binom{N}{N_+} e^{-\beta|J|(2N_+ - N)^2/N} =$$

$$= \sqrt{\frac{N}{2\pi\beta|J|}} \int_{-\infty}^{+\infty} dx \sum_{N_+=0}^{N} \binom{N}{N_+} \exp\left[-\frac{N}{4\beta|J|}x^2 + i(2N_+ - N)x\right] =$$

$$= \sqrt{\frac{N}{2\pi\beta|J|}} \int_{-\infty}^{+\infty} dx \, \exp\left[-N\left(\frac{x^2}{4\beta|J|} - \frac{1}{2}\log(4\cos^2 x)\right)\right], \tag{7.20}$$

---

[2] Notice that in this integral representation only correlations which are even under $\mathbb{Z}_2$ are well-defined and real since $\mathbb{Z}_2$ acts by complex conjugation. The natural prescription for the odd correlations sets them to zero i.e. the symmetry is unbroken, as expected for anti-ferromagnets.

Fig. 7.1 Graphical solution of Eq. (7.21). The blue curve is $y = -\tan x$; the straight lines represent the function $y = Ax$ for different values of A

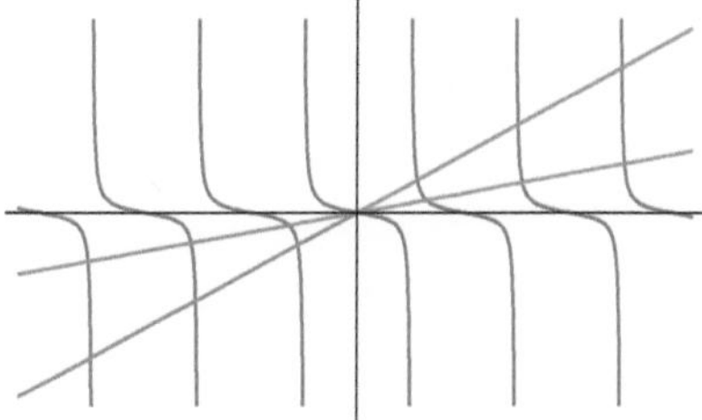

so now the saddle points are at

$$\frac{\partial}{\partial x}\left(\frac{1}{2}Ax^2 - \log(2\cos x)\right) = Ax + \tan x = 0 \tag{7.21}$$

which has infinitely many solutions for all $A > 0$ as Fig. 7.1 shows. Moreover the second derivative[3]

$$\frac{\partial^2}{\partial x^2}\left(\frac{1}{2}Ax^2 - \log(2\cos x)\right) = A + \frac{1}{\cos^2 x} \stackrel{\circ}{=} A + 1 + A^2x^2 > 0 \tag{7.22}$$

is always positive, in facts larger than 1, for $A > 0$. Hence in the thermodynamic limit we have infinitely-many local minima of the free energy. The value of the free energy at the various local minima is

$$\frac{\beta F}{N} \stackrel{\circ}{=} \frac{1}{2}\left(Ax^2 + \log(1 + A^2x^2)\right) - \log 2 \tag{7.23}$$

As $T \to 0$, $A \to 0$ and the solutions approach $x = k\pi$ for $k \in \mathbb{Z}$ with second derivative equal 1 and $\beta F/N \to -\log 2 \equiv -S/N|_{T=0}$ for all k. The function

$$\frac{\beta F}{N}(x) = \frac{1}{2}Ax^2 - \log(2\cos x) \tag{7.24}$$

is plotted in Fig. 7.2 for $A = 0.02$. This figure is typical of a frustrated system: the free energy presents *infinitely many* deep valleys separated by high barriers. The bottom of each valley is a local minimum which is a meta-stable state, while the stable (equilibrium) state corresponds to the unique absolute minimum of the free energy which is at $x = 0$. The meta-stable states decay to more stable meta-stable states, and eventually to the equilibrium state, by the thermal analogue of quantum tunneling but—given the height of the barriers between the valleys—this will (typically) require an exponential long time, so in many respects the several meta-stable states behave as equilibrium states.

---

[3] The symbol $\stackrel{\circ}{=}$ means that the equality holds only on the solutions of (7.21).

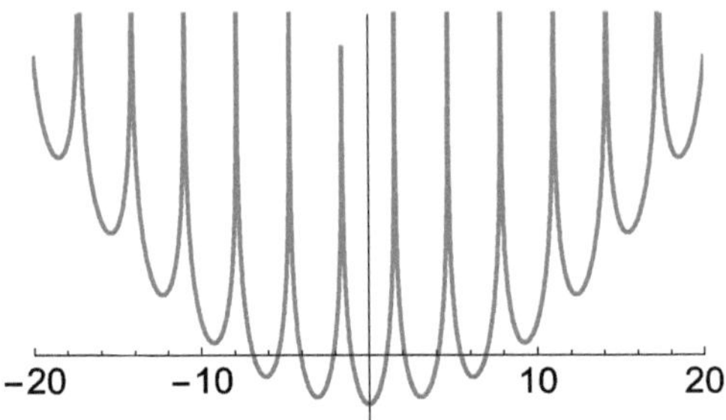

Fig. 7.2 The function (7.24) for A small (here $A = 0.02$)

For all positive temperature $T > 0$ the free energy has a unique absolute minimum at $x = 0$ which is a disordered state. Now the energy and entropy for all $T > 0$ are

$$\beta \frac{U}{N} = \beta \frac{\partial}{\partial \beta}\left(\beta \frac{F}{N}\right) = -\frac{1}{2}Ax^2 = 0 \tag{7.25}$$

$$\frac{S}{N} = \beta \frac{U}{N} - \beta \frac{F}{N} = -Ax^2 + \frac{1}{2}\log(4\cos^2 x) = \log 2, \tag{7.26}$$

i.e. the same values (7.16) as for the mean-field ferromagnetic model with $T > T_c$.

7.2 Annealed Versus Quenched Systems

We have in mind the following situation (which is realized in a long list of real-world materials[4]). For simplicity we take the microscopic degrees of freedom of our system to be standard Ising spins $\sigma_i = \pm 1$ located at the sites i of a lattice Λ. However now the lattice Λ contains also other magnetic elements, called "impurities", whose effect is to modify the couplings between nearby spins, so that we end up with a spin system governed by an Ising-like Hamiltonian of the general form

$$H[J_{ij}, \sigma_k] = -\sum_{i,j} J_{ij}\,\sigma_i \sigma_j \tag{7.27}$$

where now each coupling J_{ij} depends on the two locations i, j on the lattice as well as on the distribution of the "impurities" in the lattice Λ.

There are two different physical situations. We already considered the first one. The "impurities" may be allowed to thermalize with the spin degrees of freedom. In particular the impurities may be free to move around in the lattice, so that at

---

[4] For a partial list, see Appendix of [3].

equilibrium their distribution will be homogeneous in space, and the Ising couplings J_{ij} invariant under translations in the lattice. When this happens we say that the system is *annealed*. In this annealed situation the "impurities" are just additional degrees of freedom in the definition of the canonical ensemble, and they should be treated on the same footing as the Ising spins σ_i. The free energy F is, as usual at equilibrium, the logarithm of the canonical partition function, which is given by an expression of the schematic form

$$e^{-\beta F} = \sum_{\{\sigma_k = \pm 1\}} \mathrm{Tr}_{\mathrm{imp}}\left[e^{-\beta H_{\mathrm{imp}} - \beta H[J_{ij}(\mathrm{imp}), \sigma_k]} \right] =$$

$$= \int [DJ]\, P[J] \sum_{\{\sigma_k = \pm 1\}} e^{-\beta H[J_{ij}, \sigma_k]} = \int [DJ]\, P[J]\, Z[J]_{\mathrm{Ising}} \tag{7.28}$$

where

$$P[J] = \mathrm{Tr}_{\mathrm{imp}}\left[\exp\left(-\beta H_{\mathrm{imp}} \right) \delta\left[J_{ij} - J_{ij}(\mathrm{imp}) \right] \right] \tag{7.29}$$

is the (unnormalized) statistical distribution of the couplings J_{ij} as determined by the thermal distribution of the impurity degrees of freedom imp at equilibrium, and

$$Z[J]_{\mathrm{Ising}} = \sum_{\{\sigma_i = \pm 1\}} \exp\left[\beta \sum_{ij} J_{ij} \sigma_i \sigma_j \right] \tag{7.30}$$

is the Ising partition function for the (link dependent) couplings $\{J_{ij}\}$.

But there is a second possibility: the positions of the "impurities" are kept fixed by some external mechanism and they *do not* thermalize. Moreover the impurities are not evenly distributed through the sample of material. We can subdivide the lattice Λ into a large number L of same-size sub-lattices Λ_α which are big enough to be still macroscopic systems: each sub-lattice contains, say, M spins with

$$1 \lll M \ll N \equiv LM. \tag{7.31}$$

The free energy (per spin) of the α-th sub-lattice is given by

$$-\beta F[J^\alpha] \approx \frac{1}{M} \log \left(\sum_{\substack{\{\sigma_i = \pm 1\} \\ i \in \Lambda_\alpha}} e^{-\beta H(J^\alpha, \sigma_k)} \right) \tag{7.32}$$

where $J^\alpha \equiv (J_{ij}^\alpha)$ are the i, j-dependent couplings as determined by the particular "impurity" array present in the α-th subregion $\Lambda_\alpha \subset \Lambda$: we see $F[J^\alpha]$ as a function(al) of the coupling configuration $\{J_{ij}^\alpha\}$. Since the free energy is additive

($\equiv$ extensive), the total free energy (per spin) F of our system is

$$-\beta F = \frac{1}{N} \log \left(\sum_{\substack{\{\sigma_i = \pm 1\} \\ i \in \Lambda}} e^{-\beta H(J, \sigma_k)} \right) \approx$$

$$\approx \frac{1}{L} \sum_\alpha \frac{1}{M} \log \left(\sum_{\substack{\{\sigma_i = \pm 1\} \\ i \in \Lambda_\alpha}} e^{-\beta H(J^\alpha, \sigma_k)} \right) \equiv -\beta \frac{1}{L} \sum_\alpha F[J^\alpha], \tag{7.33}$$

where $\approx$ means that we are neglecting the $M^{-1/d}$ contributions from spins on the boundary between two sub-lattices, which is legitimate when the thermodynamic limit exists and $M \gg 1$. The distribution of "impurities" between the several subregions Λ_α then yields a probability distribution $P[J]$ for the couplings,

$$P[J][DJ] \stackrel{\text{def}}{=} \frac{1}{L} \sum_\alpha \delta[J_{ij} - J_{ij}^\alpha][DJ] \qquad L \to \infty, \tag{7.34}$$

and the RHS of Eq. (7.33) can be written simply as

$$F = \int [DJ]\, P[J]\, F[J] \equiv -\frac{1}{\beta} \int [DJ]\, P[J] \, \log Z[J]_{\text{Ising}}. \tag{7.35}$$

The main difference between this situation and the previous annealed one (7.28) is that now we average the *free energy* $F[J]$ over the couplings' probability distribution $P[J]$, whereas in the annealed case we averaged the *partition function* itself, cf. the RHS of (7.28). In this new physical situation we say that the system is *quenched* and call the RHS of (7.35) the *quenched average* (of the free energy). See the off-text box for the origin of this terminology in technology.

> **Quenched vs. Annealed in Technology**
> The two terms *annealed* and *quenched* arise from the jargon of metallurgy: *to anneal* a hot metal means to decrease its temperature very slowly, while to *quench* it means to lower its T quite rapidly. In the second case the impurities inside the metal have no time to get in thermal equilibrium with the metal, so Eq. (7.28) does not apply. The impurities get frozen in some random positions described by a probability distribution, and we have to average the thermodynamic potentials (such as the free energy) with respect to this random distribution. In metallurgy quenching and annealing are used to get metals with diverse characteristics appropriate for different technological applications.

In a quenched set-up we have two different notions of statistical average. The first one, written as $\langle O \rangle$ (or $\langle O \rangle[J]$ when we wish to emphasize the dependence on the random couplings J), is the usual canonical expectation value obtained by summing over the spin configurations at *fixed values* of the coupling $J \equiv (J_{ij})$. $\langle O \rangle[J]$ is then a functional of the coupling configuration J. The second notion is the expectation value with respect to the statistical distribution $P[J]$ of the couplings themselves: this average is denoted by an overbar

$$\overline{\mathcal{A}} \overset{\text{def}}{=} \int [DJ]\, P[J]\, \mathcal{A}[J]. \tag{7.36}$$

Averaging over both distributions gives the final expectation value $\overline{\langle O \rangle}$. In this notation Eq. (7.35) reads

$$\beta F = -\overline{\log Z} \quad \text{where} \quad Z \equiv Z[J]_{\text{Ising}}. \tag{7.37}$$

Usually one assumes that the couplings $\{J_{ij}\}$ obey a Gaussian distribution $P[J]_{\text{Gauss}}$ whose mean and variance are invariant under lattice translations.

The Replica Trick

The reader may easily convince herself that computing the quenched average (7.35) is much harder than to compute an annealed average (7.28) (i.e. a standard canonical partition function). Indeed it is typically easier to compute integrals of exponentials than integrals of logarithms. We have a big box of tools to compute integrals of exponentials: saddle-point methods, Stokes analytic methods, Ward-like identities, Schwinger-Dyson equations, etc. For integrals of logarithms our tool box is almost empty. Therefore any dirty trick which may convert quenched averages into annealed ones is very welcomed!

The *replica trick* performs the desired black magics: when it works, it allows us to compute a quenched mean using an annealed ensemble. One observes that

$$\log Z = \lim_{n \to 0} \frac{Z^n - 1}{n}. \tag{7.38}$$

When n is a positive integer, Z^n is just the canonical partition function of n identical copies of the system (called *replicas*). This suggests a three-steps procedure to compute quenched averages known as the *replica trick*. In the first step one computes

$$\int [DJ] P[J]\, Z[J]^n \equiv \sum_{\{\sigma_k^a = \pm 1\}} \mathrm{Tr}_{\text{imp}}\left[e^{-\beta H_{\text{imp}} - \beta \sum_{a=1}^{n} H[J_{ij}(\text{imp}), \sigma_k^a]} \right], \tag{7.39}$$

which is the annealed ensemble of a system consisting of n *replicas* $\{\sigma_i^1\}, \ldots, \{\sigma_i^n\}$ of our spin system $\{\sigma_i\}$ coupled to the "impurity" degrees of freedom **imp**. In the second step one analytically continues the result from n integer to n real (or complex). Finally, in the third step one takes the limit

$$\lim_{n \to 0} \frac{1}{n} \left(\int [DJ]\, P[J]\, Z[J]^n - 1 \right) = \int [DJ]\, P[J]\, \log Z[J] \tag{7.40}$$

which is the *quenched average* we wanted to compute. All very nice, except that the correct story is a bit more subtle as we are going to explain.

Subtleties with the Replica Trick
In the thermodynamic limit the actual quantity of interest is

$$-\beta \overline{F} = \lim_{N \to \infty} \frac{1}{N} \lim_{n \to 0} \frac{1}{n} \left(\overline{Z_N^n} - 1 \right) \tag{7.41}$$

where the proper order of limits is first $n \to 0$ and then $N \to \infty$. What is easy to compute (thanks to saddle-point methods[5]) is the large N asymptotics of $\overline{Z_N^n}$ for *fixed integral n*, that is, the limit computed in the *reverse order*. If the two limits commute, the computation becomes (reasonably) simple, but if they do not commute we may expect trouble and big surprises.

If the system has only one pure state (as it happens at high temperature) the limits are expected to commute, but in presence of multiple pure states it looks rather unlikely that they commute. We known that a potential source of problems with the uniqueness of the thermodynamic limit is the spontaneous breaking of a symmetry. The n-replica model has a symmetry $\mathfrak{S}_n$ given by the permutation of the n identical copies. If this symmetry is *unbroken,* for large but finite N all replicas will be in the same "approximate" phase, that is, their probability measures will be mainly concentrated near the same saddle point, and the dependence of the partition function on n for large but finite N will be "easy" to compute, the analytic continuation in n straightforward, and the limits will commute. But when the replica symmetry $\mathfrak{S}_n$ is spontaneously broken—possibly in a multitude of different patterns with a variety of possible unbroken subgroups $\mathfrak{S} \subset \mathfrak{S}_n$—for large N the several copies will be in *distinct* "approximate" phases. For n a positive integer we get a distribution function describing how many replicas are in each such phase. In this situation the dependence of the free energy from n becomes a very intricate mess, its analytic continuation rather subtle, and the limits are *not* expected to commute.

---

[5] See Sect. 2.5.

7.3   Spin Glasses: The Sherrington-Kirkpatrick Model

Spin glasses are quenched systems where frustration plays a major role.[6]

The simplest model of a spin glass is the *Sherrington-Kirkpatrick (SK) model* [5].[7] Its Hamiltonian is

$$\beta H[\sigma_i, J_{kl}] = - \sum_{1 \leq i < j \leq N} J_{ij}\, \sigma_i \sigma_j - B \sum_i \sigma_i \tag{7.42}$$

where $\sigma_i = \pm 1$ are Ising spins, and the coupling constants $\{J_{ij}\}_{i<j}$ are quenched independent random variables distributed with the same Gaussian probability $P[J_{ij}]$ of mean J_0/N and standard deviation $K/\sqrt{N}$

$$P[J_{ij}] = \left(\frac{K^2}{2\pi N} \right)^{N(N-1)/4} \exp\left[-N \frac{\sum_{i<j}(J_{ij} - J_0/N)^2}{2K^2} \right] \tag{7.43}$$

for all pairs $i < j$. The particular dependence on N of the probability measure $P[J_{ij}]$ is chosen so that the thermodynamic limit $N \to \infty$ exists and is non-trivial. $K \propto \beta$ and we may take the two equal with no loss. Applying the *replica trick* we get

$$-\beta \overline{F} = \lim_{N \to \infty} \frac{1}{N} \int [DJ_{ij}]\, P[J_{ij}] \log \sum_{\{\sigma_i = \pm 1\}} e^{-\beta H[\sigma_i, J_{kl}]} =$$
$$= \lim_{N \to \infty} \frac{1}{N} \lim_{n \to 0} \frac{1}{n} \left[\overline{Z_n} - 1 \right], \tag{7.44}$$

where for $n \in \mathbb{N}$

$$\overline{Z_n} = \int [DJ_{ij}]\, P[J_{ij}] \sum_{\{\sigma_i^a = \pm 1\}} \exp\left[\sum_{a=1}^{n} \left(\sum_{i<j} J_{ij}\, \sigma_i^a \sigma_j^a + B \sum_i \sigma_i^a \right) \right], \tag{7.45}$$

---

[6] For a review of spin glass theory with a summary of recent developments see [4].

[7] For a recent review on the developments about the SK model see [6].

the superscript $a = 1, \ldots, n$ labels the different replicas, and $i, j = 1, \ldots, N$. Performing the Gaussian integral in J_{ij} we get

$$\overline{Z_n} = \int [DJ_{ij}] \, P[J_{ij}] \sum_{\{\sigma_i^a = \pm 1\}} \exp\left(\sum_{a=1}^{n} \sum_{i<j} J_{ij} \sigma_i^a \sigma_j^a + B \sum_{a=1}^{n} \sum_{i=1}^{N} \sigma_i^a \right) =$$

$$= \sum_{\{\sigma_i^a = \pm 1\}} \exp\left[\frac{K^2}{2N} \sum_{a,b=1}^{n} \sum_{i<j} \sigma_i^a \sigma_j^a \sigma_i^b \sigma_j^b + \sum_{a=1}^{n} \left(\frac{J_0}{N} \sum_{i<j} \sigma_i^a \sigma_j^a + B \sum_{i=1}^{N} \sigma_i^a \right) \right]$$

$$\tag{7.46}$$

We have the elementary identity

$$\sum_{a,b=1}^{n} \sum_{i<j} \sigma_i^a \sigma_j^a \sigma_i^b \sigma_j^b = \sum_{a<b} \left(\sum_{i=1}^{N} \sigma_i^a \sigma_i^b \right) \left(\sum_{j=1}^{N} \sigma_j^a \sigma_j^b \right) + \frac{1}{2} n N^2 - n^2 N, \tag{7.47}$$

where we may ignore the last term since it does not contribute to the thermodynamic limit. Next we rewrite the RHS of (7.46) as a Gaussian integral in the form

$$\mathcal{N} e^{nK^2 N/4} \int [d\Phi_{ab}][dy_a] \exp\left[-N \left(\frac{1}{2K^2} \sum_{a<b} \Phi_{ab}^2 + \frac{1}{4J_0} \sum_a y_a^2 \right) \right] \times$$

$$\times \prod_{i=1}^{N} \sum_{\{\sigma_i^a = \pm 1\}} \exp\left[\sum_{a<b} \Phi_{ab} \, \sigma_i^a \sigma_i^b + \sum_a (B + y_a) \sigma_i^a \right] =$$

$$= \mathcal{N} e^{nK^2 N/4} \int [d\Phi_{ab}][dy_a] \exp\left[-N \left(\frac{1}{2K^2} \sum_{a<b} \Phi_{ab}^2 + \frac{1}{4J_0} \sum_a y_a^2 \right) \right] \times$$

$$\times \left(\sum_{\{\sigma^a = \pm 1\}} \exp\left[\sum_{a<b} \Phi_{ab} \, \sigma^a \sigma^b + \sum_a (B + y_a) \sigma^a \right] \right)^N =$$

$$= \mathcal{N} e^{nK^2 N/4} \int [d\Phi_{ab}][dy_a] \exp\left[-N A[\Phi_{ab}, y_a] \right]$$

$$\tag{7.48}$$

where $\mathcal{N}$ is an inessential normalization constant and

$$
A[\Phi_{ab}, y_a] = \frac{1}{2K^2} \sum_{a<b} \Phi_{ab}^2 - \frac{nK^2}{4} + \frac{1}{4J_0} \sum_{a=1}^{n} y_a^2 -
$$
$$
- \log \left(\sum_{\{\sigma^a = \pm 1\}} \exp \left[\sum_{a<b} \Phi_{ab}\, \sigma^a \sigma^b + \sum_{a=1}^{n} \left(B + y_a \right) \sigma^a \right] \right).
$$
$$(7.49)$$

In conclusion we have

$$
- \beta \overline{F} = \lim_{N \to \infty} \frac{1}{N} \lim_{n \to 0} \frac{1}{n} \left[\mathcal{N} \int [\mathrm{d}\Phi_{ab}][\mathrm{d}y_a] \, e^{-NA[\Phi, y]} - 1 \right].
\qquad (7.50)
$$

Note that if we take the limit $N \to \infty$ while keeping n a fixed positive integer (that is, for the n copies annealed system), we have

$$
Q_{ab} \stackrel{\mathrm{def}}{=} \frac{\Phi_{ab}}{K^2} = \overline{\langle \sigma^a \rangle \langle \sigma^b \rangle} = \lim_{N \to \infty} \frac{1}{N} \sum_{i} \overline{\langle \sigma_i^a \rangle \langle \sigma_i^b \rangle} \quad a \neq b,
\qquad (7.51)
$$

$$
m_a \stackrel{\mathrm{def}}{=} \frac{y_a}{2J_0} = \overline{\langle \sigma^a \rangle} = \lim_{N \to \infty} \frac{1}{N} \sum_{i} \overline{\langle \sigma_i^a \rangle}.
\qquad (7.52)
$$

These quantities look like order parameters. m_a is the a-th replica magnetization and $m_a \neq 0$ means that the a-th replica is in a ferromagnetic phase. $Q_{ab} \neq 0$ means that the system has "some new kind" of magnetic order. Roughly speaking the *spin glass phase* is the peculiar magnetic order where Q_{ab} is non-zero while $m_a = 0$. However to give a precise meaning to the notion of "spin glass phase" we need first to dwell with the intricacies of the $n \to 0$ limit. We shall see below that the issue of order parameters for the spin glass phase is extremely subtle.

To simplify the story we mainly focus on the basic situation $B = J_0 = 0$ leaving the general case to the diligent reader (who may consult the extensive literature on the subject where the general case is analyzed in detail [3, 5]). When needed, we switch back the magnetic field B for the sake of comparison. When $J_0 = 0$ we can set $y_a \equiv 0$ and in Eq. (7.50) the functional $A[\Phi, y]$ simplifies to

$$
A[\Phi] = -\frac{nK^2}{4} + \frac{1}{2K^2} \sum_{a<b} \Phi_{ab}^2 - \log \left(\sum_{\{\sigma^a = \pm 1\}} \exp \left[\sum_{a<b} \Phi_{ab}\, \sigma^a \sigma^b \right] \right).
\qquad (7.53)
$$

The functional $A[\Phi]$ is a function of the $n(n-1)$ parameters Φ_{ab} with $a < b$.

Internal Energy   From thermodynamics we know that

$$\overline{U} = \frac{\partial}{\partial \beta}(\beta \overline{F}) \tag{7.54}$$

Since in suitable units $\beta = K$ we get

$$\overline{U} = -\frac{K}{2} - \frac{1}{2K^3} \lim_{n \to 0} \frac{1}{n} \mathrm{Tr}\, \Phi^2. \tag{7.55}$$

Magnetic Susceptibility   We compute the magnetic susceptibility χ at $B = 0$ following [7, 8]. From linear response theory we know that

$$\frac{\chi}{\beta} = \sum_j \left[\langle \sigma_i \sigma_j \rangle - \langle \sigma_i \rangle \langle \sigma_j \rangle \right] \tag{7.56}$$

We average this equation over the distribution $P[\boldsymbol{J}]$. The terms with $i \neq j$ vanish by $\mathbb{Z}_2$ gauge invariance, and we remain with

$$\chi = \beta \left(1 - \overline{\langle \sigma_i \rangle \langle \sigma_i \rangle} \right) \quad \text{not summed over } i! \tag{7.57}$$

From Eq. (7.51) we expect

$$\chi = \beta \left(1 - \lim_{n \to 0} \frac{1}{n} \sum_{a<b} Q_{ab} \right) \tag{7.58}$$

a formula which may be rigorously established.

A **Fancy** *Variational Problem*

In the limit $N \to \infty$ the integral inside the large bracket in (7.50) can be easily computed by the saddle-point technique. Thus, ***if*** *inverting the order of the limits is legitimate,* we can easily get the exact solution of the SK model. While this inversion may look suspicious, it is expected to be valid for large enough temperatures, i.e. for small K. The result of the inversion of limits is

$$\beta \overline{F} = \lim_{n \to 0} \left[\frac{1}{n} \min_\Phi A[\Phi] \right]. \tag{7.59}$$

A moment reflection gives the expression that we would get keeping the proper order of limits

$$\beta\overline{F} = {}^{\text{``}}\min_{\Phi}\left[\lim_{n\to 0}\frac{1}{n}A[\Phi]\right]{}^{\text{''}}, \qquad (7.60)$$

that is, we have to solve the variational problem of extremizing the putative free energy *directly at* $n = 0$ instead of analytically continue down to $n = 0$ the solution of the variational problem for n positive integral which is the *naive* prescription in Eq. (7.59). We have put the RHS of (7.60) between quotation marks because we have still to give a precise meaning to it.

The function(al) $A[\Phi]$ depends on $n(n-1)/2$ parameters. As we approach $n = 0$ from the above, the number of parameters gets *negative*: that is, we are confronted with the variational problem of "minimizing" a function depending on a *negative* number of variables. More precisely: to "minimize" a function defined in a vector space of *negative dimension* $-\epsilon(1 - \epsilon)/2$. This is a fancy and very subtle problem. We can think of several *inequivalent* ways of defining the notion of "minimum" in such an esoteric set-up. *What is the meaning of the variational problem* (7.60)?

To give a definite prescription for this peculiar variational problem we need to address two issues:

Var1   the space (ensemble) $\mathfrak{E}$ of *"the* 0×0 *symmetric matrices* Φ_{ab} *with zeros along the main diagonal"* in which we have to search for the "minimum" of the functional $\widetilde{A}: \mathfrak{E} \to \mathbb{R}$ where

$$\widetilde{A}[\cdot] \stackrel{\text{def}}{=} \lim_{n\to 0}\frac{1}{n}A[\cdot]; \qquad (7.61)$$

Var2   what we really mean by the RHS of Eq. (7.60), that is: what it means that the point $\Phi_0 \in \mathfrak{E}$ realizes the "minimum" of the functional $\widetilde{A}: \mathfrak{E} \to \mathbb{R}$?

Both issues have rather surprising answers.

Var1 The ensemble $\mathfrak{E}$—which at first sight has dimension ≤ 0—is quite *big*, in facts has dimension $+\infty$. $\mathfrak{E}$ is just the *largest space* on which we can define the functional $\widetilde{A}[\cdot]$. The rationale for this definition is physically obvious: in principle any element $\mathring{\Phi} \in \mathfrak{E}$ yields a legitimate candidate free energy $\widetilde{A}[\mathring{\Phi}]/\beta$ and we have no right to rule it out *a priori:* we have to duly check whether $\widetilde{A}[\mathring{\Phi}]/\beta$ satisfies the variational conditions to be the *actual* free energy. How we work *in practice* with such a huge space? In the usual way: recall the story of the Hamilton variational principle in mechanics [9]: one gets the Euler-Lagrangian equations (and Jacobi's second variation operator) by considering *arbitrary finite-dimensional* subfamilies in the infinite-dimensional variational ensemble $\mathfrak{E}$. Therefore our task is to describe *all* finite-dimensional subfamilies of $\mathfrak{E}$ rather than $\mathfrak{E}$ *per se*. We write $\mathring{S}(n)$ for the vector space of symmetric real $n \times n$ matrices with zeros along the main diagonal.

Definition 7.1 A *k-dimensional subfamily in* $\mathfrak{E}$ is an infinite sequence of matrices

$$\mathbf{\Phi}(\boldsymbol{\phi}) \equiv \left\{ \Phi^{(n)}(\phi^1, \ldots, \phi^k) \in \mathring{S}(n), \ n \in \mathbb{N} \text{ with } n_0 \mid n \right\} \quad \text{for some } n_0 \in \mathbb{N},$$

$$(7.62)$$

which depend smoothly on k real parameters $\phi^1, \ldots, \phi^k$, with the property that all the $\mathfrak{S}_n$ invariants of $\Phi^{(n)}$ such as (for instance)

$$\text{Tr}\left(\Phi^{(n)}(\phi^1, \ldots, \phi^k) \right)^\ell = f_\ell(n; \phi^1, \ldots, \phi^k), \quad \ell \in \mathbb{N} \tag{7.63}$$

extend to analytic functions of n defined in a neighborhood of zero. If $\mathbf{\Phi}(\boldsymbol{\phi})$ is such a subfamily, we set

$$\widetilde{\text{Tr}}\left(\mathbf{\Phi}(\boldsymbol{\phi}) \right)^\ell = \lim_{n \to 0} \frac{1}{n} f_\ell(n; \phi^1, \ldots, \phi^k). \tag{7.64}$$

In particular the functional $A[\Phi]$ is a $\mathfrak{S}_n$ invariant, so each finite-dimensional subfamily $\mathbf{\Phi}(\boldsymbol{\phi})$, for which the limit

$$\widetilde{A}[\mathbf{\Phi}(\boldsymbol{\phi})] \stackrel{\text{def}}{=} \lim_{n \to 0} \left(\frac{1}{n} A[\Phi^{(n)}(\phi^1, \ldots, \phi^k)] \right) \tag{7.65}$$

exists, gives a well-defined function $\widetilde{A}[\boldsymbol{\phi}]$ of the parameters $\boldsymbol{\phi} \equiv (\phi^1, \ldots, \phi^k)$ which is a perfectly legitimate finite-dimensional family of candidates for $\beta \overline{F}$. It remains to determine which one of the several candidates yields the correct quenched free energy. We know that the actual free energy has a variational characterization. A necessary condition for $\boldsymbol{\phi}_* = (\phi_*^1, \cdots, \phi_*^k)$ to give the actual quenched free energy $\widetilde{A}[\mathbf{\Phi}(\boldsymbol{\phi}_*)]/\beta$ is that it extremizes the functional

$$\left. \frac{\partial}{\partial \Phi_{ab}} \widetilde{A}[\mathbf{\Phi}(\boldsymbol{\phi})] \right|_{\boldsymbol{\phi}=\boldsymbol{\phi}_*} = 0. \tag{7.66}$$

The meaning of this equation, however, is subtle on several counts. In addition to the new phenomena described below in issue **Var2**, we mention the following point. Given a k-dimensional family we can construct ℓ-dimensional subfamilies with $\ell < k$ by specialization of the parameters ϕ^j. Therefore we can speak of the set of families containing a given family as a specialization. Equation (7.66) should be valid not just along the particular family, but also in any bigger family containing it for the same $\boldsymbol{\phi}_*$. I.e. the derivatives along directions which are orthogonal to the particular finite-dimensional family should also vanish.

Var2 We know from Chap. 2 that, in order for the thermodynamic limit to exist, the thermodynamic potential should be convex. This is a physical requirement which generalizes the condition that the heat capacity should be non-negative (cf.

Chap. 1). However the notion of being "convex" is subtle for a function defined in a space of (formally) negative dimension. We borrow the following example from [8]. Consider the function

$$h(\Phi_{ab}) = \frac{1}{n}\text{Tr}(\Phi^2) = \frac{2}{n}\sum_{1 \le a < b \le n} \Phi_{ab}^2 \tag{7.67}$$

where $\Phi_{ab} \in \mathring{S}(n)$ are $n \times n$ symmetric matrices with 0's along the diagonal. As an operator, the Hessian ($a < b, a' < b'$)

$$\frac{\partial^2 h(\Phi_{cd})}{\partial \Phi_{ab}\, \partial \Phi_{a'b'}} = \frac{4}{n}\, \delta_{aa'}\, \delta_{bb'} \tag{7.68}$$

is a positive multiple of the identity in the vector space $\mathring{S}(n)$, hence all its eigenvalues are positive, so one would conclude that the critical point $\Phi_{ab} = 0$ is a minimum of $h(\Phi_{ab})$: this is certainly the case in *positive* dimension. But consider the matrix

$$\phi_{ab} = \begin{cases} \phi & \text{for } a \ne b \\ 0 & \text{for } a = b \end{cases} \tag{7.69}$$

In this case

$$h(\phi_{ab}) = (n-1)\phi^2 \xrightarrow{\ n \to 0\ } -\phi^2, \tag{7.70}$$

so $\phi = 0$ is actually a *maximum* of the function. The point is that the trace of the identity is negative in negative dimension! In positive dimension the two statements

(i) the critical point x_0 of the smooth function $h(x)$ is a local minimum $h(x)$
(ii) the Hessian operator of $h(x)$ has positive eigenvalues at the critical point x_0

are equivalent. In "negative" dimension they are mutually exclusive. Then which one of the two is the proper definition of "minimum" in "negative" dimension?

The eigenvalues of the Hessian—seen as a linear operator not as a bilinear form!—are the susceptibilities in the sense of linear response theory: indeed by their very physical definition the susceptibilities are linear maps between the external field and the induced order parameter. Physics requires the susceptibilities to be positive (heating up a body should increase its temperature!), and positivity of the susceptibilities is what we mean physically by "convexity". In a space of positive dimension this notion agrees with the usual mathematical concept of convexity.

This physical notion of "convexity" is our guiding principle—*Alas! We certain do not expect that heating up a spin-glass will make its temperature smaller!*—and even less that its entropy decreases. The heat capacity and the other susceptibilities should be non-negative, hence the *eigenvalues of the Hessian should be non-negative at the "minimum"* ϕ_*. As the example (7.70) shows, this means that:

At its "minimum" the function(al) $\widetilde{A}[\phi]$ *has a maximum*

or, in Parisi's own words:

> A careful analysis shows that the free energy A must be maximized with respect to all variables [10, page 322]
> Detailed arguments suggest that we should look for the maximum of the free energy as a function of the [parameters] [10, page 323]

Thus the physically correct requirement is that the free energy at the critical point should have a Hessian

$$\frac{\partial^2 \widetilde{A}}{\partial \Phi_{ab} \, \partial \Phi_{a'b'}} \tag{7.71}$$

whose eigenvalues are non-negative in *all* directions. *All* means that the eigenvalues should be computed in *any* family containing the one we used to find the critical value, so that our Hessian has *infinitely many* eigenvalues, all of which should be non-negative! Pragmatically this means the following [11]: we compute the eigenvalues of the Hessian as analytic functions of n and require that their analytic continuation to $n = 0$ is non-negative. This is a bit tricky because the multiplicities of most of these eigenvalues (typically given by polynomials of degree ≤ 2 in n) become zero or negative as $n \to 0$ since the sum of their multiplicities should be zero in the $n \to 0$ limit. Yet physical "convexity" says that a negative eigenvalue of any multiplicity, *even zero or negative,* is an instability signaling that the point ϕ_* is not the correct solution to the variational problem which computes the free energy of the model.

The (Naive) Symmetric Solution

We consider first the solution obtained by inverting the order of limits due to Sherrington-Kirkpatrick (SK) [5]. While it is not guaranteed to be the correct one, it is certainly valid at high enough temperatures, i.e. above some critical temperature T_c. Using the previous variational characterization of the correct solution, we can also read from this naive solution the correct value of the critical temperature T_c below which its ceases to yield the valid free energy.

We formulate this solution not in the original "naive" fashion, but using the fancy variational language of **Var1,Var2** for uniformity with the successive improved treatments. We change normalization of the variables and consider a family

$$Q(q) = \frac{1}{K^2}\Phi(q) \quad \text{and set } K = \beta, \tag{7.72}$$

so that our variables Q_{ab} and q_a are order parameters in the sense of (7.51).

We first look for a family $Q(q)$ which is invariant under the replica symmetry $\mathfrak{S}_n$, that is,

$$Q_{ab}^{SK} = \begin{cases} q & a \neq b \\ 0 & a = b. \end{cases} \tag{7.73}$$

The first step is to compute the expression inside the large parenthesis in Eq. (7.53)

$$\sum_{\{\sigma^c=\pm 1\}} \exp\left[\beta^2 q \sum_{a<b}\sigma^a\sigma^b\right] = \sum_{\{\sigma^c=\pm 1\}} \exp\left[\beta^2 \frac{q}{2}\left(\sum_a \sigma^a\right)^2 - n\frac{q}{2}\right] =$$

$$= e^{-n\beta^2 q/2} \sum_{n_+=0}^{n}\binom{n}{n_+} e^{\beta^2 q(2n_+-n)^2/2} \tag{7.74}$$

where n_+ is the number of *up* spins. The sum in the RHS is evaluated around Eq. (7.17)

$$e^{-n\beta^2 q/2} \sum_{n_+=0}^{n}\binom{n}{n_+} e^{\beta^2 q(2n_+-n)^2/2} = e^{-n\beta^2 q/2}\int_{-\infty}^{+\infty} \frac{dz}{\sqrt{2\pi}}\, e^{-z^2/2+n\log[2\cosh(\beta\sqrt{q}\,z)]}$$

$$\tag{7.75}$$

that is, $A[q; n]$ is equal to

$$-\frac{n\beta^2}{4} + \frac{\beta^2}{4}n(n-1)q^2 + \frac{\beta^2}{2}nq - \log\int_{-\infty}^{+\infty}\frac{dz}{\sqrt{2\pi}}e^{-z^2/2+n\log(2\cosh(\beta\sqrt{q}\,z))} =$$

$$= -\frac{n\beta^2}{4} + \frac{\beta^2}{4}n(n-1)q^2 + \frac{\beta^2}{2}nq -$$

$$- \log\left(1 + n\int_{-\infty}^{+\infty}\frac{dz}{\sqrt{2\pi}}e^{-z^2/2}\log(2\cosh(\beta\sqrt{q}\,z)) + O(n^2)\right)$$

$$\tag{7.76}$$

so that the Sherrington-Kirpatrick free energy is

$$\beta \overline{F(q)}_{\mathrm{SK}} \equiv \lim_{n \to 0} \frac{1}{n} A[q; n] = -\frac{\beta^2}{4}\left(1 - q\right)^2 - \int_{-\infty}^{+\infty} \frac{dz}{\sqrt{2\pi}} e^{-z^2/2} \log[2\cosh(\beta\sqrt{q}\,z)].$$

$$(7.77)$$

The "self-consistency" condition is obtained by setting to zero the derivative of $\overline{F(q)}_{\mathrm{SK}}$ with respect to q. We have

$$2\frac{\partial}{\partial q}\int_{-\infty}^{+\infty} \frac{dz}{\sqrt{2\pi}}\, e^{-z^2/2} \log[2\cosh(\beta\sqrt{q}\,z)] = \beta \int_{-\infty}^{+\infty} \frac{dz}{\sqrt{2\pi}} e^{-z^2/2} \frac{z}{\sqrt{q}} \tanh(\beta\sqrt{q}z)$$

$$= \beta^2 \int_{-\infty}^{+\infty} \frac{dz}{\sqrt{2\pi}} \frac{e^{-z^2/2}}{\cosh^2(\beta\sqrt{q}z)} = \beta^2 - \beta^2 \int_{-\infty}^{+\infty} \frac{dz}{\sqrt{2\pi}} e^{-z^2/2} \tanh^2(\beta\sqrt{q}z)$$

$$(7.78)$$

where in the second equality we integrated by parts (i.e. used the Gaussian Ward identity). Putting everything together we get the equation for q

$$q = \int_{-\infty}^{+\infty} \frac{dz}{\sqrt{2\pi}} e^{-z^2/2} \tanh^2\left(\beta\sqrt{q}z\right) \leq \beta^2 q, \tag{7.79}$$

so that for $\beta < \beta_c \equiv 1$, that is at high temperature, the only solution is $q = 0$ which is the ordinary paramagnetic phase. A positive solution q with a given value of $\phi \equiv \beta^2 q > 0$ to Eq. (7.79) exists at a unique temperature β^{-1}; indeed

$$\frac{1}{\beta^2} = \int_{-\infty}^{+\infty} \frac{dz}{\sqrt{2\pi}} e^{-z^2/2} \frac{\tanh^2(\sqrt{\phi}z)}{\phi} \equiv \frac{1}{\phi} - \frac{1}{\phi}\int_{-\infty}^{+\infty} \frac{dz}{\sqrt{2\pi}} \frac{e^{-z^2/2}}{\cosh^2(\sqrt{\phi}z)} \tag{7.80}$$

where the function of ϕ in the RHS is strictly monotonically decreasing from 1 at $\phi = 0$ to 0 at $\phi = \infty$. From Eq. (7.77) it is clear that—when it exists—the solution with $q > 0$ has lower free energy than the $q = 0$ one, and so the last one should be ruled out. As $\beta \to \infty$ (i.e. $T \to 0$) $q \to 1$.

Definition 7.2 In the SK model the phase where $m = 0$ but $q \neq 0$ is a new phase called the *spin glass phase*. The higher temperature phase with $m = q = 0$ is the usual *paramagnetic phase*.

Remark 7.1 Up to averaging with a Gaussian distribution, Eq. (7.79) coincides with the mean field equation for the Ising model (which is exact in infinite dimension which is the case for the SK model). In presence of a non-zero magnetic field (but still $J_0 = 0$) the only modification of Eq. (7.79) is a shift of the argument of $\tanh^2$

$$\beta\sqrt{q}z \rightsquigarrow \beta\sqrt{q}z + \beta B. \tag{7.81}$$

We compute some easy quantity in the simplified SK model. Since

$$\frac{1}{\cosh^2(\beta\sqrt{q}\,z)} \rightarrow \frac{2}{\beta\sqrt{q}}\,\delta(z) \quad \text{as } \beta \rightarrow \infty, \tag{7.82}$$

for T *small* (β large)

$$q \approx 1 - \frac{2}{\sqrt{2\pi}}\beta^{-1}. \tag{7.83}$$

Near the critical temperature q is small and we may compute the integrals in Eq. (7.80) as a power series in q (for $\beta \approx 1$)

$$q = \int\limits_{-\infty}^{+\infty} \frac{dz}{\sqrt{2\pi}}\,e^{-z^2/2}\,\tanh^2(\beta\sqrt{q}z) = \beta^2 q - 2\beta^4 q^2 + \frac{17}{3}\beta^6 q^3 + O(q^4) \tag{7.84}$$

Hence as long as the temperature is smaller than T_c, q is non-zero: for $\beta = \beta_c + \epsilon$

$$q \approx \frac{\beta^2 - \beta_c^2}{2\,\beta^4} \approx \epsilon. \tag{7.85}$$

The energy is

$$\overline{U} = \frac{\partial}{\beta}(\beta\overline{F}) = -\frac{\beta}{2}\left(1 - q^2\right) \tag{7.86}$$

so in the high temperature regime, where $q = 0$, $\overline{U} = -\beta/2$, while at $T = 0$

$$U\Big|_{T=0} = -\frac{2}{\sqrt{2\pi}} \approx -0.7979. \tag{7.87}$$

The entropy is

$$\overline{S} = K\frac{\partial}{\partial K}(\beta\overline{F}) - (\beta\overline{F}) =$$

$$= -\frac{K^2}{4} - \frac{q}{2} + \frac{3}{2}\frac{q^2}{K^2} + \int\limits_{-\infty}^{+\infty} \frac{dz}{\sqrt{2\pi}}e^{-z^2/2}\,\log[2\cosh(\sqrt{q}z)] \tag{7.88}$$

which in the higher temperature phase is simply

$$\overline{S}\Big|_{K<K_c} = -\frac{K^2}{4} + \log 2. \tag{7.89}$$

Detailed computations show that at zero temperature the entropy becomes *negative*

$$\overline{S}\Big|_{T=0} = -\frac{1}{2\pi} \tag{7.90}$$

a pathology which indicates that the solution is *not* correct for T near zero. This means the for $T < T_c$ the solution we found is a saddle point of the free energy which is not really a "minimum" (that is, a maximum). In other words: while it is a maximum in the one-parameter family we have considered, it is not a "minimum" in the full ensemble $\mathfrak{E}$, and we must enlarge the family in which we look for the variational solution.

Instability of the SK Solution

The instability of the above "naive" SK solution was demonstrated by de Almeida and Thouless [11] which computed the eigenvalues of the Hessian at the SK solution for any finite n, then analytically continued them to $n = 0$ and found that $n(n-3)/2$ of them become negative when (setting $J_0 = B = 0$ for simplicity)

$$\frac{1}{\beta^2} < \int_{-\infty}^{+\infty} \frac{dz}{\sqrt{2\pi}} \frac{e^{-z^2/2}}{\cosh^4(\beta\sqrt{q}\,z)}. \tag{7.91}$$

Note that the RHS is always ≤ 1. Thus, in particular, when $\beta < \beta_c \equiv 1$ this inequality is not satisfied and there are no negative eigenvalues, hence the SK solution is *stable* when $T > T_c$, so it describes the correct physics in the high temperature phase. Now

$$\frac{1}{\cosh^4(\beta\sqrt{q}z)} \rightarrow \frac{4}{3\,\beta\sqrt{q}}\,\delta(z) \quad \text{as } \beta \rightarrow \infty,$$

$$\Rightarrow \quad \int_{-\infty}^{+\infty} \frac{dz}{\sqrt{2\pi}} \frac{e^{-z^2/2}}{\cosh^4(\beta\sqrt{q}\,z)} \approx \frac{4}{3\beta}\frac{1}{\sqrt{2\pi q}}, \quad \beta \rightarrow \infty \tag{7.92}$$

and the inequality is certain satisfied for low T, hence the SK solution is ***not*** correct at low temperatures. Just below the critical temperature $\beta^2 \gtrsim \beta_c^2 \equiv 1$, we have

$$\int_{-\infty}^{+\infty} \frac{dz}{\sqrt{2\pi}} \frac{e^{-z^2/2}}{\cosh^4(\beta\sqrt{q}z)} = 1 - 2\beta^2 q + 7\beta^4 q^2 + O(q^3), \tag{7.93}$$

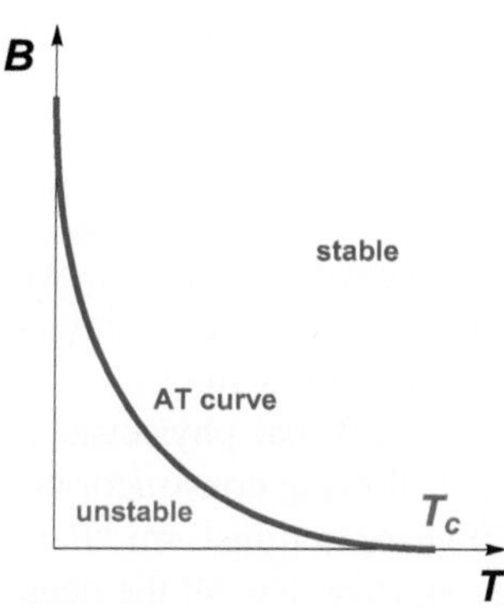

Fig. 7.3 The AT curve [11] in the T-B plane which separates the regions where the SK solution is stable from the region where it is unstable (for the case $J_0 = 0$)

while Eq. (7.84) gives (for $\beta \approx 1$)

$$\frac{1}{\beta^2} = 1 - 2\beta^2 q + \frac{17}{3}\beta^4 q^2 + O(q^3) \tag{7.94}$$

hence as soon as the temperature $T < T_c$ $q \neq 0$ and the inequality (7.91) gets satisfied. Below the critical temperature the SK solution becomes unstable, hence not the physically correct solution (however the solution remains still "qualitatively" good for temperatures as low as $\approx 0.4\,T_c$ [8]). For the more general analysis where J_0 and B are not set to zero see the original paper by de Almeida and Thouless [11]. We borrow from Ref. [11] the phase diagram in Fig. 7.3 which is valid for $J_0 = 0$ but with both B and T arbitrary.

Below the AT curve (Fig. 7.3) we are in a new phase with peculiar properties: in particular the free energy is nowhere analytic in the unstable region.

Reference [11] describes explicitly the $n(n-3)/2$ eigenvectors of the Hessian matrix (7.71) which become negative as $n \to 0$ below the AT line (i.e. for $\beta > 1$ setting $B = 0$). They correspond to infinitesimal deformations of the SK matrix Q_{ab}^{SK} in Eq. (7.73) of the form

$$Q_{ab} = Q_{ab}^{\mathrm{SK}} + \epsilon\,\Delta_{ab} \tag{7.95}$$

where Δ_{ab} satisfies the condition

$$\sum_c (\Delta_{ac} - \Delta_{bc}) = 0 \quad \forall\, a, b = 1, \dots, n. \tag{7.96}$$

To get the correct physics now we have to solve the variational problem in the ensemble of analytic families of matrices, searching for a better ansatz. The first author to achieve this goal and solve the SK model was Giorgio Parisi (he got the Nobel Prize for Physics also for this result): see [8, 10, 12, 13].

7.4   The Parisi Solution: Broken Replica Symmetry

The Parisi formula can be seen as a theorem of mathematical analysis

M. Talagrand [14]

We opened this section with a quotation stating that Parisi's result has now the logical status of a math theorem. We did that because the very physical reasoning which follows may look to some mathematically oriented reader as the kind of witchcraft that physicists so often indulge in. We wished to reassure her that all the following constructions, *however peculiar,* lay on solid rock. It took more than 30 years to transform "Parisi conjectural formula" into a theorem, and we are not going to review all the deep work that went into that. The curious reader may have a look to [14] and references therein.

According to our discussion in Sect. 7.3 we have to find a better family $Q(q)$. To have some chance to work, the family $Q(q)$ should satisfy certain conditions:

Q1 $Q(q)$ should contain a subsequence of the SK family (7.73) because the actual solution lays in this family for $T > T_c$;

Q2 $Q(q)$ should contain the directions in matrix space corresponding to the negative eigenvalues of the Hessian, since moving in these directions we increase the free energy and approach the actual "minimum" (i.e. maximum) at the given T. Comparing with Eqs. (7.95) and (7.96) we conclude that the matrices in the family better must satisfy

$$\sum_c (Q_{ac} - Q_{bc}) = 0 \quad \forall\, a, b = 1, \ldots, n. \tag{7.97}$$

Q3   the limit $n \to 0$ must exists. In particular

$$-\infty < \lim_{n \to 0} \frac{1}{n} \mathrm{Tr}\, Q^2 \le 0, \tag{7.98}$$

where the second inequality follows from the bound given by the naive solution (we want to find a greater free energy than the SK one because in "negative" dimension "minimization" is maximization).

We give a first example of a sequence of matrices which satisfy the conditions **Q1-Q3**. Let $m \mid n$ be integers and

$$
Q^{(n)}(q_0, q_1) =
\begin{pmatrix}
\boxed{Q^{(m)}(q_0)} & & & & \\
& \boxed{Q^{(m)}(q_0)} & & q_1 & \\
& & \boxed{Q^{(m)}(q_0)} & & \\
& q_1 & & \ddots & \\
& & & & \boxed{Q^{(m)}(q_0)}
\end{pmatrix}
\tag{7.99}
$$

where the blocks along the diagonal are $m \times m$ SK matrices $Q^{(m)}(q_0)_{ab} \equiv (1 - \delta_{ab})q_0$ and the entries outside the diagonal blocks are all equal q_1. Written in components, this matrix reads

$$
Q_{ab}^{(n)} =
\begin{cases}
(1 - \delta_{ab})q_0 & \text{if } \lceil a/m \rceil = \lceil b/m \rceil \\
q_1 & \text{if } \lceil a/m \rceil \neq \lceil b/m \rceil
\end{cases}
\tag{7.100}
$$

where $\lceil x \rceil$ is the *ceiling function* (the smallest integer no smaller than x). Clearly the sums of all matrix columns are equal and

$$
\widetilde{\mathrm{Tr}}(Q^{(n)})^2 \equiv \lim_{n \to 0} \frac{1}{n} \mathrm{Tr}(Q^{(n)})^2 = -mq_1^2 + (m - 1)q_0^2
\tag{7.101}
$$

which satisfies the condition (7.98) provided we take $m < 1$ while performing the analytic continuation. The matrix $Q^{(n)}(q_0, q_1)$ corresponds to the pattern of spontaneous breaking of the replica symmetry

$$
\mathfrak{S}_n \to \mathfrak{S}_{n/m} \ltimes (\mathfrak{S}_m)^{n/m}.
\tag{7.102}
$$

Now the candidate free energy becomes (see [8])

$$
\widetilde{A}(q_0, q_1, m) = -\frac{\beta^2}{4} - \frac{\beta}{4}\left(mq_1^2 - (m - 1)q_0^2\right) -
$$
$$
- \int \frac{dz}{\sqrt{2\pi}} e^{-z^2/2} \frac{1}{m} \log\left(\int \frac{dy}{\sqrt{2\pi}} e^{-y^2/2} \cosh^m[\beta\sqrt{q_1}z + \beta\sqrt{q_0 - q_1}y]\right)
\tag{7.103}
$$

Next we have to find the extremum in q_0, q_1 and $m \in (0, 1)$. We are not going to do that; we just state that the solution one gets shows very good agreement with the numerical simulations: the entropy at $T = 0$, that in the SK symmetric solution was

$S(0) = -1/2\pi \approx -0.1591$, now has the value $S(0) \approx -0.01$. While still negative, the situation has definitely improved. Clearly we are on the right track, and we just have to keep improving our variational family of "0×0 symmetric matrices with zeros on the main diagonal".

In view of the criteria **Q1–Q3** it is obvious what we have to do: repeat recursively the same construction. In the third step we take integers $m_0 \mid m_1$ with $m_1 \mid n$ and consider the matrix

$$
Q^{(n)}(q_0, q_1, q_2) =
\begin{pmatrix}
\boxed{Q^{(m_1)}(q_0, q_1)} & & & q_2 \\
 & \boxed{Q^{(m_1)}(q_0, q_1)} & & \\
 & & \ddots & \\
q_2 & & & \boxed{Q^{(m_1)}(q_0, q_1)}
\end{pmatrix}
\tag{7.104}
$$

where in the blocks along the diagonal we put the matrix $Q^{(m_1)}(q_0, q_1)$ of the previous step with $m = m_0$. Going on with the recursion procedure, at the k-th step we have a sequence of integers m_i [8]

$$
m_1, \; m_2, \; \ldots, \; m_{k+1} \equiv n \qquad m_i \mid m_{i+1}
\tag{7.105}
$$

and we set

$$
Q_{ab} = q_i \quad \text{iff} \quad \lceil a/m_i \rceil \neq \lceil b/m_i \rceil \;\; \text{and} \;\; \lceil a/m_{i+1} \rceil = \lceil b/m_{i+1} \rceil
\tag{7.106}
$$

Then

$$
-\widetilde{\mathrm{Tr}}\, Q^2 = \sum_{i=1}^{k} (m_i - m_{i+1}) q_i^2
\tag{7.107}
$$

and Eq. (7.98) requires that after the analytic continuation to $n = 0$ we take

$$
0 \leq m_{i+1} \leq m_i \leq 1.
\tag{7.108}
$$

Increasing k yields a richer family satisfying conditions **Q1–Q3**. It is natural to take the limit $k \to \infty$. To keep track of the parameters q_i, m_i one introduces the function $q(x)$ defined in the interval $[0, 1]$ [8]:

$$
q(x) = q_i \quad \text{for} \;\; m_{i+1} < x < m_i.
\tag{7.109}
$$

In the $k \to \infty$ limit $q(x)$ becomes a function in $L^2([0, 1])$. *The function $q(x)$ plays the role of order parameter for spin glasses.* Thus spin glasses has an infinite number of order parameters as it is natural for a frustrated system where the number of ground states is potentially infinite (in facts exponentially large in the size of the system). One can show that $q(x)$ is a non-decreasing function of x. Now

$$- \tilde{\mathrm{Tr}}\, Q^2 = \int_0^1 q(x)^2 \, \mathrm{d}x \tag{7.110}$$

so that the internal energy is

$$\overline{U} = -\frac{\beta}{2} \left(1 - \int_0^1 q(x)^2 \mathrm{d}x \right) \tag{7.111}$$

The susceptibility is

$$\chi = \beta \int_0^1 \mathrm{d}x \, (1 - q(x)) \tag{7.112}$$

One can write an expression for the free energy as a functional of the function $q(x)$ which is however not explicit since it depends on a specific solution to a certain non-linear partial differential equation [8]. Now one can show that all eigenvalues of the Hessian are non-negative (some are zero) [15], so the Parisi solution appears to be correct. The k-th step Parisi free energy can be written as follows (we restore the external magnetic field B). We have two sequences of real numbers

$$0 = m_0 \leq m_1 \leq \cdots \leq m_{k-1} \leq m_k = 1 \tag{7.113}$$

$$0 = q_0 \leq q_1 \leq \cdots \leq q_k \leq q_{k+1} = 1. \tag{7.114}$$

We define recursively the $(k+1)$ quantities $\Psi_{k+1}, \Psi_k \cdots, \Psi_1, \Psi_0$:

$$\Psi_{k+1} = \log \cosh \left(\sum_{0 \leq s \leq k} \beta z_s \sqrt{q_{s+1} - q_s} + B \right) \tag{7.115}$$

$$\Psi_\ell = \begin{cases} \displaystyle\int_{-\infty}^{+\infty} \frac{\mathrm{d}z_\ell}{\sqrt{2\pi}} \, e^{-z_\ell^2/2} \exp\left[m_\ell \, \Psi_{\ell+1} \right] & m_\ell \neq 0 \\[2ex] \displaystyle\int_{-\infty}^{+\infty} \frac{\mathrm{d}z_\ell}{\sqrt{2\pi}} \, e^{-z_\ell^2/2} \, \Psi_{\ell+1} & m_\ell = 0 \end{cases} \tag{7.116}$$

and the k-step Parisi free energy

$$\mathcal{F}_k(\beta, B, \boldsymbol{m}, \boldsymbol{q}) = \frac{\beta^2}{4} \sum_{s=1}^k m_s(q_{s+1}^2 - q_s^2) - \log 2 - \log \Psi_0(\beta, B, \boldsymbol{m}, \boldsymbol{q}). \tag{7.117}$$

This formula is correct in the following sense.

Theorem 7.1 (Talagrand [14]) [8] *One has*

$$- \lim_{N \to \infty} \frac{1}{N} \overline{\log Z_N} = \sup_{k,\boldsymbol{m},\boldsymbol{q}} \mathcal{F}_k(\beta, B, \boldsymbol{m}, \boldsymbol{q}) \tag{7.118}$$

where

$$Z_N = \sum_{\{\sigma_k = \pm 1\}} \exp\left[-\frac{\beta}{\sqrt{N}} \sum_{ij} J_{ij}\, \sigma_i \sigma_j \right] \tag{7.119}$$

and the J_{ij} are Gaussian random variables with mean 0 and covariance 1.

Note that, indeed, a "minimization" is replaced by taking a *supremum* in agreement with Parisi's insight (cf. gray-box on page 342). The proof of this theorem is purely in the language of mathematical statistics, no replicas or other physicists' dirty tricks are involved.

It is convenient to introduce the probability distribution

$$P(q) = \lim_{n \to 0} \frac{1}{n(n-1)/2} \sum_{a<b} \delta(Q_{ab} - q) = \sum_{s=0} (m_s - m_{s-1})\, \delta(q - q_{s-1}) \tag{7.120}$$

which is non-negative and with total mass 1 by Eq. (7.113), clearly it contains the same information as $q(x)$. In the limit $k \to \infty$ the function $q(x)$ is expected to become strictly monotonically increasing: in this case we have

$$\mathrm{d}x = P(q)\, \mathrm{d}q. \tag{7.121}$$

7.5   Physical Interpretation of the Parisi Solution

The replica trick is powerful, but gives little clue about the physical meaning of the results. For instance, what is the physical interpretation of the functional order parameter $q(x)$? In this final section we say a few words, mostly at a heuristic level, following ideas better described in [10] and references therein.

At low temperature we expect the spin glasses to have a huge number of pure equilibrium states (*pure* in the sense of Chap. 4). All equilibrium states have the same free energy density in the thermodynamic limit, and the states with higher free energy density are meta-stable.

---

[8] Michel Talagrand got the Abel prize for this result. For more details see the two-volume set [16].

We know that a general *mixed* phase is a convex linear combination of pure ones, in the sense that the expectation values $\langle \cdots \rangle$ in any such mixed phase can be written

$$\langle O \rangle = \sum_\alpha w_\alpha \langle O \rangle_\alpha, \qquad 0 \le w_\alpha \le 1, \qquad \sum_\alpha w_\alpha = 1, \tag{7.122}$$

where $\langle \cdots \rangle_\alpha$ is the thermal expectation value in the α-th pure (equilibrium) state. A blind application of the formalism of statistical mechanics, without selecting specific boundary conditions and/or switching on infinitesimal external fields, will generically produce a mixed state. We can easily understand what the weights w_α are for such a "blind" state. Consider a system whose size N is large but finite. To define it we need to specify boundary conditions; we take them periodic, so that the finite system is translational invariant. Before the thermodynamic limit, the notion of "phase" is not well defined, but nevertheless when $N \ggg 1$ we do have density matrices ϱ_α which behave *approximatively* as they were pure states. There may be several such matrices ϱ_α which have a free energy of the form

$$N F_\alpha \equiv -\log \mathrm{Tr}(\varrho_\alpha) = N\beta F_{\min} + O(1) \tag{7.123}$$

where

$$N F_{\min} = \inf_\alpha \left(-\log \mathrm{Tr}\, \varrho_\alpha \right) \tag{7.124}$$

is the lowest possible free energy. As $N \to \infty$ all states α have the same free energy per site, and all are equilibrium states. Before taking the limit we had approximatively

$$\langle O \rangle_\alpha \approx \sum_\alpha e^{-N(F_\alpha - F_{\min})} \langle O \rangle_\alpha \tag{7.125}$$

and as $N \to \infty$ we expect this formula to become exact. Clearly only states with $(F_\alpha - F_{\min}) = O(1/N)$ contribute to the sum in the thermodynamic limit. In particular the meta-stable ones drop out. This discussion holds for any statistical system: what is peculiar of the spin-glasses is that the number of terms contributing to the limit sum (7.125) now may be *infinite*.

To orient ourselves in this infinite set of pure states, we have first to determine their *geography*: we need ways to distinguish them, that is, observables to measure how much different any two given pure states are. A natural proposal is to look for a *distance* between pure states. Let

$$m_\alpha^i = \langle \sigma^i \rangle_\alpha \tag{7.126}$$

be the average magnetization of the i-th spin in state α. A natural notion of square-distance is

$$d^2_{\alpha\beta} \overset{\text{def}}{=} \frac{1}{N} \sum_i (m^i_\alpha - m^i_\beta)^2 = 2(q_{\text{EA}} - q_{\alpha\beta}) \tag{7.127}$$

where the *overlap* between two states is defined as

$$q_{\alpha\beta} \overset{\text{def}}{=} \frac{1}{N} \sum_i m^i_\alpha \, m^i_\beta. \tag{7.128}$$

On can argue on physical grounds that the overlap of a state α with itself is independent of the particular state and equal to the *Edwards-Anderson* (EA) *order parameter*

$$q_{\alpha\alpha} = q_{\beta\beta} \equiv q_{\text{EA}}. \tag{7.129}$$

Given the large number of pure phases, it is natural to introduce a probability distribution for their overlaps

$$P(q) \overset{\text{def}}{=} \sum_{\alpha,\beta} w_\alpha w_\beta \, \delta(q_{\alpha\beta} - q). \tag{7.130}$$

Now we can state the physical interpretation of the functional order parameter $q(x)$ which we found to describe the breaking pattern of the replica symmetry:

Claim (Parisi [17]) *The overlap probability* (7.130) *is equal to the function defined in* (7.121) *and yields the same information as the functional order parameter* $q(x)$.

This explains physically why $q(x)$ is monotonically increasing and shows that $q(1) = q_{\text{EA}}$, while $q(0)$ is the minimum overlap between two pure states.

We conclude with a final observation [18, 19]: it turns out that the metric defined in Eq. (7.127) is *ultrametric*, that is, it satisfies the strong form of the triangular inequality

$$d_{\alpha\beta} \leq \max(d_{\alpha\gamma}, d_{\beta\gamma}) \quad \forall\, \alpha, \beta, \gamma, \tag{7.131}$$

i.e. all triangles are isosceles with the base the short side, and two distinct spheres of radius R have zero overlap. Equivalently: the pure phases are organized in a hierarchy of clusters, superclusters, etc. If the distance spectrum is $0 = d_0 < d_1 < d_2 \cdots$ the states at distance $\leq d_1$ belong to the same cluster, those at distance $\leq d_2$ to the same supercluster, those at distance $\leq d_3$ to the same super-supercluster, and so on.

This result was to be expected for the following reason: the free energy landscape has several valleys, with secondary valleys inside them, separated by high barriers. We thus have a hierarchy of valleys inside valleys inside valleys ... an ordering which is expected to be reflected in the hierarchy of clusters of states inside clusters.

References

1. G.H. Wannier, Antiferromagnetism. The triangular Ising net, Phys. Rev. **79**, 357–364 (1950). ERRATUM: Phys. Rev. **B7**, 5017 (1973)
2. S. Galam, P.-V. Koseleff, Solving the triangular Ising antiferromagnet by simple mean field, Eur. Phys. J. **B28**, 149155 (2002)
3. D. Chowdhury, *Spin Glasses and Other Frustrated Systems* (Princeton University Press, 1986)
4. A. Altieri, M. Baity-Jesi, An introduction to the theory of spin glasses in *Encyclopedia of Condensed Matter Physics*, (Elsevier, 2024), pp. 361–370
5. D. Sherrington, S. Kirkpatrick, Solvable model of a spin-glass. Phys. Rev. Lett. **35**, 1792–1796 (1975)
6. D. Panchenko, The Sherrington-Kirkpatrick model: an overview. J. Stat. Phys. **149**, 362–383 (2012), arXiv:1211.1094
7. K.H. Fisher, Static properties of spin glasses. Phys. Rev. Lett. **34**, 1438 (1976)
8. G. Parisi, An introduction to the Statistical Mechanics of amorphous systems, in *Recent Advances in Field Theory and Statistical Mechanics*, ed. by J.B. Zuber, R. Stora (North-Holland, 1984), pp. 474–523
9. S. Cecotti, *Analytic Mechanics. A Concise Textbook* (Springer, 2024)
10. G. Parisi, Spin glass theory, in *Time Dependent Effects in Disordered Matherials*, ed. by R. Pynn, T. Riste (Plenum Press, 1987), pp. 317–329
11. J.R.L. de Almeida, D.J. Thouless, Stability of the Sherrigton-Kirpatrick solution of a spin glass model. J. Phys. **A11**, 983–990 (1978)
12. G. Parisi, Spin glasses and optimization problems without replicas, in *Chance and Matter*, ed. by J. Souletie et al. (North-Holland, 1988)
13. M. Mézard, G. Parisi, M. Virasoro, *Spin Glass Theory and Beyond* (World Scientific, 1987)
14. M. Talagrand, The Parisi formula. Ann. Math. **163**, 221–263 (2006)
15. I. Kondor, C. De Dominicis, The spectrum of fluctuations around Sompolinsky's mean field solution for a spin glass. J. Phys. **A16**, L73 (1983)
16. M. Talagrand, *Mean-field Models for Spin Glasses,* vols. I and II. Modern Surveys in Mathematics, vols. 54, 55 (Springer, 2011)
17. G. Parisi, Order parameter for spin-glasses. Phys. Rev. Lett. **50**, 1946 (1983)
18. G. Parisi, On the emergence of tree-like structures in complex systems, in *Perspectives on Biological Complexity*, ed. by O.T. Solbrig, G. Nicolis (IUBS, 1991)
19. G. Parisi, F. Ricci-Tersenghi, On the origin of ultrametricity. J. Phys. **A33**, 113 (2000)

Index